S0-CFC-624

STUDENT COMPANION TO ACCOMPA

FUNDAMENTALS OF
BIOCHEMISTRY
LIFE AT THE MOLECULAR LEVEL

Third Edition

Akif Uzman
University of Houston

Joseph Eichberg
University of Houston

William Widger
University of Houston

Donald Voet
University of Pennsylvania

Judith G. Voet
Swarthmore College

Charlotte W. Pratt
Seattle, Washington

WILEY

JOHN WILEY & SONS, INC

Background Photo Cover Credit: Lester Lefkowitz/Getty Images

Inset Photo Credits: Based on X-ray structures by left to right Thomas Steitz, Yale University; Daniel Koshland, Jr., University of California at Berkeley; Emmanual Skordalakis and James Berger, University of California at Berkeley; Nikolaus Grigorieff and Richard Henderson, MRC Laboratory of Molecular Biology, U.K.; Thomas Steitz, Yale University.

To order books or for customer service call 1-800-CALL-WILEY (225-5945)

Copyright © 2009 John Wiley & Sons, Inc. All rights reserved.

No part of this publication may be reproduced, stored in a retrieval system or transmitted in any form or by any means, electronic, mechanical, photocopying, recording, scanning, or otherwise, except as permitted under Sections 107 or 108 of the 1976 United States Copyright Act, without either the prior written permission of the Publisher, or authorization through payment of the appropriate per-copy fee to the Copyright Clearance Center, Inc., 222 Rosewood Drive, Danvers, MA 01923, or on the web at www.copyright.com. Requests to the Publisher for permission should be addressed to the Permissions Department, John Wiley & Sons, Inc., 111 River Street, Hoboken, NJ 07030-5774, (201) 748-6011, fax (201) 748-6008, or online at http://www.wiley.com/go/permissions.

To order books or for customer service, please call 1-800-CALL-WILEY (225-5945).

ISBN-13 978- 0-470-22842-5

Printed in the United States of America

10 9 8 7 6 5 4 3

Printed and bound by Bind-Rite Graphics, Inc.

Preface

Welcome to biochemistry! You are about to become acquainted with one of the most exciting scientific disciplines. The biotechnology industry, with its roots in molecular genetics, is one of the most visible manifestations of the explosion of biochemical knowledge that has occurred during our lifetime. Drug design and novel approaches such as gene therapy rely on the fundamental knowledge of the chemistry of biological molecules, particular proteins. Our most common diseases (e.g., diabetes and heart disease) have pleiotropic multifaceted physiological effects that are best understood in terms of biochemistry. You will soon discover that biochemistry's impact on our lives cannot be over-emphasized. We are excited to bring you an ever-expanding understanding of this magnificent subject!

Learning biochemistry is not easy but it can be fun! Most students discover that biochemistry is a synthetic science, merging knowledge of general chemistry, organic chemistry, and biology. Hence, a more mature and creative kind of thinking is required to gain a deep understanding of biochemistry. In addition to a solid foundation in chemistry and biology, you will need to recognize and assimilate some general principles from other disciplines within biology, including physiology, genetics, and cell biology. In this respect, biochemistry is not all that different from nonscientific pursuits that require some degree of "lateral thinking" across disciplines.

This Student Companion accompanies *Fundamentals of Biochemistry Third Edition* by Donald Voet, Judith G. Voet, and Charlotte W. Pratt. It is designed to help you master the basic concepts and exercise your analytical skills as you work your way through the textbook. Each chapter of the Student Companion is divided into four parts, beginning with a general summary reminding you of the topics covered in that chapter. This is followed by a section called Essential Concepts, which provides an overview of the main facts and ideas that are essential for you understanding of biochemistry. This can be regarded as a set of brief notes for each chapter, alerting you to the key facts you need to commit to memory and to the concepts you need to master. You will soon notice that biochemical knowledge is cumulative: new concepts often rely on a solid understanding of previously presented concepts. Hence, one of the key goals of this Companion is to help you gain this understanding. The third and last section is the Questions. These are organized in a manner to help you gain a firm understanding of each section of a chapter. Some questions ask you to recall essential facts while others exercise your problem-solving skills. Answers to all the questions are provided. However, you do yourself a great disservice by turning to them too soon. Don't know the answer right away? Keep trying! Go back to the text to find clues for yourself. Use the answers to check yours, not to fill in a temporary void in your understanding.

"How should I study biochemistry beyond reading the textbook and working in this Companion?" is a likely question from students. The phrase, "if you don't use it, you lose it" applies here. It is pointless to simply read your biochemistry textbook over and over. Unless you are actively engaged in working with the material, you become a passive reader. Active engagement includes using your hands to work problems, drawing pictures, or writing an outline or flow chart. As you read, ask yourself questions and seek answers from your text and your instructor. By doing these things, you use the material, and it becomes more efficiently transferred to your long-term memory.

"How often should I study biochemistry?" is also a common question. Most instructors agree that frequent short study sessions—even daily—will pay greater dividends than a single long session once a week. Because short-term memory lasts just a few minutes, take a few minutes after every class to review you notes. Similarly, stop reading your text, and review on paper with diagrams, word charts, flow diagrams, what you just read. Talk biochemistry with anyone who will listen. Form a study group to enrich your knowledge and test your memory. All these activities will result in the transfer of knowledge from short-term memory to long-term memory. In other words, the more you use biochemistry, the better you know it and the more fun you will have with it.

One of the truly most satisfying ways to learn biochemistry is to apply its principles and findings to problems that integrate your knowledge. To this end, Dr. Kathleen Cornely has developed numerous case studies that test your analytical skills. You will find these case studies in the *Fundamentals of Biochemistry 3e* WileyPLUS course (**www.wileyplus.com**). Topics for the case studies were chosen to cover a range of interesting areas relevant to biochemistry. The cases themselves are based on data from research reports as well as clinical studies. The prerequisites include material that you are likely to be studying in biochemistry class, but occasionally include concepts from genetics and immunology on a level likely to be encountered in a first-year general biology course. The answers to the questions posed in these case studies can be obtained from your instructor.

We have many people to thank for helping us get this Companion to you. First and foremost, we would like to acknowledge Assistant Editor Ms. Aly Rentrop and

Ms. Kelly Tavares who worked on the design and lay-out, and Production Editor Ms. Sandra Dumas. In addition, we would like to acknowledge Ms. Petra Recter, Associate Publisher for Physics and Chemistry at John Wiley and Sons for her support of this project. We are also indebted to Caroline Breitenberger of The Ohio State University and Laura Mitchell of St. Joseph's University for their much-appreciated reviews of our original draft. We would also like to thank our students at Swarthmore College, the University of Houston, and the University of Houston-Downtown for pointing out errors and ambiguities in earlier drafts of this work. It is still possible that errors persist, and so we would greatly appreciate being alerted to them. Please forward you comments to Akif Uzman (uzmana@uhd.edu).

Akif Uzman

Joseph Eichberg

William Widger

Charlotte W. Pratt

Donald Voet

Judith G. Voet

Contents

1

Introduction to the Chemistry of Life

This chapter introduces you to life at the biochemical and cellular level. It begins with a discussion of the chemical origins of life and its early evolution. This discussion continues into ideas and theories about the evolution of organisms, followed by a brief introduction to taxonomy and phylogeny viewed from a molecular perspective. The chapter concludes with an introduction to the basic concepts of thermodynamics and its application to living systems. Biochemistry, like all other sciences, is a based on the measurement of observable phenomena. Hence, it is important to become familiar with the conventions used to measure energy and mass. Box 1-2 presents the essential biochemical conventions that we will encounter throughout *Fundamentals of Biochemistry*.

Essential Concepts

The Origin of Life

1. Living matter consists of a relatively small number of elements, of which C, N, O, H, Ca, P, K, and S account for ~98% of the dry weight of most organisms (which are 70% water). These elements form a variety of reactive functional groups that participate in biological structure and biochemical reactions.

2. The current model for the origin of life proposes that organisms arose from the polymerization of simple organic molecules to form more complex molecules, some of which were capable of self-replication.

3. Most polymerization reactions involving the building of small organic molecules into larger more complex ones occur by the formation of water. This is called a condensation reaction.

4. A key development in the origin of life was the formation of a membrane that could separate the critical molecules required for replication and energy capture from a potentially degradative environment.

5. Complementary surfaces of molecules and macromolecules provide a template for biological specificity (e.g., macromolecular assembly, enzyme activity, and expression and replication of the genome).

6. Modern cells can be classified as either prokaryotic or eukaryotic. Eukaryotic cells are distinguished by a variety of membrane-bounded organelles and an extensive cytoskeleton.

Organismal Evolution

7. Prokaryotes show a limited range of morphologies but very diverse metabolic capabilities.

8. Phylogenetic evidence based on comparisons of ribosomal RNA genes have been used by Woese and colleagues to group all organisms into three domains: archaea, bacteria, and eukarya.

9. The evolution of sexual reproduction marks an important step of the evolution of organisms, because it allows for genetic exchanges that lead to an increase in the adaptability of a population of organisms to changing environments.

10. Eukaryotes contain several membrane-bounded structures, such as mitochondria and chloroplasts, which may be descended from ancient symbionts.

11. Archaea represent a third domain or branch of life in the three-domain system of classification. While they outwardly resemble bacteria, their genomes and the proteins encoded in them more closely resemble those of eukaryotes.

12. Biological evolution is not goal-directed, requires some built-in sloppiness, is constrained by its past, and is ongoing.

13. Natural selection directs the evolution of species.

Thermodynamics

14. The first law of thermodynamics states that energy (U) is conserved; it can neither be created nor destroyed.

15. Enthalpy (H) is a thermodynamic function that is a sum of the energy of the system and the product of the pressure and the volume (PV). Since biochemical processes occur at constant pressure and have negligible changes in volume, the change of energy of the system is nearly equivalent to the change in enthalpy ($\Delta U \approx \Delta H$).

16. The second law of thermodynamics states that spontaneous processes are characterized by an increase in the entropy of the universe, that is, by the conversion of order to disorder.

17. The spontaneity of a process is *determined by its free energy change ($\Delta G = \Delta H - T\Delta S$).* Spontaneous reactions have $\Delta G < 0$ (exergonic) and nonspontaneous reactions have $\Delta G > 0$ (endergonic).

15. For any process at equilibrium, the rate of the forward reaction is equal to the rate of the reverse reaction, and $\Delta G = 0$.

18. Energy, enthalpy, entropy, and free energy are state functions; that is, they depend only on the state of the system, not its history. Hence, they can be measured by considering only the initial and final states of the system and ignoring the path by which the system reached its final state.

19. The entropy of a solute varies with concentration; therefore, so does its free energy. The free energy change of a chemical reaction depends on the concentration of both its reactants and its products.

20. For the general reaction

$$aA + bB \leftrightharpoons cC + dD$$

the free energy change is given by the following relationship:

$$\Delta G = \Delta G^\circ + RT \ln \left(\frac{[C]^c [D]^d}{[A]^a [B]^b} \right)$$

21. The equilibrium constant of a chemical reaction is related to the standard free energy of the reaction when the reaction is at equilibrium ($\Delta G = 0$) as follows:

$$\Delta G^\circ = -RT \ln K_{eq}$$

where K_{eq} is the equilibrium constant of the reaction:

$$K_{eq} = \frac{[C]^c_{eq} [D]^d_{eq}}{[A]^a_{eq} [B]^b_{eq}}$$

The equilibrium constant can therefore be calculated from standard free energy data and *vice versa*.

22. The equilibrium constant varies with temperature by the relationship:

$$\ln K_{eq} = \frac{-\Delta H^\circ}{R} \left(\frac{1}{T} \right) + \frac{\Delta S^\circ}{R}$$

where ΔH° and ΔS° represent enthalpy and entropy in the standard state. A plot of K_{eq} versus $1/T$, known as a van't Hoff plot, permits the values of ΔH° and ΔS° (and therefore ΔG° at any temperature) to be determined from measurements of K_{eq} at two (or more) temperatures.

23. The biochemical standard state is defined as follows: The temperature is 25°C, the pH is 7.0, and the pressure is 1 atm. The activities of reactants and products are taken to be the total activities of all their ionic species, except for water, which is assigned an activity of 1. $[H^+]$ is also assigned an activity of 1 at the physiologically relevant pH of 7. These conditions are different than the chemical standard state, so that the biochemical standard free energy is designated as $\Delta G^{\circ\prime}$ and in the chemical standard state is ΔG°. We assume that activity equals molarity for dilute solutions.

24. An isolated system cannot exchange matter or energy with its surroundings. A closed system can exchange only energy with its surroundings. A closed system inevitably reaches

equilibrium. Open systems exchange both matter and energy with their surroundings and therefore cannot be at equilibrium. Living organisms must exchange both matter and energy with their surroundings and are thus open systems. Living organisms tend to maintain a constant flow of matter and energy, referred to as the steady state.

25. Living systems can respond to slight perturbations from the steady state to restore the system back to the steady state.

26. The recovery of free energy from a biochemical process is never total, and some energy is lost to the surroundings as heat. Hence, while the system becomes more ordered, the surroundings experience an increase in entropy.

27. Enzymes accelerate the rate at which a biochemical process reaches equilibrium. They accomplish this by interacting with reactants and products to provide a more energetically favorable pathway for the biochemical process to take place.

Key Equations

Be sure to know the conditions for which the following thermodynamics equations apply and be able to interpret their meaning.

1. $\Delta U = q - w$

2. $H = U + PV$

3. $\Delta H = q_P - w + P\Delta V$

4. $S = k_B \ln W$

5. $\Delta S \geq \dfrac{q}{T} = \dfrac{\Delta H}{T}$

6. $\Delta G = \Delta H - T\Delta S$

7. $\Delta G = \Delta G° + RT \ln \left(\dfrac{[C]^c [D]^d}{[A]^a [B]^b} \right)$

8. $\Delta G° = -RT \ln K_{eq}$

9. $K_{eq} = \dfrac{[C]^c_{eq} [D]^d_{eq}}{[A]^a_{eq} [B]^b_{eq}}$

10. $\ln K_{eq} = \dfrac{-\Delta H°}{R} \left(\dfrac{1}{T} \right) + \dfrac{\Delta S°}{R}$

Questions

The Origin of Life

1. What are the elements that account for 98% of the dry weight of living cells?

2. In what critical ways was the atmosphere of the primitive earth different than the earth's current atmosphere?

3. What is the rationale and significance of the Miller and Urey experiments?

4. What was, and continues to be, the source of atmospheric oxygen?

5. Examine the reaction shown below for the condensation and hydrolysis of lactose (two covalently linked sugars, a disaccharide). Circle the functional groups that will form water during the condensation.

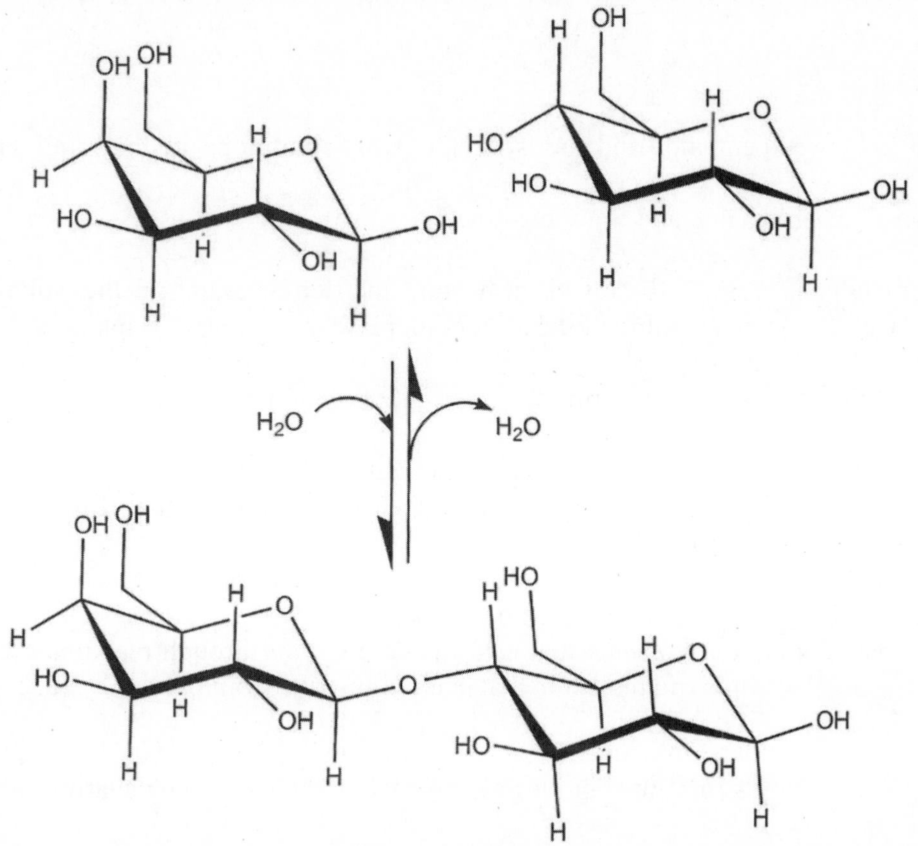

6. The condensation of two functional groups can result in the formation of another common functional group, which can be referred to as a compound functional group. Examine the functional groups in Table 1-2. Which compound functional groups are the combination of two other functional groups found in that figure? Show how these compound functional groups form.

Cellular Architecture

7. Draw a schematic diagram of an eukaryotic cell showing its principal organelles.

8. Give the principal distinguishing feature(s) of each pair of terms:
    (a) Prokaryote and eukaryote
    (b) Cytosol and cytoplasm
    (c) Endoplasmic reticulum and cytoskeleton

9. Living organisms are classified into three domains: _____ ,

    _____ , and _____ .

10. What is the most compelling evidence that mitochondria and chloroplasts represent descendents of symbiotic bacteria that lived inside of ancient eukaryotic cells?

11. What is the relationship between mutation and genetic variation in a population of organisms? Of what significance is it to evolution?

Thermodynamics

12. Distinguish between enthalpy and energy. Under what conditions are they equivalent?

13. What does it mean when q and w are positive?

14. When crystalline urea is dissolved in water, the temperature of the solution drops precipitously. Does the enthalpy of the system increase or decrease? Explain.

15. List and define the four major thermodynamic state functions.

16. Which of the following pairs of states has higher entropy?
    (a) Two separate beakers of NaCl and KCH_3COO in solution and a beaker containing a solution of both salts.
    (b) A set of dice in which all the dice show 6 dots on the top side and a set of dice in which the 6's show up on one of the side faces.
    (c) A small symmetric molecule that can form a polymer through reaction at either end, and a small asymmetric molecule that can polymerize from only one end.

17. Rationalize the temperature dependence of Gibbs free energy changes when both the enthalpy and entropy terms are positive values or when they are both negative values.

18. Hydrogen gas combines spontaneously with oxygen gas to form water

$$2\,H_2 + O_2 \rightarrow 2\,H_2O$$

Which term, enthalpy or entropy, predominates in the equation for the Gibbs free energy? How are the surroundings affected by this reaction?

19. Evaluate the following statement: Enzymes accelerate the rate of a reaction by increasing the spontaneity of the reaction.

20. Based on your reading in this chapter, suggest simple criteria for a reasonable definition of life.

21. The breakdown of glucose to pyruvate utilizes the free energy of 2 ATP molecules; however, the synthesis of glucose from pyruvate requires 4 ATP. Using the basic ideas of thermodynamics presented in this chapter, provide a thermodynamic explanation. *Hint*: Think about the relationship $\Delta G = \Delta H - T\Delta S$ in your argument.

Answers to Questions

1. Carbon, nitrogen, oxygen, hydrogen, phosphorus, calcium, potassium, and sulfur account for 98% of the dry weight of living cells.

2. The primitive earth's atmosphere was a relatively reducing atmosphere lacking appreciable amounts of O_2. Much recent controversy revolves around just how much NH_3 and CH_4 existed in the young earth's atmosphere. Relatively low amounts would indicate that the atmosphere was both nonreducing and nonoxidizing. In either case, the absence of significant levels of O_2 indicates that the atmosphere was nonoxidizing compared to the present atmosphere.

3. The Miller and Urey experiments were designed to ask whether biological molecules could be generated by mimicking the atmospheric conditions of the early earth. The early atmosphere was thought then to consist of H_2O, CH_4, NH_3, SO_2, and possibly H_2. Miller and Urey subjected these molecules, except for SO_2, to electrical discharges that were meant to simulate discharges believed to be prevalent in the earth's early atmosphere. The generation of diverse organic acids showed that the precursors for larger biological molecules could be spontaneously produced in the early atmosphere, paving the way for subsequent chemical evolution.

4. Photosynthesis.

5. Shown below is the condensation of galactose and glucose to form lactose. The atoms involved in the elimination of water are circled.

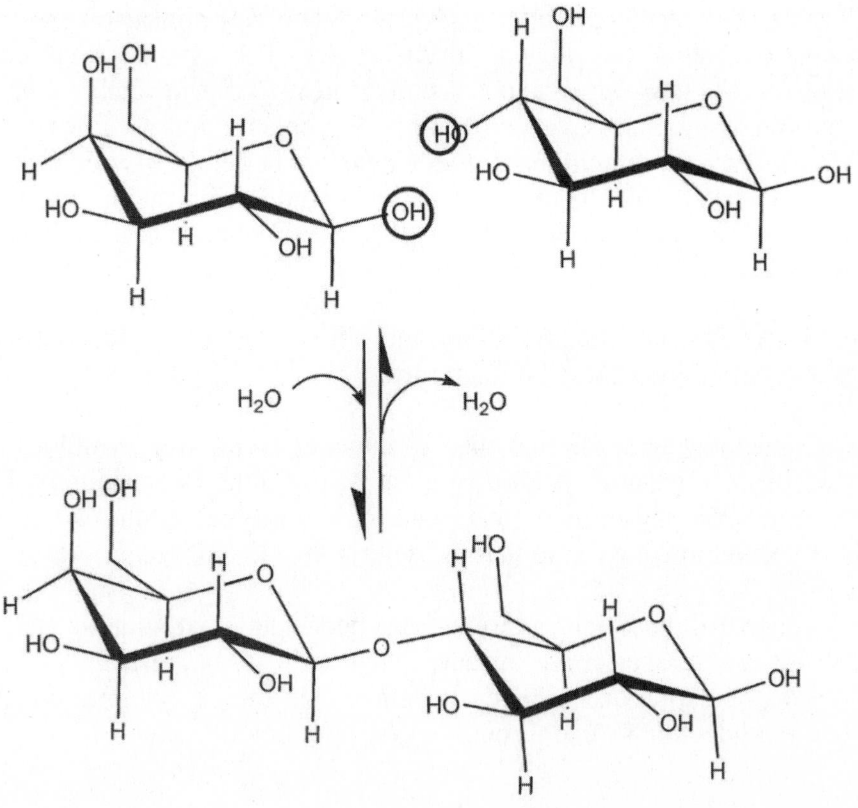

6. Shown below are condensation reactions of two pairs of functional groups found in Table 1-2. The atoms involved in the elimination of water are circled.

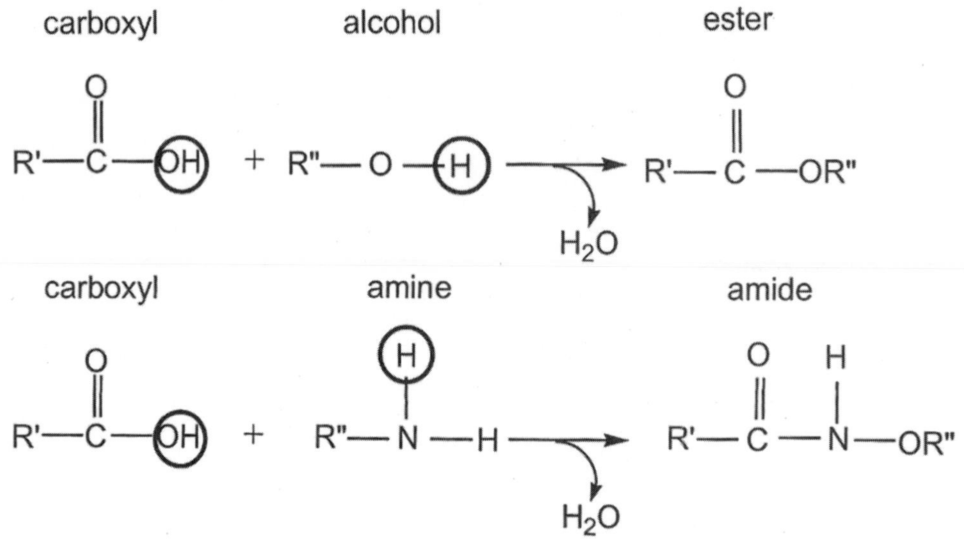

7. See Figure 1-8.

8. (a) Eukaryotes are defined by their elaborate internal membrane system and the enclosure of genomic DNA inside a double-membrane compartment, the nucleus, which is lacking in prokaryotes.
   (b) The cytoplasm is the contents of a prokaryotic cell and also refers to the cellular contents outside of the nucleus in eukaryotes. The cytosol is the cellular contents minus all of the membranous compartments of the cell (e.g., mitochondria, endoplasmic reticulum, Golgi apparatus, chloroplasts, and vacuoles).
   (c) The endoplasmic reticulum is the extensive network of internal membranes in eukaryotic cells (which is topologically continuous with the nuclear membrane). The cytoskeleton is the extensive network of protein filaments in the cytosol of eukaryotic cells.

9. All living organisms can be classified into three domains: archaea (archaebacteria), bacteria (eubacteria), and eukarya (eukaryotes).

10. The most compelling evidence that these organelles were once symbiotic bacteria is the presence of distinct genetic material and protein synthesis machinery inside of these organelles. The RNA and proteins that make up the protein synthesis machinery of these organelles is much more similar to that of bacteria than that of eukaryotes (see Chapter 27).

11. Genetic variation results from mutations that persist in a population and have not been eliminated by natural selection because they did not significantly decrease fitness. Evolution of a population occurs when variations that increase an individual's chances for survival and reproduction spread throughout the population.

12. Enthalpy is defined as $H = U + PV$. Hence $\Delta H = \Delta U + \Delta(PV)$. The enthalpy of a system is equal to its energy change when $\Delta(PV) = 0$, which occurs at the constant pressure and volume conditions typical of living things.

13. If q is positive, then heat has been transferred *to* the system, increasing its internal energy. If w is positive, then work has been done *by* the system, decreasing its internal energy.

14. In this example, urea and the water are the system, and the vessel and beyond are the surroundings. As urea dissolves, the enthalpy increases (an endothermic process). Heat is absorbed into the interactions between the urea and water, making the solution cooler.

15. **Energy** is measured as the heat absorbed by a system minus the work done by the system on its surroundings. **Enthalpy**, or heat content, is the amount of heat generated or absorbed by a system when a process occurs at constant pressure, as in biological systems, and no work is done other than the work of expansion or contraction (ΔV) of the system. **Entropy** is a measure of the heat absorbed or generated by a system at constant temperature and reflects the number of equivalent ways of arranging a system with no change in its internal energy. **Gibbs free energy** is the energy available to do work; it is a combination of enthalpy and entropy ($\Delta H - T\Delta S$) and an indicator of the spontaneity of a process at constant pressure and temperature.

16. (a) The beaker containing the solution with both salts has higher entropy. The molecules in this solution can be arranged in many more ways with respect to each other than each salt solution alone.
    (b) The set of dice with the 6's showing on the side faces has more entropy since there are many more ways to obtain this configuration.
    (c) The symmetric molecule has greater entropy, since it can polymerize by joining reactions involving either end. Note that the information content of the asymmetric molecule is higher, however. All biological polymers are formed from asymmetric subunits.

17. When enthalpy and entropy are both positive, ΔG decreases with increasing temperature, and the temperature at which the reaction occurs spontaneously must be high enough that the $T\Delta S$ term is larger than the ΔH term in the equation $\Delta G = \Delta H - T\Delta S$. For instance, dissolving crystalline urea in water is endothermic; however the process is spontaneous when it is carried out at room temperature. When enthalpy and entropy are both negative, ΔG decreases with decreasing temperature, and for the reaction to be spontaneous, the temperature must be low enough that the $T\Delta S$ term is not more negative than the ΔH term.

18. The expression for the Gibbs free energy is $\Delta G = \Delta H - T\Delta S$. In this reaction, the entropy decreases since the number of molecules decreases and the product is therefore more ordered than the reactants. Hence, in order for this reaction to be spontaneous the enthalpy term (ΔH) must be more negative than the entropy term ($T\Delta S$). In many cases such as this, the enthalpy of the reaction is negative and the reaction releases heat to the surroundings.

19. This statement is incorrect. An enzyme does not alter the spontaneity of a reaction; rather, it increases the rate at which a reaction reaches equilibrium.

20. Catalysis, replication, and mutability have been argued to be the minimum criteria for life. In addition, an organism must be able to maintain a novel chemical environment that is not in equilibrium with its surroundings and to resist the environmental fluctuations that might disturb its ability to carry out the other three essential functions of living systems—this steady state condition is called homeostasis.

21. Since the synthesis of glucose requires more ATP input than the amount produced during glucose degradation into pyruvate, we know that more chemical work is required for glucose synthesis than its degradation. Hence, glucose synthesis is endergonic (non-spontaneous) and its degradation is exergonic (spontaneous).

    The figure below shows one representation of this process ($P_i = HPO_4^{2-}$)

    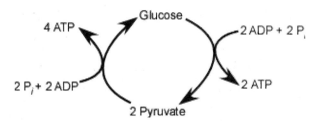

    Remember that the second law of thermodynamics tells us that every transformation of energy results in an increase in entropy of the universe, so that all the free energy released during glucose degradation to pyruvate cannot supply the total energy needed for the reverse process. We can see this in the relationship $\Delta G = \Delta H - T\Delta S$, which, when rearranged, gives $\Delta H = \Delta G + T\Delta S$. Hence, the total energy contained in the bonds of two pyruvate molecules is less than that of the bonds in glucose, so that that an additional input of energy is required. This is a biochemical example of the idea that one cannot have a perpetual motion machine.

2
Water

This chapter introduces you to the unique properties of water and to acid–base reactions. The discussion of water begins with a look at its structure and how its polarity provides a basis for understanding its powers as a solvent. You are then introduced to the hydrophobic effect, osmosis, and diffusion. The chemical properties of water are then described, beginning with the ionization of water, which sets the stage for a discussion of acid–base chemistry and the behavior of weak acids and buffers. This discussion includes the Brønsted–Lowry definition for acids and bases, the definition of pH, and the derivation of the Henderson–Hasselbalch equation. As we shall see in subsequent chapters, a solid understanding of acid–base equilibria is fundamental to understanding key aspects of amino acid biochemistry, protein structure, enzyme catalysis, transport across membranes, energy metabolism, and other metabolic transformations.

Essential Concepts

1. Water is essential to biochemistry because:
    (a) biological macromolecules assume specific shapes in response to the chemical and physical properties of water.
    (b) biological molecules undergo chemical reactions in an aqueous environment.
    (c) water is a key reactant in many reactions, usually in the form of H^+ and OH^- ions.
    (d) water is oxidized in photosynthesis to produce molecular oxygen, O_2, as part of the process that converts the sun's energy into usable chemical form. Expenditure of that energy under aerobic conditions leads to the reduction of O_2 back to water.

Physical Properties of Water

2. The structure of water closely approximates a tetrahedron with its two hydrogen atoms and the two lone pairs of electrons of its oxygen atom "occupying" the vertices of the tetrahedron.

3. The high electronegativity of oxygen relative to hydrogen results in the establishment of a permanent dipole in water molecules.

4. The polar nature of water results in negative portions of the molecule being attracted to the positive portions of neighboring water molecules by a largely electrostatic interaction known as the hydrogen bond.

5. Hydrogen bonds are represented as D—H$\cdots$A, where D—H is a weakly acidic compound so that the hydrogen atom (H) has a partial positive charge, and A is a weakly basic group that bears lone pairs of electrons. A is often an oxygen atom or a nitrogen atom (occasionally sulfur).

6. Water is strongly hydrogen bonded, with each water molecule participating in four hydrogen bonds with its neighbors; two in which it donates and two in which it accepts. Hydrogen bonds commonly form between water molecules and the polar functional groups of biomolecules, or between the polar functional groups themselves.

7. The strongly hydrogen bonded character of water is responsible for many of its characteristic properties, most notably:
   (a) a high heat of fusion, which allows water to act as a heat sink, such that greater heat loss is necessary for the freezing of water compared to other substances of similar molecular mass.
   (b) a high heat of vaporization, such that relatively more heat must be input to vaporize water compared to other substances of similar molecular mass.
   (c) an ability to dissolve most polar compounds.
   (d) an open structure that makes ice less dense than liquid water, thereby making ice float, insulating the water beneath it, and inhibiting total freezing of large bodies of water.

8. A variety of weak electrostatic interactions are critical to the structure and reactivity of biological molecules. These interactions include, in order of increasing strength, London dispersion forces, dipole–dipole interactions, hydrogen bonds, and ionic interactions (see Table 2-1).

9. Water is an excellent solvent of polar and ionic substances due to its property of surrounding polar molecules and ions with oriented shells of water, thereby attenuating the electrostatic forces between these molecules and ions.

10. The tendency of water to minimize its contact with nonpolar (hydrophobic) molecules is called the hydrophobic effect. This effect is largely driven by the increase in entropy caused by the necessity for water to order itself around nonpolar molecules. This causes the nonpolar molecules to aggregate, thereby reducing the surface area that water must order itself about. Consequently, nonpolar substances are poorly soluble in aqueous solution.

11. Many biological molecules have both polar (or charged) and nonpolar functional groups and are therefore simultaneously hydrophilic and hydrophobic. These molecules are said to be amphiphilic or amphipathic.

12. Osmosis is the movement of solvent across a semipermeable membrane from a region of lower concentration of solute to a region of higher concentration of solute. Osmotic pressure of a solution is the pressure that must be applied to the solution to prevent an inflow of solvent. Hence, an increase in solute concentration results in an increase in osmotic pressure.

13. Diffusion is the random movement of molecules in solution (or in the gas phase). It is responsible for the movement of solutes from a region of high concentration to a region of low concentration.

Chemical Properties of Water

14. Water is a neutral, polar molecule that has a slight tendency to ionize into H^+ and OH^-. However, the proton is never free and binds to a water molecule to form H_3O^+ (hydronium ion).

15. The ionization of water is described as an equilibrium between the unionized water (reactant) and its ionized species (products)

$$H_2O \leftrightharpoons H^+ + OH^-$$

In which

$$K = \frac{[H^+][OH^-]}{[H_2O]}$$

Since in dilute aqueous solution, the concentration of water is essentially constant (55.5 M), the concentration of H_2O is incorporated into the value of K, which is referred to as K_w, the ionization constant of water.

$$K_w = [H^+][OH^-]$$

16. The values of both H^+ and K are inconveniently small; hence, their values are more conveniently expressed as negative logarithms, so that

$$pH = -\log[H^+]$$

$$pK = -\log K$$

17. According to the Brønsted–Lowry definition, an acid is a substance that can donate a proton, and a base is a substance that can accept a proton. The strength of a weak acid is proportional to its dissociation constant, which is expressed as

$$K = \frac{[H^+][A^-]}{[HA]}$$

18. The pH of a solution of a weak acid is determined by the relative concentrations of the acid and its conjugate base. The equilibrium expression for the dissociation of a weak acid can be rearranged to

$$pH = pK + \log\frac{[A^-]}{[HA]}$$

This relationship is known as the Henderson–Hasselbalch equation. When the concentration of a weak acid is equal to the concentration of its conjugate base, pH = pK. Hence, the stronger the acid, the lower its pK (see Table 2-4).

19. Solutions of a weak acid at pH's near its pK resist large changes in pH as OH^- or H^+ is added. Added protons react with the weak acid's conjugate base to reform the weak acid; whereas added OH^- combines with the acid to form its conjugate base and water. A solution of a weak acid and its conjugate base (in the form of a salt) is referred to as a buffer. Buffers are effective within 1 pH unit of the pK of the component acid.

Key Equation

Know how the use the Henderson–Hasselbalch equation:

$$pH = pK + \log \frac{[A^-]}{[HA]}$$

Questions

Physical Properties of Water

1. Draw a 3-dimensional structure of a water molecule, including any lone pairs of electrons, and indicate the dipole moment. What is the name of the geometrical figure you have drawn?

2. Which gas in each of the following pairs would you expect to be more soluble in water? Why?

    (a) oxygen and carbon dioxide
    (b) nitrogen and ammonia
    (c) methane and hydrogen sulfide

3. Why do bottles of beer break when placed in a freezer?

4. Distinguish between hydrophilic, hydrophobic, and amphipathic substances.

5. Mixing olive oil with vinegar creates a salad dressing that is an emulsion, a mixture of vinegar with many tiny oil droplets. However, in a few minutes, the olive oil separates entirely from the vinegar. Describe the changes in entropy that occur during the initial mixing and the subsequent separation of the olive oil and vinegar.

6. Shown on the next page is a beaker that contains a solution of 0.1 M NaCl. Floating inside is a cellulose bag that is permeable to water and small ions like Na^+ and Cl^- but is impermeable to protein. The bag contains 1 M NaCl and 0.1 M protein. Describe the movement of solutes and solvent.

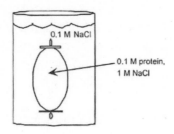

Chemical Properties of Water

7. In the most general definition of an acid and a base, a Lewis acid is a compound that can accept an electron pair, and a Lewis base is a compound that can donate an electron pair. Which of the following compounds can be classified as Brønsted–Lowry acids and bases and which as Lewis acids and bases?

$$CO_2, BF_3, \ NH_3, \ Zn^{2+}, \ HCOOH, \ NH_4^+, \ OH^-, \ Cl^-, \ H_2O$$

8. Define pK and pH, and write the Henderson–Hasselbalch equation that relates the two.

9. The pK's of trichloroacetic acid and acetic acid are 0.7 and 4.76, respectively. Which is the stronger acid? What is the dissociation constant of each?

10. What is the $[OH^-]$ in a 0.05 M HCl solution? What is the pH?

11. A solution of 0.1 M HCl has a pH of 1. A solution of 0.1 M acetic acid has a pH of 2.8. How much 1 M NaOH is needed to titrate a 100 mL sample of each acid to its respective equivalence point? Hint: The OH^- from strong bases such as NaOH reacts completely with the protons from both strong acids such as HCl and weak acids such as acetic acid.

12. A 0.01 M solution of a weak acid in water is 0.05% ionized at 25°C. What is its pK?

13. What is a buffer? How does it work? What compounds act as buffers in cells?

14. What is the pH of a 0.1 M solution of cacodylic acid (pK = 6.27)?

15. How would the pH of a 0.1 M solution of acetic acid be affected by the addition of a 4.5 M solution of sodium acetate (NaOAc)? NaOAc is a relatively strong base (weak acids have strong conjugate bases).

16. A 0.02 M solution of lactic acid (pK = 3.86) is mixed with an equal volume of a 0.05 M solution of sodium lactate. What is the pH of the final solution?

17. You need a KOAc solution at pH 5, which is 3 M in K^+. Such a solution is used in bacterial plasmid DNA isolation. How many moles of KOAc and acetic acid (HOAc) do you need to make 500 mL of this solution?

18. What are the predominant phosphate ions in a neutral pH phosphate buffer?

19. A beaker of pure distilled water sitting out on your lab bench is slightly acidic. Explain. *Hint*: See Box 2-1.

20. What is the pH of a solution of 10^{-12} M HCl? Explain.

21. Mammalian erythrocytes contain the enzyme carbonic anhydrase, which catalyzes the equilibrium $CO_2 + H_2O \leftrightharpoons HCO_3^- + H^+$ with no carbonic acid intermediate. Hence, we have the following acid–base equilibrium:
$$CO_2 + H_2O \leftrightharpoons HCO_3^- + H^+$$

    (a) Describe how this equilibrium is affected in the lungs. What happens to blood pH in the lungs?

    (b) What process decreases pH in the peripheral tissues?

Answers to Questions

1.

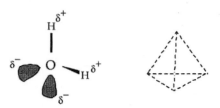

    The molecule forms a tetrahedron.

2. (a) Carbon dioxide (O=C=O), which is more polarizable than oxygen (O=O). (b) Ammonia (NH_3), which is more polar than nitrogen (N≡N). (c) Hydrogen sulfide (HS), which is more polar than methane (CH_4).

3. Water expands upon freezing. Full bottles of beer have little room for expansion and the frozen liquid expands beyond the boundaries of the glass and cap. Similarly, bottles of wine lose their corks in the freezer.

4. Hydrophilic substances are polar compounds that are readily soluble in water. Hydrophobic substances are nonpolar compounds that are insoluble in water. Amphipathic substances have both polar and nonpolar segments; they form micelles or bilayers in aqueous solution in which the polar groups face the water and the nonpolar groups exclude water and face each other (see Figure 2-11).

5. When the olive oil and vinegar (which is an aqueous solution) are mixed, the entropy of the solution decreases (the action of mixing is an input of energy that drives this process). Mechanical mixing results in an ordering of water around the thousands of oil droplets. Once the mixing has stopped, the system moves toward equilibrium, which maximizes entropy. As the oil droplets fuse, the surface-to-volume ratio of all of the oil decreases, thereby decreasing the amount of ordered water. Consequently, the entropy of the solution increases.

6. Na^+ and Cl^- ions diffuse out of the bag down their concentration gradient across the cellulose membrane. Water moves into the bag via osmosis since its concentration inside the bag is relatively lower due to the presence of the protein that cannot traverse the membrane (and initially due to the higher [NaCl] inside the bag—which eventually equalizes with the [NaCl] outside the bag). The bag expands as water enters. This setup is called dialysis and is used to remove or add salts to solutions of macromolecules.

7. CO_2, BF_3, Zn^{2+}, $HCOOH$, NH_4^+, and H_2O are Lewis acids; however, only $HCOOH$, NH_4^+, and H_2O are Brønsted–Lowry acids. NH_3, OH^- and Cl^- are considered bases under both definitions.

8. pK is the negative logarithm of the dissociation constant of a weak acid. pH is the negative logarithm of the [H^+]. The Henderson–Hasselbalch equation is

$$pH = pK + \log \frac{[\text{conjugate base}]}{[\text{acid}]}$$

9. pK is the negative logarithm of the dissociation constant, K, so the smaller the value of pK, the larger the value of K. The larger the value of K, the greater the dissociation of the acid in water. Therefore, trichloroacetic acid is the stronger acid since it has a lower pK than acetic acid.

For trichloroacetic acid, p$K = -\log K$

Therefore, $K = 10^{-0.7} = 0.2$

Similarly for acetic acid,

$$\log K = -4.76$$
$$K = 10^{-4.76} = 1.74 \times 10^{-5}$$

10. HCl is a strong acid and completely dissociates in water, so [HCl] = [H⁺] = 0.05 M. The
     pH is given by

$$pH = -\log [H^+]$$
$$= -\log 0.05$$
$$= 1.3$$

Since $[H^+][OH^-] = K_w$,

$$[OH^-] = pK_w / [H^+]$$
$$= 10^{-14} / 0.05$$
$$= 2 \times 10^{-13} \text{ M}$$

11. The amount of NaOH needed to titrate equivalent amounts of all monovalent acids is the
     same for both acid solutions. In both cases, 10 mL of 1 M NaOH will bring the titration
     of the acid to its end point. (Note: 10 mL of 1 M NaOH provides 0.01 moles of OH⁻ to
     neutralize the 0.01 moles of H⁺.)

12. The equilibrium constant for a weak acid is

$$K = \frac{[H^+][A^-]}{[HA]}$$

The concentration of each of the dissociated ions is 0.05% of 0.01 M, or
0.0005×0.01 M = 5×10^{-6} M. Because only 0.05% of the acid is ionized, we can assume
that [HA] does not change. We then calculate

$$K = \frac{(5 \times 10^{-6})(5 \times 10^{-6})}{0.01} = 2.5 \times 10^{-9}$$

$$pK = -\log (2.5 \times 10^{-9}) = 8.6$$

13. A buffer is a solution of a weak acid and its conjugate base. A buffer works by
     "absorbing" base or acid equivalents within about one pH unit of its pK. For example,
     when a small amount of OH⁻ is added, it reacts with HA to form water and A⁻, with little
     change in pH. Similarly, a small amount of H⁺ reacts with A⁻ to form HA. In biological
     systems, protons produced during catabolic reactions must be buffered by the cell.
     Phosphate and carbonate ions, as well as proteins, nucleic acids, and fatty acids, serve as
     buffers in cells.

14. Since $pK = 6.27$, $K = 10^{-6.27} = 5.37 \times 10^{-7}$. For the dissociation of cacodylic acid

$$\text{cacodylic acid} \leftrightharpoons H^+ + \text{cacodylate}^-$$

we can approximate [H⁺] = [cacodylate⁻] = x.

We approximate [cacodylic acid] by $0.1 - x$. Therefore,

$$K = \frac{[H^+][cacodylate^-]}{[cacodylic\ acid]}$$

$$5.37 \times 10^{-7} = \frac{x^2}{0.1 - x}$$

$$0 = x^2 + (5.37 \times 10^{-7})x - (5.37 \times 10^{-8})$$

Solve for x using the quadratic equation

$$x = \frac{-b \pm \sqrt{b^2 - 4ac}}{2a}$$

$$= \frac{-5.47 \times 10^{-7} \pm \sqrt{(5.47 \times 10^{-7})^2 + [4(5.47 \times 10^{-8})]}}{2}$$

$$x = -2.32 \times 10^{-4}\ or\ 2.31 \times 10^{-4}$$

The result can only be 2.31×10^{-4} since there is no such thing as a negative concentration.

Therefore, $x = [H^+] = 2.31 \times 10^{-4}$ M and

pH = 3.64

Note that for weak acids with $pK > 5$, one can approximate the pH by ignoring the change in acid concentration due to its dissociation. Thus, [cacodylic acid] = 0.1 M and

$$5.37 \times 10^{-7} = \frac{x^2}{0.1}$$

Therefore, $x = 2.32 \times 10^{-4}$ M

15. The added acetate anion reacts with the H^+ present in the solution to form undissociated acid, thereby causing a decrease in $[H^+]$ and an increase in pH.

16. The final concentrations are [lactic acid] = 0.01 M and [lactate] = 0.025 M. Use the Henderson–Hasselbalch equation to calculate the pH:

$$pH = pK + \log\frac{[lactate]}{[lactic\ acid]}^{[K15]}$$
$$= 3.86 + \log 2.5$$
$$= 3.86 = 0.398$$
$$= 4.26$$

17. Use the Henderson–Hasselbalch equation to calculate the concentration of acetic acid (x) necessary to obtain a solution of pH 5 that contains 3 M K^+ (3 M KOAc). The pK of acetic acid is 4.76 (Table 2-4).

$$pH = pK + \log\frac{[KOAc]}{[HOAc]}$$

$$5 = 4.76 + \log\frac{3}{x}$$

$$_{[K17]}0.24 = \log 3 - \log x$$

$$0.24 - 0.48 = -\log x$$

$$0.24 = \log x$$

$$x = 1.73 \text{ M}$$

Therefore, for a 500 mL solution, the amount of acetic acid added should be 1.73/2 or 0.86 moles. The amount of KOAc should be 3/2 or 1.5 moles.

$$x = \frac{3}{100.24}$$

$$= 1.73 \text{ M}$$

18. At pH 7, $H_2PO_4^-$ is in equilibrium with HPO_4^{2-} at nearly a 1:1 ratio since pH 7 is near pK_2 (6.82). See Figure 2-18 and Table 2-4.

19. Atmospheric carbon dioxide reacts with the distilled water in the beaker to form carbonic acid. The carbonic acid dissociates into protons and bicarbonate with a pK of about 6. These protons make the distilled water slightly acidic.

$$CO_2 + H_2O \leftrightarrows H_2CO_3$$

$$H_2CO_3 \leftrightarrows HCO_3^- + H^+$$

20. The pH is 7. Note here that the dissociation of water ($H_2O \leftrightarrows OH^- + H^+$), produces a concentration of 10^{-7} M H^+, compared to the added 10^{-12} H^+ from HCl. Hence, HCl at this concentration contributes negligibly little H^+.

21. (a) The low ambient partial pressure of CO_2 in air coming into the lung alveoli provides a steep concentration gradient across the endothelium of the alveolar and vascular lining for the incoming blood, where the partial pressure of CO_2 is close enough to its solubility limit to favor diffusion of CO_2 into the lung air space. The loss of CO_2 from the left-hand side of this equilibrium will favor the reaction shifting to the left to re-establish equilibrium. Hence, the $[H^+]$ will decline and the pH of the blood will rise (from 7.2 to about pH 7.4).

    (b) Respiration in the tissues produces CO_2, such that the elevation of CO_2 in the capillaries will drive this equilibrium to the right to re-establish equilibrium, thereby decreasing the pH of the blood (back to 7.2).

3
Nucleotides, Nucleic Acids, and Genetic Information

This chapter introduces you to the structure and function of nucleotides and their polymers, ribonucleic acid (RNA) and deoxyribonucleic acid (DNA). The chapter begins with a discussion of the various kinds of nucleotides and the large variety of their functions in cellular processes. The nucleic acid polymers, RNA and DNA, are the primary players in the storage, transmission, and decoding of the genetic material. Scientists use a variety of powerful techniques to characterize and manipulate DNA from any organism. This chapter discusses the sequence-specific cleavage of DNA by restriction endonucleases; DNA sequencing; amplification of DNA by cloning in unicellular organisms such as bacteria and yeast; and the *in vitro* amplification of DNA by the polymerase chain reaction.

Essential Concepts

Nucleotides

1. Of the four major classes of biological molecules (amino acids, sugars, lipids, and nucleotides), nucleotides are the most functionally diverse. They are involved in energy transfer, catalysis, and signaling within and between cells, and are essential for the storage, decoding, and transmission of genetic information.

2. Nucleotides are composed of a nitrogenous base linked to a ribose sugar to which at least one phosphate group is attached. The eight common nucleotides, which are the monomeric units of RNA and DNA, contain the bases adenine, guanine, cytosine, thymine, and uracil.

3. The best known nucleotide is the energy transmitter adenosine triphosphate (ATP), which is synthesized from adenosine diphosphate (ADP). Transfer of one or two of the phosphoryl groups of ATP is an exergonic process whose free energy can be used to drive an otherwise nonspontaneous process.

Introduction to Nucleic Acid Structure

4. Nucleic acids are polymers of nucleotides in which phosphate groups link the 3′ and 5′ positions of neighboring ribose residues. This linkage is called a phosphodiester bond because the phosphate is esterified to the two ribose groups. The phosphates are acidic at biological pH and so the polynucleotide is a polyanion.

5. Nucleic acids are inherently asymmetric, so that one end (with a 5′ phosphate) is different form the other end (with a 3′ hydroxyl). This asymmetry, or polarity, is critical for the information storage function of nucleic acids. In fact, all linear biological molecules show this kind of polarity.

6.   The structure of DNA was determined by Francis Crick and James Watson in 1953. Key information used to build their model included the following:
     (a) DNA has equal numbers of adenine and thymine residues and equal numbers of cytosine and guanine residues (called Chargaff's rules).
     (b) X-Ray diffraction studies (principally by Rosalind Franklin) revealed that the polymer is helical with a uniform width.
     (c) Structural studies had indicated that the nitrogenous bases should assume the keto tautomeric form.
     (d) Chemical evidence indicated that the polymer was linked by phosphodiester bonds between the 3′ and the 5′ carbons of adjoining ribose units.

7.   The Watson–Crick model has the following major features:
     (a) Two polynucleotide chains wind around a central axis to form a right-handed helix.
     (b) The polynucleotide chains are antiparallel to each other.
     (c) The nitrogenous bases occupy the core of the double helix, while the sugar–phosphate chains are the backbones of DNA, running along the outside of the helical structure.
     (d) Adenine is hydrogen bonded to thymine to form a planar base pair (like the rung of a ladder). In a similar fashion, cytosine is hydrogen bonded to guanine.

8.   The complementary strands of DNA immediately suggest that each strand of DNA can act as a template for the synthesis of its complementary strand and so transmit hereditary information across generations.

9.   The DNA of an organism (its genome) is unique to each organism and, in general, increases in amount in rough proportion to the complexity of the organism. In eukaryotes, genomic DNA occurs in discrete pieces called chromosomes.

10. While single-stranded DNA is uncommon in cells, RNA (ribonucleic acid) is usually single-stranded. However, by complementary base pairing, RNA can form intrastrand double helical sections, which bend and fold these molecules into unique three-dimensional shapes.

Overview of Nucleic Acid Function

11. The link between DNA and proteins is RNA. DNA replication, its transcription into RNA, and the translation of messenger RNA into a protein constitute the central dogma of molecular biology: DNA makes RNA makes protein.

12. RNA has many functional forms. The product of DNA transcription that encodes a protein is called messenger RNA (mRNA) and is translated in the ribosome, an organelle containing ribosomal RNA (rRNA). In the ribosome, each set of three nucleotides in mRNA base pairs with a transfer RNA (tRNA) molecule that is covalently linked to a specific amino acid. The order of tRNA binding to the mRNA determines the sequence of amino acids in the protein. It follows that any alteration to the genetic material (mutation) manifests itself in a new nucleotide sequence in mRNA, which often results in the pairing of a different RNA and hence a new amino acid sequence.

13. Some RNAs have catalytic capabilities. In cells, the joining of two amino acids in the ribosome is catalyzed by rRNA.

Nucleic Acid Sequencing

14. Nucleic acid sequences provide the following:
     (a) Information on the probable amino acid sequence of a protein (but not on any posttranslational modifications to these amino acids).
     (b) Clues about protein structure and function (based on comparisons of known proteins).
     (c) Information about the regulation of transcription (DNA sequences adjoining amino acid coding sequences are usually involved in regulation of transcription).
     (d) Discoveries of new genes and regulatory elements in DNA.

15. The overall strategy for sequencing DNA or any other polymer is
     (a) cleave the polymer into specific fragments that are small enough to be fully sequenced.
     (b) sequence the residues in each segment.
     (c) determine the order of the fragments in the original polymer by repeating the preceding steps using a different set of fragments obtained by a different degradation procedure.

16. Nucleic acid sequencing became relatively easy with the following:
     (a) The development of recombinant DNA techniques, which allows the isolation of large amounts of specific segments of DNA.
     (b) The discovery of restriction endonucleases, which cleave DNA at specific sequences.
     (c) The development of the chain-termination method of sequencing.

17. Gel electrophoresis is a technique in which charged molecules move through a gel-like matrix under the influence of an electric field. The rate a molecule moves is proportional to its charge density, its size, and its shape. Nucleic acids have a relatively uniform shape and charge density, so their movement through the gel is largely a function of their size.

18. The most common method used to sequence DNA is called the chain-terminator or dideoxy method. This technique relies on the interruption of DNA replication *in vitro* by small amounts of dideoxynucleotides. These lack the 3′-hydroxyl group to which another nucleotide would normally be added during polymerization. The DNA fragments, each ending at a position corresponding to one of the bases, are separated according to size by electrophoresis, thereby revealing the sequence of the DNA.

19. Automated DNA sequencing has allowed scientists to obtain over 500,000 base pairs of DNA sequence information per day. This technology utilizes a different fluorescent dye that is attached to a primer in each of four different reaction mixtures of the chain-termination procedure. The four reaction mixtures are combined for gel electrophoresis, wherein each lane of the gel is scanned by a spectrophotometer that records the fluorescence along the lane from top to bottom. Each band, which corresponds to an oligonucleotide terminated by either ddATP, ddCTP, ddGTP, or ddTTP, is identified by one of the four characteristic fluorescence emissions (indicated by the green, red, black, and blue curves in Fig. 3-22).

20. Phylogenetic relationships can be evaluated by comparing sequences of similar genes in different organisms. The number of nucleotide differences roughly corresponds to the extent of evolutionary divergence between the organisms.

21. The human genome appears to contain ~23,000 protein-encoding genes, compared to ~6000 in yeast, ~13,000 in *Drosophila*, ~18,000 in *C. elegans*, and ~26,000 in *Arabidopsis*. Most proteins encoded in the human genome occur in all the other organisms sequenced to date. Up to 60% of the human genome is transcribed, and 50% of it consists of various repeated sequences, a few of which are repeated >100,000 times.

Manipulating DNA

22. Recombinant DNA technology makes it possible to amplify and purify specific DNA sequences that may represent as little as one part per million of an organism's DNA.

23. Cloning refers to the production of multiple genetically identical organisms from a single ancestor. DNA cloning is the production of large amounts of a particular DNA segment through the cloning of cells harboring a vector containing this DNA. Cloned DNA can be purified, after which it can be sequenced. If the DNA of interest is flanked by the appropriate regulatory sequences (which themselves can be introduced into the vector), then the cells harboring the cloned DNA will transcribe and translate the cloned DNA to produce large amounts of protein. The protein can be purified and used for a variety of analytical or biomedical purposes.

24. Organisms such as *E. coli* and yeast are commonly used to carry cloning vectors, which are small, autonomously replicating DNA molecules. The most frequently used vectors are circular DNA molecules called plasmids, which occur as one or hundreds of copies in the host cell. Other important vectors for DNA cloning are bacteriophage λ in *E. coli* and yeast artificial chromosomes (YACs) and bacterial artificial chromosomes (BACs). Both YACs and BACs can accommodate much larger pieces of foreign DNA than can plasmids.

25. The following strategy is typically used to clone a segment of DNA:
    (a) A fragment of DNA is obtained using restriction endonucleases that generate sticky ends. The fragment is then isolated for subsequent ligation to a vector that has been cut with the same restriction endonuclease. The vector contains two selectable genes: (i) one that allows for the selection of transformed bacteria (usually via antibiotic resistance); and (ii) another that allows for the identification of recombinant vectors (vector plus inserted DNA) among a population of transformed colonies.
    (b) The fragment is ligated to the vector using an enzyme called DNA ligase, which catalyzes the formation of phosphodiester bonds.
    (c) The recombinant DNA molecule is introduced into cells by a method called transformation. Cells containing the recombinant DNA are usually selected for by their ability to resist specific antibiotics.
    (d) Cells containing the desired DNA are identified by a second selectable marker, often via the interruption of an easily assayable gene such as β-galactosidase. These identified recombinant clones are then isolated and stored or allowed to grow in order to obtain large quantities of the recombinant DNA.

26. In most cases, the desired DNA exists in just a few copies per cell, so all the cell's DNA is cloned at once by a method called shotgun cloning to produce a genomic library. The challenge is then to use a screening technique to find the desired clone among thousands to millions in the library. Colony hybridization is a common screening technique in which cells harboring recombinant DNA are transferred from a master plate to a filter that traps the DNA after the cells have been lysed. The filters are then probed using a gene fragment or mRNA that binds specifically to the DNA of interest.

27. If some sequence information is available, a specific DNA molecule can be amplified using a powerful technique called the polymerase chain reaction (PCR). In PCR, a DNA sample (often quite impure) is denatured into two strands and incubated with DNA polymerase, deoxynucleotides, and two oligonucleotides flanking the DNA of interest that serve as primers for DNA replication by DNA polymerase. Twenty rounds of DNA replication and DNA denaturation (all of which are automated in simple computerized incubators) can increase the amount of the target sequence by a million-fold.

28. Individuality in humans and other organisms derives from their high degree of genetic polymorphism. For example, homologous human chromosomes differ in the number of tandemly repeated sequences of 2 to 7 bp known as short tandem repeats (STRs). A number of STR sites with multiple alleles in the human population are the basis of DNA fingerprinting. In this technique, the allelic forms at several sites are identified following electrophoresis of PCR-amplified DNA, with results that can identify an individual with high confidence.

29. DNA cloning can be used to obtain large amounts of protein, referred to as recombinant protein, in bacteria or other organisms (although posttranslational modifications unique to eukaryotes cannot be obtained in bacteria). Another important application is the ability to mutate the amino acid sequence in the DNA of interest at specific sites by a technique called site-directed mutagenesis. Another exciting application of DNA cloning is the ability to introduce modified or novel genes into an intact animal or plant, creating a transgenic organism. The transplanted gene is called a transgene. When this technique produces a therapeutic effect, it is referred to as gene therapy. Gene therapy also refers to the introduction of recombinant DNA into selected cells to obtain a therapeutic effect.

30. Recombinant DNA technology raises ethical considerations that must be evaluated before its use in medicine, forensics, and agriculture.

Key Equation

$$P = 1 - (1 - f)^N$$

Questions

Nucleotides

1. Indicate which of the following are purine or pyrimidine bases, nucleosides, or nucleotides:
   (a) uracil
   (b) deoxythymidine

(c) guanosine monophosphate
(d) adenosine
(e) cytosine
(f) guanylic acid
(g) UMP
(h) guanine

Introduction to Nucleic Acid Structure

2.   What key sets of data were used to build the Watson–Crick model of DNA?

3.   Define or explain the following terms: (a) antiparallel; (b) complementary base pairing.

4.   The base compositions of samples of genomic DNA from several different animals are given below. Which samples are likely to come from the same species?
(a) 27.3% T
(b) 29.5% G
(c) 13.1% C
(d) 36.9% A
(e) 22.7% C
(f) 19.9% T

Nucleic Acid Sequencing

5.   A certain gene in one strain of *E. coli* can be cleaved into fragments of 3 kb and 4 kb using the restriction endonuclease *Pst*I, but it is unaffected by *Pvu*II. In another strain of bacteria, the same gene cannot be cleaved with *Pst*I, but treatment with *Pvu*II yields fragments of 3 kb and 4 kb. Explain the probable genetic difference between the two strains.

6.   You have inserted a 5-kb piece of eukaryotic DNA into a 3.5-kb plasmid at the *Bam*HI site. When you digest the recombinant DNA with various restriction enzymes, you obtain the fragments listed below. Draw a circular map of the recombinant plasmid showing all the restriction sites. Is it possible to determine the orientation of the insert in the plasmid? If so, what information reveals this? If not, what restriction digest would you perform to determine the orientation of the insert?

Restriction enzyme	*Size of DNA fragments (kb)*
*Eco*RI and *Bam*HI	1, 1.25, 2.75, 3.5
*Pst*I and *Bam*HI	1.75, 3.25, 3.5
*Eco*RI, *Pst*I, and *Bam*HI	0.75, 1, 1.25, 2, 3.5
*Sal*I	8.5
*Sal*I and *Bam*HI	0.5, 3, 5
*Bam*HI	3.5, 5

7.   In order to separate different-sized fragments of purified DNA by agarose gel electrophoresis at pH 7.5, should one allow the DNA to migrate from the cathode (the – electrode) or from the anode (the + electrode)? Explain.

8.   T7 DNA polymerase is less sensitive to dideoxynucleotides than the *E. coli* DNA polymerase I fragment described in Section 3-4C.

(a)   For a given concentration of dideoxynucleotides, which enzyme is more likely to give a longer "ladder" on the sequencing gel?

(b) Which enzyme is best used for sequencing DNA close to the primer?

9.   You are interested in the sequence of a 15-base segment of a gene. Using the chain-terminator procedure, you obtain the following results after gel electrophoresis. What is the sequence of the gene segment?

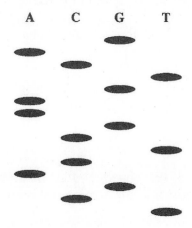

10.   An automated DNA sequencing system using chain-terminator sequencing technology can identify ~ 550,000 bases per day. A sequencing center has 15 such systems running simultaneously. How many days would it take to obtain a single read of human DNA on both strands (assuming no maintanence is required)?

11.   Assume that the average gene size in a prokaryotic genome (listed in Table 3-3) is about 1000 bases pairs in the smaller prokaryotic genomes and 1400 base pairs in the larger bacterial genomes. What is the average number of genes in the smallest and largest prokaryotic genomes shown in the table if all of the DNA in these organisms is part of a gene?

12.   Using the data in Table 3-3, compute the average chromosome length in humans. Assuming that the length of ten base pairs of DNA found in one turn of the helix is 34Å, what is the average length of a human chromosome in centimeters? Does your calculation lead to a surprising result?

Manipulating DNA

13.   Outline a procedure for identifying *E. coli* cells that contain a recombinant pUC18 plasmid in which a foreign gene has been inserted at the polylinker site.

14. The size of the human haploid genome is about 3.2×10^6 kb. Using human sperm DNA and bacteriophage λ as a cloning vector, you have made a human genome library containing 10^5 clones. Each clone has a 10-kb insert on average. What is the probability that a particular 10-kb single-copy gene is present in your library? Are you confident that you can find your gene with this library? Does a library with 10 times more clones significantly improve your chances? If not, why not?

15. The amplification of DNA by PCR is exponential for many rounds and then becomes linear. What factors are likely to become rate limiting so that rate of DNA accumulation becomes linear?

16. During PCR, which cycle of DNA replication results in a double-stranded, unit-length DNA product? How does the rate of accumulation of this product differ from that of the DNA product wherein only one strand is of unit length?

17. Many eukaryotic peptide hormones cloned in and isolated from bacteria have no biological activity. Why?

18. DNA synthesized using RNA as a template is known as cDNA. In the technique called subtractive hybridization cloning, small amounts of cDNA from all the mRNA of tissue A is hybridized to all the RNA from tissue B. The unhybridized cDNA is then cloned. What has been cloned?

19. Carefully examine Figure 3-30, showing the binding of primers during *in vitro* mutagenesis. During DNA replication *in vitro*, two populations of DNA will be produced. What are they? What problem is faced by the molecular biologist once she transforms bacteria with this DNA?

Answers to Questions

1. (a) Pyrimidine base; (b) pyrimidine nucleoside; (c) purine nucleotide; (d) purine nucleoside; (e) pyrimidine base; (f) purine nucleotide; (g) pyrimidine nucleotide; (h) purine base.

2. (i) Chargaff's rules, (ii) knowledge that nitrogenous bases are predominantly in the keto tautomeric form, (iii) X-ray crystallographic data indicating that DNA has a double-helical structure, (iv) X-ray crystallographic data suggesting that the planar aromatic bases form a stack parallel to the fiber axis.

3. (a) The nucleotides of a DNA strand are linked between the 3′ and 5′ ribose carbons by a phosphate group, thereby giving the strand polarity. The two strands of DNA form a duplex in which the strands run in opposite directions, that is, in an antiparallel arrangement.
   (b) Each nitrogenous base of one strand is hydrogen bonded to a different specific nitrogenous base on the opposing strand to form a planar base pair. This qualitatively specific hydrogen bonding between base pairs is called complementary base pairing. In the Watson–Crick structure, adenine specifically pairs with thymine, and guanine pairs with cytosine.

4.   Samples (a) and (e), since by Chargaff's rules, 27.3% T implies 27.3% A, so that G + C = 45.4% and C = 22.7%. Samples (c) and (d), since 13.1% C implies 13.1% G and 36.9% each of A and T.

5.   The gene in the first strain contains the sequence CTGCAG, which is recognized and cleaved by *Pst*I but not by *Pvu*II. In the second strain, the site has mutated so that it cannot be recognized by *Pst*I. Its sequence must be CAGCTG, since it is recognized and cleaved by *Pvu*II.

6.   Shown below is one of two possible restriction maps of the recombinant plasmid (in the other possible map, the *Sal*I site would be 0.5 kb from the other *Bam*HI site). To determine the orientation of the insert, it would be necessary to perform a restriction enzyme digest with *Sal*I and *Pst*I (or *Sal*I and *Eco*RI) to reveal the location of the *Sal*I site. For the map shown below, the *Sal*I/*Pst*I digestion products would be 2.25 kb and 6.25 kb.

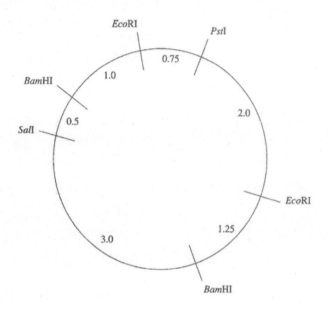

7.   At pH 7.5, purified DNA is negatively charged; hence, the DNA fragments will migrate from the cathode toward the anode.

8.   (a)   T7 DNA polymerase is likely to give a longer ladder, since it is less likely to terminate DNA chain elongation.
      (b)   The *E. coli* polymerase I fragment is better suited for obtaining sequences close to the primer, since it terminates DNA chain elongation more often and hence produces more short fragments than does T7 DNA polymerase.

9.   The bands on the gel identify the 3′ ends of DNA fragments synthesized by the DNA polymerase in the presence of dideoxynucleotides. Their sequence, read from bottom to top, is

5′-T-C-G-A-C-T-C-G-A-A-G-T-C-A-G- 3′

This is complementary to the DNA of interest, which has the sequence

$$5'\text{- C-T-G-A-C-T-T-C-G-A-G-T-C-G-A- }3'$$

10. The human genome contains 3.2×10^9 base pairs (bp), which means we need sequence information for 6.4×10^9 bases. So, with 15 systems producing 550,000 bases per day, we have 8.2×10^6 bases read every day.

    Therefore, it would take

    $$\frac{6.4 \times 10^9 \text{ bases in human genome}}{8.25 \times 10^6 \text{ bases per day}} = 775.75 \text{ days} = 2.13 \text{ years}$$

11. The smallest prokaryotic genome in Table 3-3 is *Mycoplasma genitalium* with 580,000 base pairs; the largest shown in the table is *Escherichia coli* with 4,649,000 base pairs. Dividing each of these genome lengths by 1000 base pairs or 1400 base pairs per gene, respectively, shows that one can expect about 580 genes in *M. genitalium* and 3,321 genes in *E. coli*.

12. The average chromosome length is

    $$\frac{3.2 \times 10^9 \text{ bp in human genome}}{23 \text{ chromosomes}} = 1.39 \times 10^8 \text{ bp per chromosome}$$

    The length of the DNA is

    $$(1.39 \times 10^8 \text{ bp})(34 \times 10^{-10} \text{ m/turn})/10 \text{ bp/turn} = 0..47 \text{ m} = 47 \text{ cm } !!$$

    So, try to imagine how 23 chromosomes with an average length of 47 cm of DNA are packaged into a human cell nucleus with an average diameter of only ~20 μm. Molecular biologists do not yet have a satisfying answer to this enormous packaging problem.

13. Bacteria containing the plasmids will be resistant to ampicillin, since the pUC18 vector contains the amp^R gene, and so can grow in the presence of ampicillin. Bacteria lacking the vector will not grow in the presence of ampicillin. Cells containing the foreign gene at the polylinker site, which interrupts the *lacZ* gene, will not produce any β-galactosidase and so can be detected as white colonies in the presence of a β-galactosidase substrate that turns blue when hydrolyzed. Cells containing the vector without the foreign DNA will appear blue because they express functional β-galactosidase.

14. The frequency (f) of the gene of interest is $10/(3.2 \times 10^6) = 3.12 \times 10^{-6}$. The probability ($P$) of finding this gene is

$$P = 1 - (1 - f)^N$$

where N is the number of clones. Hence, $P = 1 - (1 - 3.12 \times 10^{-6})^{100,000} = 1 - 0.73 = 0.27$. This low probability inspires no confidence that a single-copy gene can be easily found. The probability may be even lower, since not all genomic DNA fragments will be incorporated into the cloning vector with equivalent efficiency. Using the relationship above, the probability increases to 0.96 when the library contains 10^6 clones. Most investigators will shoot for a probability of 0.999 when constructing a recombinant DNA library to ensure the best chance of identifying a single-copy gene.

15. The finite amounts of nucleotides, primers, and enzyme will become rate-limiting. In the early rounds of PCR, these elements are at or near saturating levels (that is, their concentrations are much higher than that of the DNA template). Consequently, the rate of DNA production is directly proportional to the amount of template, which accumulates in an exponential fashion. But as the template DNA accumulates, its concentration approaches that of the enzyme, primers, and nucleotides. This slows the rate at which these elements come together to generate additional DNA, so the rate of DNA accumulation becomes linear.

16. Looking at the products shown in the second cycle (Figure 3-29), one sees a "hybrid-length" DNA in which one strand is of unit length and the other is much longer (and of varying length). This DNA molecule will dissociate into two strands in the third round, producing two products: a new "hybrid-length" DNA molecule and another DNA molecule in which each strand is unit length. This latter unit-length DNA duplex will accumulate exponentially in subsequent rounds. The "hybrid-length" molecule will accumulate in a linear fashion as a new pair is generated with the replication of each strand of the original strands of DNA.

17. These hormones require posttranslational processing that cannot be carried out in the bacteria.

18. What has been cloned is the cDNA corresponding to the mRNA sequences of tissue A that are absent from tissue B. The hybridization reaction eliminates the mRNA sequences shared by the tissues, including mRNAs encoding proteins required for common metabolic activities. The cloned cDNA therefore represents the genes that are active (i.e., that are transcribed into mRNA) in tissue A but not tissue B.

19. During in vitro DNA replication, one half of the population of DNA molecules will be the mutated version and another will be the original DNA sequence (the so-called wild-type sequence). Upon transformation, the replicating bacteria will form three populations (as shown in the rectangles below), where the fractions on the right show their contribution to the population. Note that only 50% of the population will carry the desired mutation.

Hence, the molecular biologist must collect many colonies and sequence them, with the anticipation that one-half will be the desired mutated form, or she must devise a vector or host that will selectively allow only the transformed bacteria with the mutated form to be easily recognized or survive. (Note that the heterozygous plasmid will express the mutant protein only if the DNA sequence is inserted in the proper orientation with respect to transcription.)

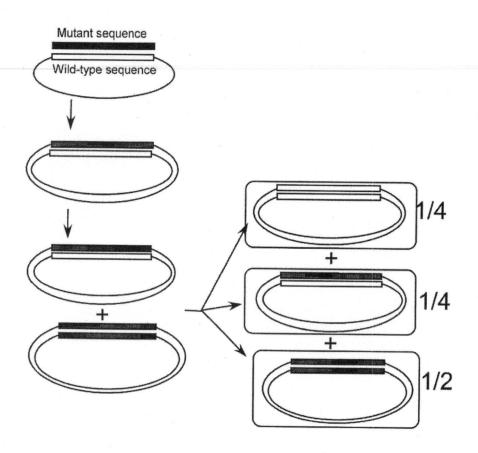

4

Amino Acids

This chapter introduces you to the structure and chemistry of amino acids. The chapter begins by discussing the zwitterionic character of amino acids at physiological pH, followed by a brief introduction to the amide linkage known as the peptide bond. Amino acids are categorized by the chemistry of their unique R groups and by their mode of biosynthesis. The standard amino acids are specified by the genetic code; nonstandard amino acids include the D stereoisomers and modified forms of the standard amino acids. This chapter also introduces you to the conventions used to describe stereoisomers, namely the Fischer convention and the *RS* system. Amino acids contain ionizable groups, so this chapter discusses the acid–base properties of amino acids and introduces you to the concept of the isoelectric point. A clear understanding of the acid–base properties of amino acids is critical for appreciating the structural and catalytic behavior of proteins. The chapter closes with a brief discussion of the biological significance of some amino acid derivatives.

Essential Concepts

Amino Acid Structure

1. All proteins are composed of 20 standard amino acids, which are specified by the genetic code.

2. The standard amino acids are called α-amino acids because they have a primary amino group and a carboxyl group bound to the same carbon atom (the α carbon). Only proline has a secondary amino group attached to the α carbon, but it is still commonly referred to as an α-amino acid.

3. The generic structure of an amino acid at pH 7 is shown below.

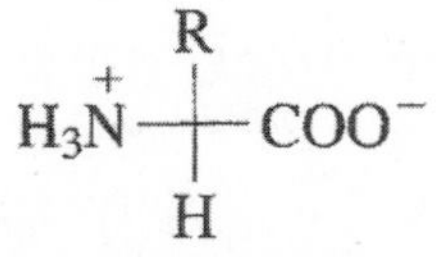

$$\overset{\displaystyle R}{\underset{\displaystyle H}{\overset{+}{H_3N} - \!\!\!\!\!\!\!\!\!\!- COO^-}}$$

    At pH 7, the amino acid is a zwitterion, or dipolar ion. A unique side chain, or R group, characterizes each amino acid.

4. Amino acids are polymerized by condensation reactions to form a chain called a polypeptide. Each polypeptide is polarized: One end has a free amino group and the other end has a free carboxyl group, referred to as the N-terminus and the C-terminus, respectively.

5.   The R groups of the standard 20 amino acids are classified into three categories based on their polarities and charge at pH 7: the nonpolar amino acids, the polar uncharged amino acids, and the charged amino acids (see Table 4-1).

6.   Among the nonpolar amino acids, glycine (shorthand Gly or G) has a hydrogen atom as its R group. Alanine (Ala; A), valine (Val; V), leucine (Leu; L), isoleucine (Ile; I), and methionine (Met; M) have aliphatic chains as R groups (Met has a sulfur rather than a methylene group). Tryptophan (Trp; W) and phenylalanine (Phe; F) contain bulky indole and phenyl groups, respectively.

7.   The polar uncharged amino acids include asparagine (Asn; N), glutamine (Gln; Q), serine (Ser; S), threonine (Thr; T), tyrosine (Tyr; Y), and cysteine (Cys; C). Amide functional groups occur in Gln and Asn. Alcoholic functional groups occur in Ser and Thr. Tyrosine and cysteine are characterized by a phenolic group and a thiol group, respectively.

8.   Among the charged amino acids, aspartate (Asp; D) and glutamate (Glu; E) contain carboxylic groups in their R groups. Lysine (Lys; K), arginine (Arg; R), and histidine (His; H) contain a butylammonium group, a guanidino group, and an imidazole group, respectively.

9.   The pK values of ionizable groups depend on the electrostatic influences of nearby groups. In proteins, the pK values of ionizable R groups may shift by several pH units from their values in the free amino acids.

Stereochemistry

10.   Except for glycine, the standard amino acids have asymmetric structures and rotate the plane of polarized light; thus, they are optically active. These molecules cannot be superimposed on their mirror images. Such nonsuperimposable pairs of molecules are called enantiomers. The asymmetric atom of an optically active molecule is called the chiral center and the molecule is said to have the property of chirality.

11.   Fischer projections are used to represent the absolute configuration of substituents around a chiral center. In the Fischer convention, horizontal lines extend above the plane of the paper, while vertical lines extend below the surface of the paper. The α-amino acid shown above is a Fischer projection of an L-amino acid. The specific arrangement of substituents around the chiral carbon is related to that of L-glyceraldehyde.

12.   A molecule with n chiral centers has 2^n different possible stereoisomers. Molecules with two or more chiral centers are better described by the *RS* system, in which each substituent bound to the chiral center is prioritized according to its atomic number. Hence, the exact molecular arrangement of a molecule can be unambiguously described.

13.   Biochemical reactions almost invariably produce pure stereoisomers, in large part due to the precise arrangement of chiral groups of enzymes, which restricts the geometry of the reactants.

Amino Acid Derivatives

14. Nonstandard amino acids in polypeptides arise from posttranslational modifications of the R groups of amino acid residues of the polypeptide. These modifications have critical roles in the structure and function of proteins. Many unpolymerized nonstandard amino acids are synthesized by chemical modifications of one of the standard amino acids. Cells use many of these amino acids as signaling molecules, particularly in the central nervous system. Among the nonstandard amino acids are also the D isomers of the standard amino acids; many of these occur in bacterial cell walls and bacterially produced antibiotics. The tripeptide glutathione is a cellular reducing agent.

Key Equation

$$pI = \tfrac{1}{2}(pK_i + pK_j)$$

Questions

Amino Acid Structure

1. Without consulting the text, draw a generic L-α-amino acid using the Fischer convention.

2. Examine the amino acids below. Assume that the pK of the carboxyl group attached to the α carbon is 2.0 and that of the primary amine attached to the α carbon is 9.5 The pK's of ionizable R groups are shown below each structure. (a) Categorize these amino acids as nonpolar, uncharged polar, or charged polar at pH 7. (b) Which of the structures cannot exist as shown at any pH in aqueous solution? (c) Name each of the amino acids.

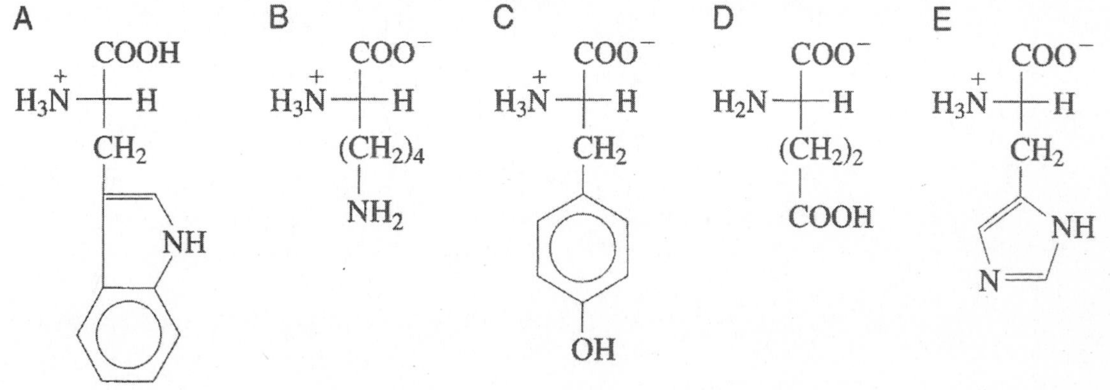

3. Which of the amino acids in Table 4-1 can be converted to another amino acid by mild hydrolysis that liberates ammonia?

4.   Which of the amino acids in Table 4-1 can generate a new amino acid by the addition of a hydroxyl group?

5.   The ionic characters of which amino acids are likely to be sensitive to pH changes in the physiological range? Explain.

6.   Circle the functional groups that are eliminated in the formation of a peptide bond between the amino acids shown below. Draw the structure of the dipeptide.

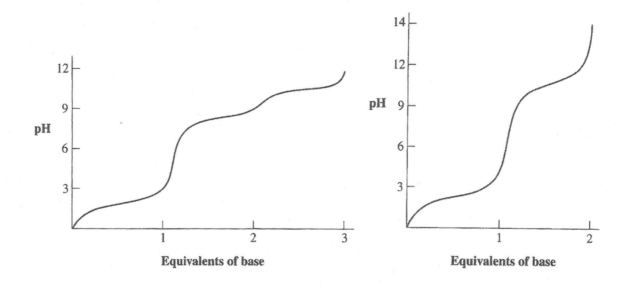

7.   What percentage of the histidine imidazole group is protonated at pH 7.2?

8.   What percentage of the cysteine sulfhydryl group is deprotonated at pH 7.6?

9.   In proteins, the imidazole groups of histidine play key roles in catalysis involving reversible protonation/deprotonation events. The pK of the imidazole group is influenced by surrounding amino acids and in many enzymes is near 7.
     (a)   What is the significance of this apparent pK in catalysis?
     (b)   Some unprotonated imidazole groups participate in hydrogen bonding a substrate to an enzyme. Do you expect the pK of these imidazole groups also to be near 7?

10.   Using the pK values in Table 4-1, identify the amino acids that have the titration curves shown below.

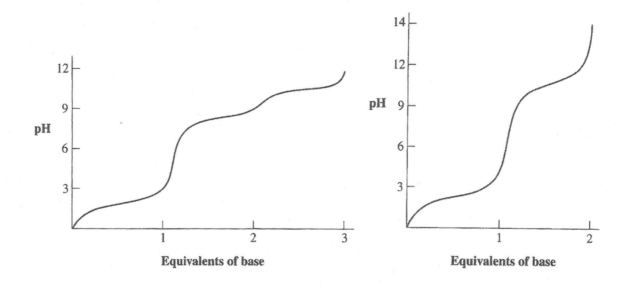

11. Which of the following tripeptides contain peptide bonds NOT commonly found in proteins? Provide a name for tripeptide B.

A

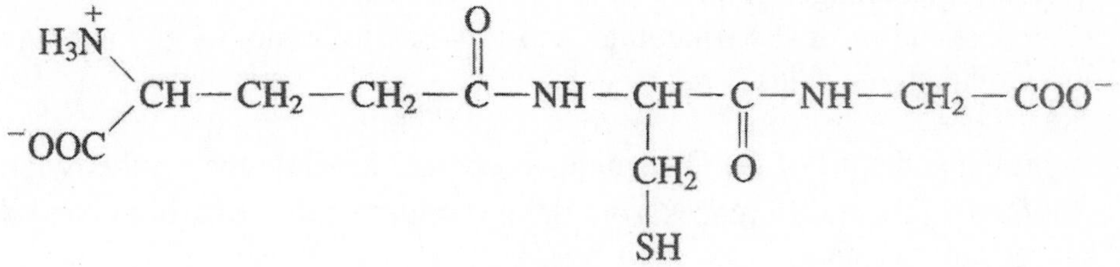

$$H_3\overset{+}{N}$$

$$CH-CH_2-CH_2-\overset{\overset{\displaystyle O}{\|}}{C}-NH-CH-\overset{\overset{\displaystyle O}{\|}}{C}-NH-CH_2-COO^-$$

$$^-OOC$$

$$CH_2$$

$$SH$$

glutathione

B

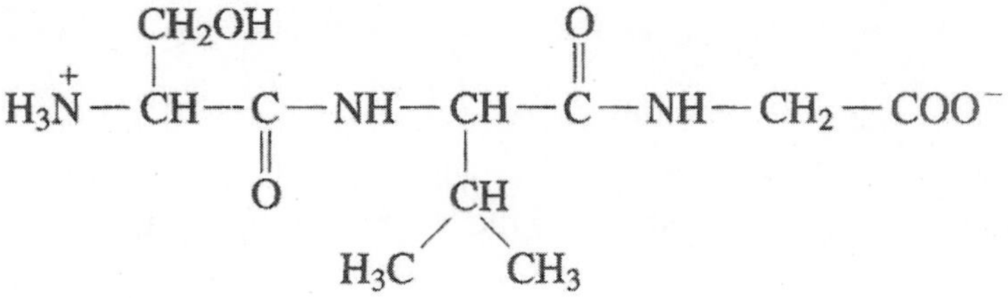

$$CH_2OH$$

$$H_3\overset{+}{N}-CH-\overset{\overset{\displaystyle}{|}}{C}-NH-CH-\overset{\overset{\displaystyle O}{\|}}{C}-NH-CH_2-COO^-$$

$$O$$

$$CH$$

$$H_3C \quad CH_3$$

C

$$COO^-$$

$$H_3\overset{+}{N}-CH_2-CH_2-\overset{\overset{\displaystyle}{|}}{C}-NH-CH-CH_2-C=CH$$

$$O$$

$$HN \quad N$$

$$CH$$

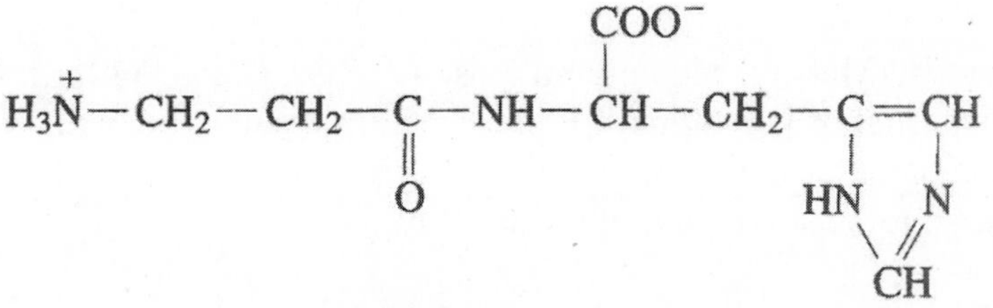

carnosine

D

$$H_3\overset{+}{N}-CH-CH_2-CH_2-CH_2-NH-CH-CH_2-SH$$

$$COO^-$$

$$C=O$$

$$NH$$

$$CH_2$$

$$COO^-$$

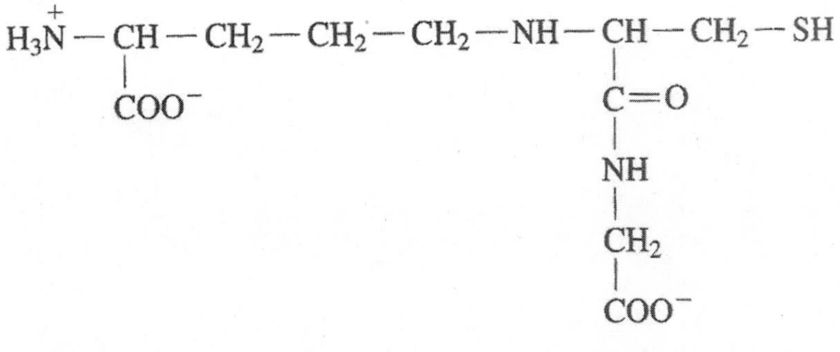

12. Why is the pK of the carboxyl group of glycine (pK = 2.3) less than that for acetic acid (pK = 4.76)?

13. Calculate the pI's of aspartate, lysine and serine.

14. Calculate the pI of the tripeptide Asp-Lys-Ser. Is it simply the average of the pI's of the individual amino acids?

15. In proteins, the pK of the C-terminus is about 3.8, while that of the N-terminus is about 7.8. Rationalize these differences from the pK values of the α-carboxyl and the α-amino groups in free amino acids.

16. Is a protein as good a *cellular* buffer at physiological pH as its constituent amino acids would be if they were present as free amino acids in proportional concentrations in the cell? Explain.

17. Based on your rationale in the previous question, describe the difference in the dissociation constants of the α-COOH of GABA (see page 88) and glutamate (Table 4-1), and the differences in the dissociation constants of the amino group in each of these amino acids.

18. 100 mg of anhydrous powder of lysine is not completely soluble in 10 mL of water but dissolves completely when base is added. Explain.

19. Examine the paper by McCoy, Meyer, and Rose (*J. Biol. Chem.* **112,** 283–302, 1935) at the Web site for the Journal of Biological Chemistry (http://www.jbc.org).
    (a) What was the name of Unknown II? What is its name today?
    (b) What remained to be solved about its structure?

20. Shown below is a titration curve for glutamic acid. Draw the structure of the species that predominates at each labeled point.

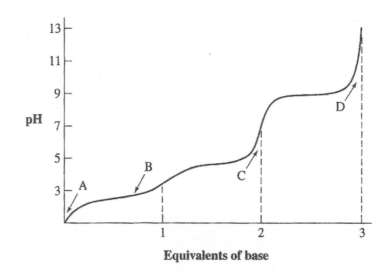

Equivalents of base

Stereochemistry

21. Which amino acid found in proteins has no optical activity?

22. Draw Fischer projections of L-aspartate and L-cysteine and show why L-Asp is (S)-Asp and L-Cys is (R)-Cys. (Refer to Box 4-2.)

23. Which amino acids in Table 4-1 have two or more prochiral centers? A prochiral center can be made chiral by substituting a different group for one of the two identical groups attached to it.

Amino Acid Derivatives

24. Which of the nonstandard amino acids below cannot occur in the interior of a polypeptide chain? Explain.

4-hydroxyproline   **N-formylmethionine**   **N,N,N-trimethylalanine**

25. From which of the standard amino acids are the following physiologically active amines derived? What modifications gave rise to these products? (See Figure 4-15 for structures.)
     (a) GABA
     (b) histamine
     (c) thyroxine
     (d) dopamine

26. How many conjugated double bonds occur in the fluorescent tyrosine derivative of green fluorescent protein?

27. Study the functional groups of the amino acids in Table 4-1 and devise a Venn diagram showing amino acids in the following categories: nonpolar, polar, negatively charged, positively charged, aliphatic, small R groups, and bulky R groups. (In a Venn diagram, objects with a given common property are enclosed by an oval. A set of objects may have several fully or partially overlapping ovals that represent different properties or the ovals may not overlap.)

Answers to Questions

1.

$$
\underset{R}{\overset{COO^-}{H_3\overset{+}{N}\!-\!\!\!-\!\!\!-H}}
$$

or

$$
\underset{H}{\overset{R}{H_3\overset{+}{N}\!-\!\!\!-\!\!\!-COO^-}}
$$

Remember, in Fischer projections, horizontal bonds extend out from the paper while vertical bonds extend behind the paper.

2. (a) Nonpolar: **A**; uncharged, polar: **C** and **E** (because **E** has a pK of 6.0, it is largely uncharged at pH 7); charged, polar: **B** and **D** (at pH 7, the ε-amino group of **B** is positively charged, and the γ-carboxyl group of **D** is negatively charged).
   (b) **B** and **D** cannot exist at any pH in aqueous solution. In **B**, the ε-amino group will become deprotonated at lower pH values than the α-amino group since its pK is lower. Similarly, the γ-carboxyl group of **D** is more acidic than the α-amino group.
   (c) **A** is tryptophan; **B** is lysine; **C** is tyrosine; **D** is glutamate; and **E** is histidine.

3. Glutamine and asparagine can be converted to glutamate and aspartate, respectively.

4. Alanine gives rise to serine, and phenylalanine gives rise to tyrosine.

5. Histidine and cysteine are the most likely to be sensitive in the physiological pH range since the pK values of their R groups are 6.04 and 8.37, respectively.

6.

7. Use the Henderson–Hasselbalch equation and rearrange terms to find the percentage of protonated species:

$$pH = pK + \log \frac{[A^-]}{[HA]} \text{[K11]}$$

$$\frac{[A^-]}{[HA]} = 10^{(pH - pK)} \text{[K12]}$$

$$= 10^{(7.2 - 6.04)}$$
$$= 10^{1.16} = 14.45$$

To obtain the percentage of the protonated species,

$$\% \, HA = \frac{[HA]}{[HA] + [A^-]} \times 100$$

$$= \frac{1}{1 + 14.45} \times 100$$

$$= 6.47\%$$

Hence, 6.47% is in the protonated form.

8. Use the Henderson–Hasselbalch equation and rearrange terms to find the percentage of deprotonated species:

$$pH = pK + \log \frac{[A^-]}{[HA]}$$

$$\frac{[A^-]}{[HA]} = 10^{(pH - pK)}$$

$$= 10^{(7.6 - 8.37)}$$
$$= 10^{-0.77} = 0.17$$

$$\% A^- = \frac{[A^-]}{[HA] + [A^-]} \times 100$$

$$= \frac{0.17}{1 + 0.17} \times 100$$

$$= 14.53\%$$

Therefore, about 14.53% of the sulfhydryl groups are deprotonated.

9. (a) The intracellular pH is also near 7, so about half the imidazole groups will be protonated. In other words, the imidazole groups can easily abstract a proton or give one up, as is required for catalysis, and remain unchanged at the end of the reaction.
   (b) It is more likely that the pK of these groups is near the pK of free histidine (6.04) so that they are available to form hydrogen bonds with potential substrates.

10. (a) Lysine; tyrosine is similar but its pK_2 is closer to 9.2.
    (b) Proline; it has only two pK's, one at about 2 and the other well above 10. The only other amino acid with a pK_2 above 10 is cysteine, but it has three pK's.

(b)

(a)

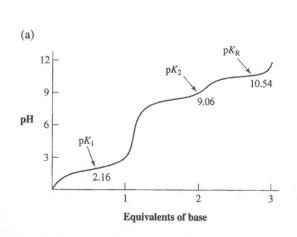

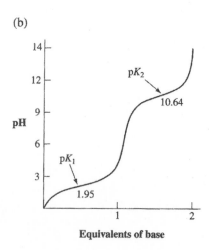

11.   A, C, and D contain peptide bonds not commonly found in proteins. In A, the γ-carboxylate group of the first amino acid participates in an amide linkage (the second and third amino acids are linked by a conventional peptide bond). C contains a linkage between the carboxylate of a β-amino acid and the amino group of an α-amino acid. In D, the side chain of the first amino acid is linked to the amino group of the second. Tripeptide B can be referred to as serylvalylglycine.

12.   The positively charged amino group of glycine electrostatically stabilizes the nearby carboxylate ion. Hence, the carboxylic acid group of glycine dissociates more readily to form the carboxylate ion than does the carboxylic acid group of acetic acid.

13.   The pI is approximately midway between the pK's of the two ionizations involving the neutral species. In amino acids with ionizable side chains, the relevant pK's may be pK_1 and pK_R or pK_R and pK_2. For aspartate, the neutral species results from ionization of the α-COOH (pK_1) and the β-COOH (pK_R):

pI = (pK_1 + pK_R)/2 = (1.99 + 3.90)/2 = 2.95.

For serine, which has no ionizable R group, pI = (pK_1 + pK_2)/2 = (2.19 + 9.21)/2 = 5.7.

For lysine, the pI lies between pK_2 and pK_R:

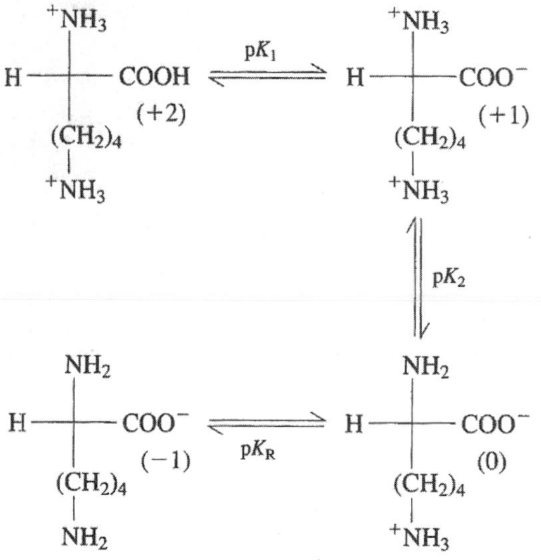

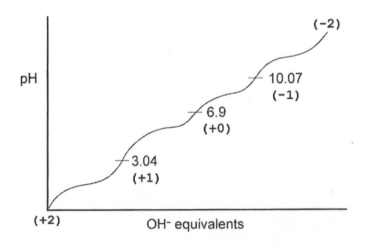

Hence, pI = (pK_2 + pK_R)/2 = (9.06 + 10.54)/2 = 9.8.

14. The pI of the tripeptide Asp-Lys-Ser cannot be precisely predicted and must be measured. However, one can use the pK values in Table 4-1 to make a good estimate. The pI of the tripeptide is 6.9, while the average of the three pI's of the individual amino acids is 6.15. The graph below shows the titration of this peptide, including the predicted charge at the beginning of the titration, when all ionizable groups are protonated, and at each equivalence point. Each equivalence point is estimated as the average between adjacent pK's. As you can see, when the peptide is fully protonated, it carries a net charge of +2. The complete ionization of the second carboxyl group in the side chain of the aspartyl residue occurs at pH 6.9, resulting in a net charge of 0.

15. In free amino acids, the neighboring α-carboxyl and α-amino groups affect each other's pK values. For instance, the negatively charged carboxyl group stabilizes the protonated amino group, making it a weaker acid and thus raising its pK value. Similarly, the positively charged amino group stabilizes the anionic carboxyl group, making it a stronger acid and thereby lowering its pK value. In proteins, the N-terminal and C-terminal residues participate in peptide bonds rather than having a free carboxyl group and a free ammonium group, respectively. Neutral peptide groups have little effect on the pK values of nearby ionizable groups.

16. Yes. The formation of the peptide bond between each amino acid in the intact protein eliminates two ionizable groups. However, at physiological pH (pH 5–9), the α-carboxyl and the α-amino groups of amino acids are poor buffers, because their respective pK values are far from the physiological pH range. However, at physiological pH, a protein containing a number of histidines and cysteines, for instance, may have a buffering capacity similar to that of an equimolar solution of its constituent amino acids; the side groups of these amino acids are free to ionize regardless of whether they are free or in a protein. Other ionizable R groups have pK's too far from physiological pH to be effective buffers.

17. The α-COOH of GABA does not come under the stabilizing influence of its amino group as it does in glutamate. Hence, its pK can be expected to be greater than that of the α-COOH of glutamate. Similarly, the amino group of GABA does not experience the stabilizing effect of the ionized form of its α-COOH and so has a lower pK than the cognate amino group in glutamate.

18. Lysine solid is of necessity neutral and therefore has the structure

$$
\begin{array}{c}
NH_2 \\
| \\
H - \!\!\!\!-\!\!\!\!-\!\!\!\!-\!\!\!\!-\!\!\!\!-\ COO^- \\
| \\
(CH_2)_4 \\
| \\
{}^+NH_3
\end{array}
$$

Neutral species are not as soluble as charged species. Adding base causes lysine's ε-NH_3^+ group to be converted to the neutral ε-NH_2, giving the molecule a net negative charge and higher solubility.

19. (a) α-Amino-β-hydroxybutyric acid (later named threonine)
     (b) The chirality of the β-carbon.

20.

A

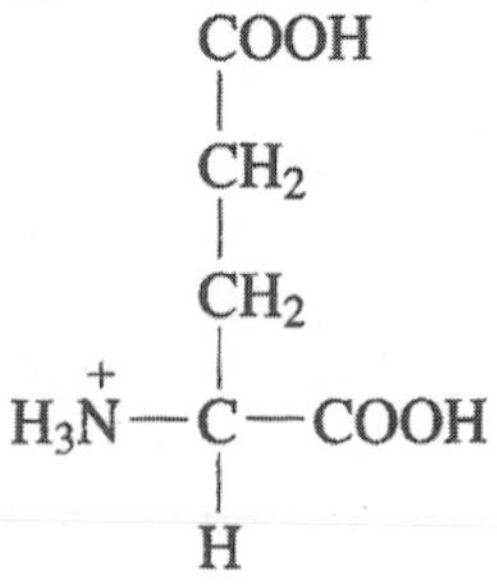

$$\begin{array}{c} COOH \\ | \\ CH_2 \\ | \\ CH_2 \\ | \\ H_3\overset{+}{N}-C-COOH \\ | \\ H \end{array}$$

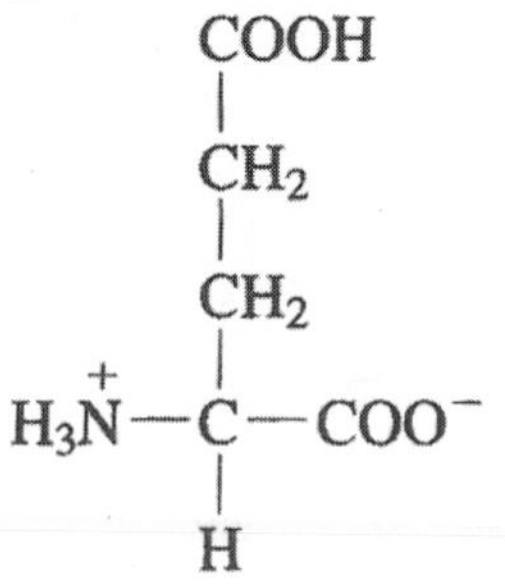

B

$$\begin{array}{c} COOH \\ | \\ CH_2 \\ | \\ CH_2 \\ | \\ H_3\overset{+}{N}-C-COO^- \\ | \\ H \end{array}$$

C

$$\begin{array}{c} COO^- \\ | \\ CH_2 \\ | \\ CH_2 \\ | \\ H_3\overset{+}{N}-C-COO^- \\ | \\ H \end{array}$$

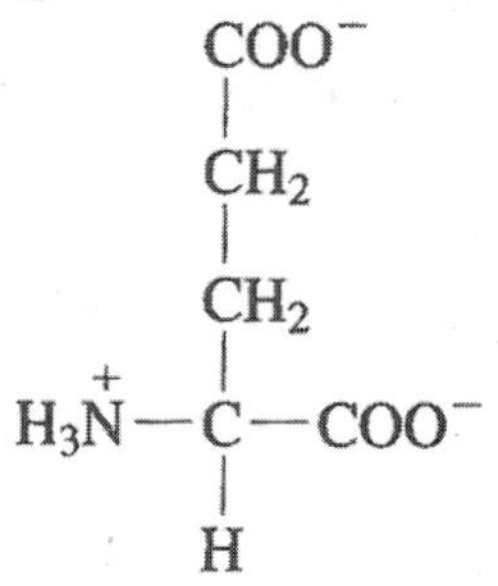

D

$$\begin{array}{c} COO^- \\ | \\ CH_2 \\ | \\ CH_2 \\ | \\ H_2N-C-COO^- \\ | \\ H \end{array}$$

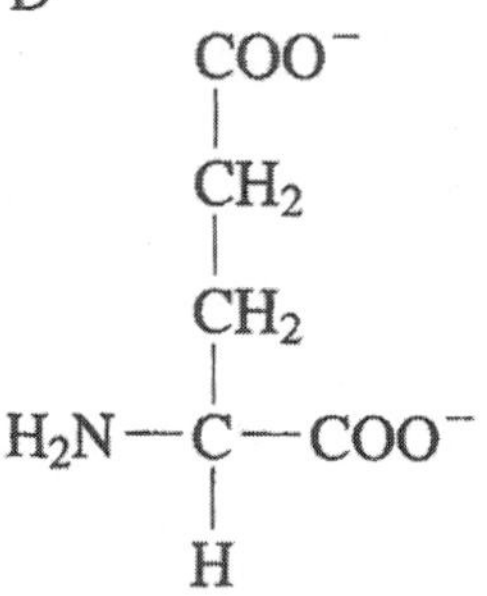

21.   Glycine; its C_α has two identical substituents, H.

22.   In a Fischer projection, the vertical bonds by convention point into the paper while the horizontal bonds point out from the paper. When the carbon groups are aligned vertically as drawn below, the amino group is on the left in an L-amino acid; it is on the right in a D-amino acid. To identify R or S configurations, swing the α-carbon hydrogen behind the α carbon. Then assign a priority number based on the atomic mass of the atom or group bound to the α carbon. The configuration is R or S according to whether the groups with decreasing priority have a clockwise or counterclockwise orientation, respectively. If two substituent atoms are the same (C, for example), the other atoms attached to them are used to assign priority. Note that for cysteine, sulfur has a higher atomic mass than oxygen, which reverses the priority of the groups. (See figure on next page.)

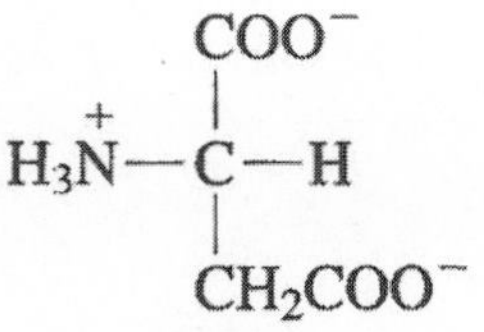

L-aspartate

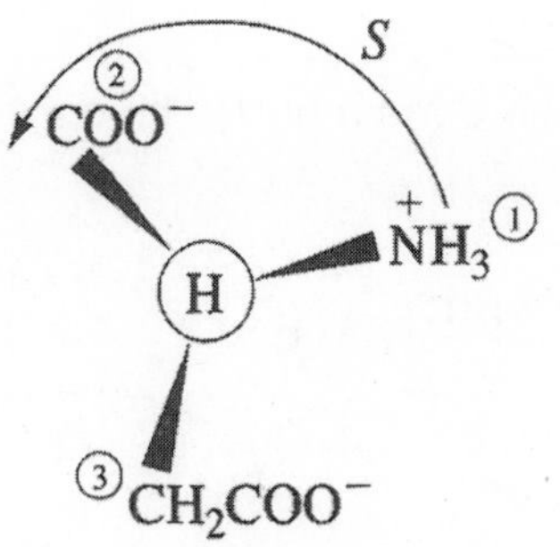

(S)-aspartate

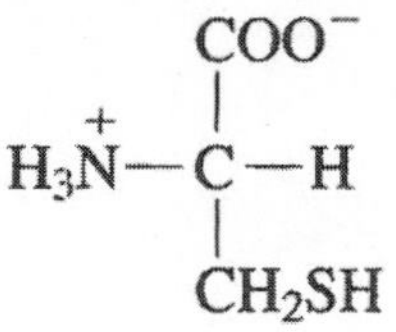

L-cysteine

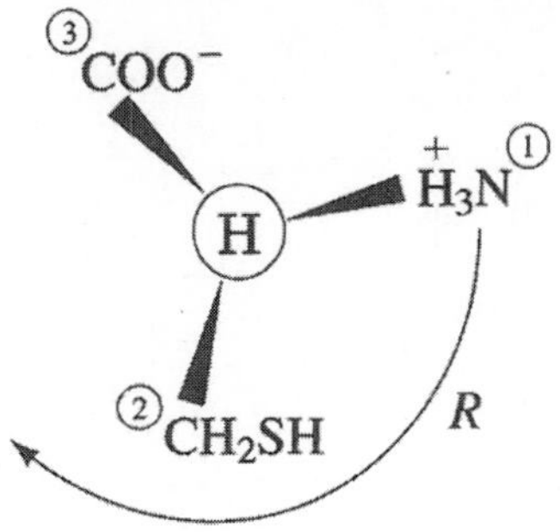

(R)-cysteine

23. The amino acids with two or more prochiral centers are Arg, Gln, Glu, Leu, Lys, Met, and Pro.

24. *N*-formylmethionine and *N,N,N*-trimethylalanine cannot occur in the interior of a polypeptide, because they lack free amino groups to form peptide bonds.

25. (a) GABA is derived from glutamate via decarboxylation of the α-carboxyl group.
    (b) Histamine is derived from histidine also by decarboxylation of the α-carboxyl group.
    (c) Thyroxine is derived from tyrosine by addition of a phenyl group and iodination.
    (d) Dopamine is derived from tyrosine via hydroxylation of the phenyl group and decarboxylation of the α-carboxyl group.

26. Six (see Box 4-3).

27.   Below is one possible Venn diagram showing the chemical and structural relationships between the R groups of the 20 genetically encoded amino acids in Table 4-1. Note that glycine is shown here as an outlier to all of the groups.

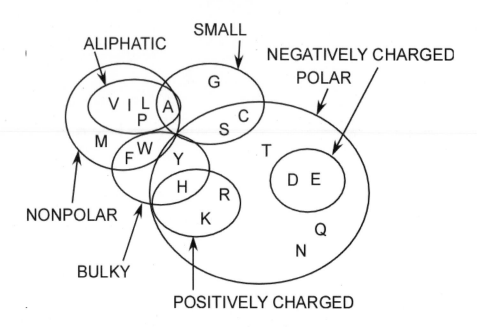

5

Proteins: Primary Structure

This chapter covers protein purification and primary structure. In this chapter you will learn how a protein's size, charge, and general shape can be analyzed and used to develop procedures to purify the protein. Many of the principles you learned in the preceding chapters about thermodynamics, aqueous solutions, and acid–base chemistry can be applied to the isolation of proteins. This chapter also includes a discussion of the strategies and chemical methods for determining the primary structure of proteins, focusing on Edman degradation and introducing mass spectrometry as a powerful technique for determining polypeptide mass and sequence. The chapter concludes with a section discussing protein evolution. Here you will see how comparisons of primary structures have led biochemists to categorize proteins into families and identify specific modules or motifs involved in specific functions.

Essential Concepts

Polypeptide Diversity

1. The primary structure of a protein is the amino acid sequence of its polypeptide chain (or chains if the protein contains more than one polypeptide).

2. Proteins are synthesized in cells by the stepwise polymerization of amino acids in the order specified by the sequence of nucleotides in its gene. The 5′ end of the messenger RNA corresponds to the amino terminus of the polypeptide, which is the end that contains a free amino group bound to the α carbon.

3. Although the theoretical possibilities for polypeptide composition and length are unlimited, polypeptides found in nature are limited somewhat in size and composition. Most polypeptides contain between 100 and 1000 residues and do not necessarily include all 20 genetically encoded amino acids. Leu, Ala, Gly, Ser, Val, and Glu are the most abundant amino acids in proteins, while Trp, Cys, Met, and His are the least common.

Protein Purification and Analysis

4. A general approach to protein purification requires the following:
   (a) A rapid and efficient method to disrupt cells so that the contents of the lysed cells (lysate) can be quickly stabilized in a buffer of appropriate pH and ionic strength.
   (b) Consideration of factors that affect the stability of the desired protein, such as pH, temperature, presence of degradative enzymes, adsorption to surfaces, and solvent conditions for long term storage.
   (c) Appropriate separation techniques to effectively purify the desired protein from the total proteins of the lysate.
   (d) A test or assay to easily assess the activity of the desired protein and hence measure its concentration at each step in the purification.

5.   A protein contains multiple charged groups, so its solubility in aqueous solution varies with the concentrations of dissolved salts (ionic strength), pH, and temperature. When salt is added to a protein at low ionic strength, the protein's solubility increases with increasing ionic strength (called salting in). At higher salt concentrations, the solubility of the protein decreases (called salting out). This variable solubility can be exploited to selectively precipitate desired or undesired proteins.

6.   Chromatography, which is used for analysis as well as purification, relies on the ability of a substance dissolved in a solvent (mobile phase) to interact with a solid matrix (stationary phase) as the solvent percolates through it. The chromatographic matrix is often in a column so that the solvent that elutes (exits) from it can be collected and assayed for the desired substance.

7.   Ion exchange chromatography takes advantage of a molecule's ionizable groups. The binding affinity of a substance for a cation or anion exchanger depends on its net charge and the pH and ionic strength of the solvent. The bound substance is eluted by increasing the salt concentration or changing the pH of the solvent.

8.   Hydrophobic interaction chromatography, which uses a matrix substituted with nonpolar groups, takes advantage of the exposed hydrophobic regions of a protein. The bound protein is eluted with a solvent that weakens the hydrophobic effects that cause the protein to bind to the matrix.

9.   Gel filtration chromatography separates molecules according to their size and shape. The stationary phase consists of beads containing pores that span a relatively narrow size range. Smaller molecules spend more time inside the beads than larger molecules and therefore elute later (after a larger volume of mobile phase has passed through the column). Within the molecular mass range that is fractionated by the specific stationary phase used, there is a linear relationship between a substance's elution time and the logarithm of its molecular mass.

10.   Affinity chromatography exploits a protein's specific ligand-binding behavior. The protein binds to a ligand that is covalently attached to the stationary phase and can be eluted with a solution containing a high salt concentration or excess ligand, which competes with the bead-bound ligand for binding sites on the protein.

11.   Electrophoresis separates molecules according to size and net charge. The molecules move through a polyacrylamide or agarose gel under the influence of an electric field. Proteins can be visualized by soaking the gel in a dye that binds to proteins or by electroblotting the proteins to nitrocellulose paper and probing with an antibody that is specific for the desired protein (Western blotting).

12.   In SDS polyacrylamide gel electrophoresis (SDS-PAGE), proteins are denatured with the detergent sodium dodecyl sulfate (SDS), which binds to a polypeptide and imparts a large negative charge by virtue of its negatively charged sulfate group. The net charge of an SDS-polypeptide is proportional to its length, giving all polypeptides the same charge density, so SDS-PAGE separates polypeptides almost entirely by sieving effects. Smaller

polypeptides migrate faster since they are less impeded by the cross-linked polyacrylamide and hence polypeptides are separated according to their molecular mass.

Protein Sequencing

13. Knowledge of a protein's amino acid sequence (its primary structure) is important for:
    (a) Determining its three-dimensional structure and elucidating its molecular mechanism of action.
    (b) Comparing sequences of analogous proteins from different species to gain insights into protein function and evolutionary relationships among the proteins and the organisms that produce them.
    (c) Developing diagnostic tests and effective therapies for inherited diseases that are caused by single amino acid changes in a protein.

14. Determining a protein's sequence may require several steps:
    (a) End group analysis reveals the number of different polypeptides or subunits in a protein. Either dansyl chloride or phenylisothiocyanate (Edman degradation) can be used to identify the N-terminal residue of a polypeptide.
    (b) Disulfide bonds within and between polypeptide chains are cleaved by reducing agents such as 2-mercaptoethanol, and the free sulfhydryl groups are then alkylated, e.g., with iodoacetate, to prevent re-formation of the disulfide bonds.
    (c) The different polypeptides of a multisubunit protein are separated so that each can be sequenced.
    (d) Polypeptides longer than 100 residues cannot be sequenced directly and must therefore be cleaved into smaller fragments using endopeptidases such as trypsin or chymotrypsin or chemical agents such as cyanogen bromide.
    (e) Each fragment is isolated and then sequenced using repeated cycles of the Edman degradation.
    (f) The order of sequenced fragments in the intact polypeptide is established by cleaving the polypeptide with a different reagent to obtain a second set of fragments whose sequences overlap those of the first.
    (g) The positions of disulfide bonds are identified by cleaving the polypeptide before it has been reduced. Each fragment containing a disulfide bond is reduced, and the resulting two peptides are separated and sequenced.

15. In Edman degradation, phenylisothiocyanate (PITC) reacts exclusively with the N-terminal amino group of a polypeptide chain. Treating the resulting PTC polypeptide with trifluoroacetic acid releases the N-terminal residue as a thiazolinone derivative, which is converted to a phenylthiohydantoin–amino acid that can be identified by chromatography. The procedure is then repeated for the newly exposed N-terminal residue of the polypeptide. This procedure has been automated and refined to the point where up to 100 residues can be sequenced with as little as 5 to 10 picomoles of protein (< 0.1 μg)!

16. Mass spectrometry is a powerful tool for polypeptide sequencing. In particular, electrospray ionization (ESI) allows for direct sequencing of short polypeptides (< 25 amino acid residues) by measuring the mass-to-charge ratio of ionic gas-phase peptides.

17.   Protein sequences are deposited in a number of public databases, most notably at the National Center for Biotechnology Information (NCBI; http://www.ncbi.nih.gov) or at the European site Expert Protein Analysis System (ExPASy: http://www.expasy.org).

Protein Evolution

18.   Mutations that alter a protein's primary structure may or may not affect the protein's function. Mutations that are deleterious to the organism are less likely to be passed on to the next generation. Mutations that enhance an organism's ability to survive and reproduce tend to propagate quickly. This is the essence of Darwinian evolution.

19.   Because related species have evolved from a common ancestor, the genes specifying each of their proteins have also evolved from a corresponding gene in that ancestor. Hence, the more closely related the two species, the more similar are the primary structures of their proteins. A phylogenetic tree can therefore be constructed by comparing the sequence divergence of proteins in different organisms.

20.   Comparisons of the primary structure of evolutionarily related proteins (homologous proteins) may indicate which amino acid residues are essential to the protein's function: Essential positions tend to contain invariant residues; less essential positions tend to contain chemically similar (conserved) residues; and positions that do not significantly affect protein structure or function can accommodate a variety of residues.

21.   The rates of evolution of proteins vary considerably. This reflects the ability of a given protein to accept amino acid changes without compromising its function.

22.   Large families of proteins such as the immunoglobulins or globins arose by gene duplication and subsequent primary structure divergence among the copies. Hence, the new copies are capable of evolving new functions.

23.   Many proteins contain amino acid sequence modules of about 40–100 residues that may have unique functions. Some modules are repeated many times in a single polypeptide. Hence, shuffling of modules is potentially a mechanism for rapidly generating new proteins with new functions.

24.   Most primary structures of proteins are known from sequencing the genes that encode them. However, DNA sequencing cannot provide information about the locations of disulfide bonds or posttranslational modifications of amino acids. In addition, some newly synthesized polypeptides are trimmed by site-specific cleavage to generate the mature functional protein. In practice, protein sequences are used to confirm nucleic acid sequences and *vice versa.*

Questions

Protein Purification and Analysis

1.   Discuss the disadvantages of each of the following in the handling of proteins:
     (a) Low concentrations of protein
     (b) Sudsing of protein solutions

(c) Lack of sterile conditions
(d) Absence of protease inhibitors
(e) Warm environments
(f) pH's far from neutrality

2.   List possible ways to track the presence of a protein during its purification.

3.   You have several cell lysates and want to determine which of them contain protein X in its phosphorylated form (with phosphorylation occurring on a tyrosine residue). You have two antibodies: One recognizes phosphotyrosine and the other recognizes protein X. Describe how you would analyze the cell lysates for the presence of phosphorylated protein X using just these two antibodies.

4.   The salting out procedure outlined in Fig. 5-5 results in significant protein purification by eliminating many less- and more-soluble proteins. What is its other advantage?

5.   You wish to separate two proteins in solution by selectively precipitating one of them. What method could you use other than salting out?

6.   The pI of pepsin is < 1.0, whereas that of lysozyme is 11.0. What amino acids must predominate in each protein to generate such pI's?

7.   Evaluate the solubility of the following peptides at the given pH. For each pair, which peptide is more soluble and why? *Hint:* Evaluate the ionization states of the amino acid functional groups.
     (a) pH 7: Gly_{20} and $(Glu-Asp)_{10}$
     (b) pH 5: $(Cys-Ser-Ala)_5$ and $(Pro-Ile-Leu)_5$
     (c) pH 7: Leu_3 and Leu_{10}

8.   Would you use an anion or cation exchange column to purify bovine histone at pH 7.0? (See Table 5-2.)

9.   Why is a DEAE ion exchange column ineffective above pH 9?

10.   When immunoaffinity chromatography is used to purify a protein, the cell lysate is often subjected to one or more purification steps before the material is applied to the immunoaffinity column. Why is this necessary?

11.   Describe two methods for determining the molecular mass of a polypeptide. Which is more accurate and why?

12.   Shown below are the biochemical fractionation and purification of an enzyme. Fill in the table on the next page with the values for specific activity and fold purification, and answer the accompanying questions. (nkat, nanokatal; one katal is the amount of enzyme that catalyzes the transformation of one mol of substrate per second.)
     (a) Which step provides the greatest purification?
     (b) Is the enzyme pure? How can purity be assessed?

	Total Protein (mg)	Activity (nkat)	Specific Activity (nkat/mg)	Fold Purification
Crude extract	50,000	10,000,000		
Ammonium sulfate precipitation	5,000	3,750,000		
DEAE-cellulose, stepwise KCl gradient	500	500,000		
DEAE-cellulose, KCl linear gradient	250	500,000		
Gel filtration	25	250,000		
Substrate affinity chromatography	1	100,000		

Protein Sequencing

13.   Match the terms below with the accompanying statements. (More than one term may apply to a statement.)

A. dansyl chloride   D. 2-mercaptoethanol
B. phenylisothiocyanate   E. cyanogen bromide
C. iodoacetate

_____   An alkylating agent that reacts with Cys residues.
_____   A reducing agent that carries out reductive cleavage of disulfide bonds.
_____   A reagent used to identify N-terminal residues.
_____   A reagent that cleaves polypeptides into smaller fragments.
_____   A reagent used in sequencing polypeptides from the N-terminus.

14.   How many fragments can be obtained by treating bovine insulin (Fig. 5-1) with endopeptidase V8? Does this number depend on whether the protein is first reduced and alkylated?

15.   You have a protein whose molecular mass is about 6.8 kD according to SDS-PAGE. However, SDS-PAGE of the protein in the presence of 2-mercaptoethanol reveals two bands of 2.3 kD and 4.5 kD. Are these polypeptides too large for direct sequencing by an automated sequencer using Edman degradation? What steps must you take to determine the complete sequence of this protein?

16. You have a polypeptide that has been degraded with cyanogen bromide and trypsin. The sequences of the fragments are listed below (using one-letter codes). Is there enough information to derive the complete amino acid sequence of the polypeptide? What further treatments might support the deduced sequence?

    Cyanogen bromide treatment:
      (1) **K**
      (2) **RFM**
      (3) **MLYCRGM**
      (4) **NIKGLM**

    Trypsin treatment:
      (5) **FMK**
      (6) **GMNIK**
      (7) **GLMR**
      (8) **MLYCR**

17. You want to know the amino acid sequence of a purified polypeptide (1341 D). Your new technician has performed the analyses outlined below. Do you have enough data to determine the amino acid sequence?
    (a) Partial acid hydrolysis yields tripeptides with the following amino acid sequences as determined by manual Edman degradation: **HSE, EGT, DYS, TSD, FTS, YSK.**
    (b) Dansyl chloride treatment yields **H** and **K**.
    (c) Brief carboxypeptidase A treatment yields one **K** per mole peptide (a carboxypeptidase removes one residue at a time from the C-terminus of a polypeptide).
    (d) Chymotrypsin treatment yields two peptides (A and B) with molecular masses of 659 D and 467 D, respectively, and a dipeptide composed of **S** and **K**.
        Peptide A
            (1) Dansyl chloride treatment yields **H**.
            (2) Brief carboxypeptidase A treatment yields **F** and **T**.
        Peptide B
            (3) Dansyl chloride treatment yields **T**.
            (4) Brief carboxypeptidase A treatment yields **Y** and **D**.
    (e) Trypsin treatment results in no fragmentation of the polypeptide.

18. Compare the molecular mass (M) of horse heart apomyoglobin calculated from the two adjacent peaks at 1884.7 and 1696.3, and 893.3 and 848.7 (Figure 5-16b). How do these results compare with those in Sample Calculation 5-1?

Protein Evolution

19. Explain why single-nucleotide mutations in DNA occur at a constant rate, but the rate of protein evolution varies.

20. Rank the following residue positions in cytochrome c in order from least to most conserved: 56, 61, 74, 85, and 89.

21. Which protein(s) shown in Figure 5-23 would be most useful for assessing the phylogenetic relationships between mammalian species (note that mammals diverged from reptiles about 300 million years ago)? Explain.

22. The sequences of two cAMP regulatory transcription factors were compared using the BLAST alignment program from NCBI. The results below show the sequences from

Escherichia coli (the Query sequence) and *Methylobacterium extorquens* (the Sbjct sequence). The BLAST "hit" compares residues 190–369 of the *E. coli* protein and residues 64–242 of the *M. extorquens* protein. Amino acid sequence identity is revealed by letters in the middle row; mismatches are spaces between the rows; and substitutions of similar amino acids are shown as a "+" in the middle row. Take a look at these "positives" as they are called in the BLAST output. Rationalize these positives based on your knowledge of R-group chemistry (recall the Venn diagram you made in Chapter 4).

```
Query   190   SGFACRHKLRASGARQINAYLLPGDLCDLDVALLDEMDHTITTLSTCTVVRLAPELIADL   249
              GFACR+KLR +GARQI AYL+PGD+CDLD   L+ MDHT+ TLSTC V R+ PE  A
Sbjct    64   EGFACRYKLRENGARQIMAYLVPGDVCDLDNGALNRMDHTVGTLSTCRVARIMPE-TARE   122


Query   250   LTHHPQIARALRKNTLVDEATLREWLMNVGRRSSVERMAHLFCELLLRFRAVGLAPEDSY   309
              L  HP +AR LRK  L+DEATLREWLMN+GRRS+VER+AHLFCELL+R RAVGL   +SY
Sbjct   123   LRQHPALARGLRKAALIDEATLREWLMNIGRRSAVERLAHLFCELLVRLRAVGLTSGNSY   182


Query   310   PLPLTQADLADTTGITSVHVSRSLKELRQGGLIELQGGRLKILDYARLRALAEFRANYLH   369
              LP+TQ DLADTTG+TSVHV+RSL EL++ GLIE +  RL + D ARL  +AEFR +YLH
Sbjct   183   ELPITQLDLADTTGMTSVHVNRSLGELKREGLIERKSKRLTLCDPARLAEVAEFRPDYLH   242
```

23   Comparisons between amino acid sequences can be used to understand which areas of a polypeptide are important for its function. A large family of proteins in bacteria includes a transcriptional regulator called the cAMP regulatory or activator protein (CRP or CAP).

   (i)   Go to the NCBI site at *http://www.ncbi.nlm.nih.gov/*

   (ii)   Retrieve the amino acid sequences for the following CRP/CAP family proteins by selecting the Protein database in the Search menu and typing the underlined accession number into the adjacent box:
   - *E. coli* catabolite gene activator (cAMP receptor protein), P0ACJ8
   - *Yersinia pestis* cyclic AMP receptor protein, AAM87500
   - *Klebsiella pneumoniae* cAMP receptor protein, A44903
   - *Shewanella* sp. MR-4 Crp/Fnr family transcriptional regulator, YP_734144
   - *Aeromonas hydrophila* fumarate and nitrate reduction regulatory protein YP_856826

   (iii)   For each retrieval, select "FASTA" from the display menu then copy and paste the amino acid sequence into a Word document, giving each sequence a recognizable byline in FASTA format: ">byline".

   (iv)   Remove the numbers from the amino acid sequence

   (v)   Go to the EMBL Clustal W site (http://www.ebi.ac.uk/Tools/clustalw/index.html), which will allow you to compare up to 10 different amino acid sequences.

   (vi)   Copy and paste the five sequences in FASTA format consecutively into the open box and hit RUN.

(vii) In a few moments, you will see an output that shows all five sequences lined up according to the best matches between them.

(a) Which parts of this protein show the most amino acid divergence?
(b) Which parts of this protein appear to be the most conserved?
(c) Go to http://www.ncbi.nlm.nih.gov/Structure/cdd/wrpsb.cgi at NCBI. Paste in the *E. coli* sequence to see what domains appear. Do the same with the *A. hydrophila* sequence. What areas of the polypeptide are identified with the two smaller, common domains (colored in red and blue)? Comparing these two domains between these species, which domain appears more divergent? Does it make sense relative to the information given about each domain (scroll the arrow over each domain to retrieve information)? The e-value is a measure of similarity; similar sequences yield low e-values.

Answers to Questions

1.
(a) Proteins at low concentrations can be adsorbed by glass or some plastic surfaces, thus significantly decreasing their recovery.
(b) Sudsing of protein solutions indicates increased contact with the air–water interface, which can denature the proteins.
(c) Proteins can be eaten or degraded by contaminating bacteria and fungi.
(d) Protease inhibitors are essential for limiting protein degradation by proteases released from lysosomes during cell lysis.
(e) Many proteins are only marginally stable and slowly denature over time at temperatures above 25°C.
(f) Protein structure as well as enzymatic activity are strictly dependent on pH.

2.
(a) If the protein is an enzyme, incubate it with a substrate that generates an easily detected (e.g., colored) product or a product that can be converted to a detectable product by a second enzyme.
(b) If an antibody to the protein is available, the protein can be detected using an immunoassay such as an RIA or ELISA.

3. Immunoaffinity chromatography, immunoassay, and Western blotting could work. One protocol would be to coat a surface with antibodies to protein X, then incubate it with a cell lysate to allow any protein X present to bind to the immobilized antibodies. Wash away unbound proteins, then incubate the surface-bound antibody–protein X complex with the anti-phosphotyrosine antibody to which a readily assayed enzyme has been attached. Wash away unbound antibody and assay the enzyme. The appearance of the enzyme's reaction product indicates the presence of phosphorylated protein X. Unphosphorylated protein X will not bind the second antibody, and lysates that contain no protein X will not cause the second antibody to bind. See Figure 5-3.

4. Following salting out, the precipitated protein can be redissolved in a very small volume, thus concentrating it. A highly concentrated protein solution may be necessary for a subsequent gel filtration step or other analytical procedure. Concentrated protein solutions are also easier to handle, and they minimize protein loss by adsorption to surfaces.

5. Adjust the pH of the solution so that it approaches the pI of one of the proteins. As that protein's net charge approaches 0, it will become less soluble and may precipitate, whereupon it can be eliminated by centrifugation or filtration. Alternatively, add a water-soluble organic solvent to selectively precipitate one of the proteins.

6. In pepsin, aspartate and glutamate must predominate because they have acidic R groups. Lysine, arginine, and tyrosine, all with basic R groups, must predominate in lysozyme.

7.

   (a) At pH 7.0, the side chains of Asp and Glu are ionized. Although both peptides contain ionized N- and C-terminal groups, $(Glu\text{-}Asp)_{10}$ contains more onized groups than Gly_{20} and is therefore the more soluble peptide.

   (b) Although Cys would be ionized at higher pH, neither peptide has ionized R groups at pH 5.0. However, $(Cys\text{-}Ser\text{-}Ala)_5$ has more polar groups and is therefore more soluble than $(Pro\text{-}Ile\text{-}Leu)_5$.

   (c) Leu_3 is less hydrophobic since it has two charged groups (its N- and C-termini) per three residues, compared to two charged groups per 10 residues in Leu_{10}. The shorter peptide is more soluble since it requires less disruption of water structure to dissolve.

8. Bovine histone is highly basic (pI = 10.8) and therefore positively charged at pH 7.0. It would therefore bind to the anionic matrix of a cation exchanger (e.g., carboxymethyl cellulose).

9. Above pH 9, the amino groups of the DEAE matrix lose their protons and are therefore unable to bind anionic substances.

10. Some pre-purification before immunoaffinity chromatography may be necessary because the cell lysate may contain other proteins that compete with the desired protein for binding to the immobilized antibody. Although such proteins would bind with lower affinity to the antibody, their greater numbers would diminish the amount of desired protein that binds specifically to the antibody.

11. Two methods for determining molecular mass are gel filtration chromatography and SDS-PAGE. SDS-PAGE is more accurate. Gel filtration is sensitive to the native shape of a protein, whereas SDS-PAGE is performed on an SDS-denatured polypeptide so that its folded shape is not an issue. Note that SDS-PAGE reveals the molecular masses of individual polypeptide chaina; gel filtration is appropriate for determining the mass of a protein consisting of multiple subunits.

12.

(a) The step that provides the greatest purification is the step that gives the largest fold purification from the previous step, or the greatest increase in specific activity from the previous step. In this example, the step is substrate affinity chromatography, with a 10-fold purification from the previous step.

(b) The enzyme may be pure, or there may yet be other proteins present. SDS-PAGE provides the most powerful technique for ascertaining whether the final protein preparation contains a single polypeptide. However, if SDS-PAGE reveals equivalent amounts of more than one polypeptide, it is possible that the native enzyme is a multimer of nonidentical subunits.

13.

C	An alkylating agent that reacts with Cys residues.
D	A reducing agent that carries out reductive cleavage of disulfide bonds.
A, B	A reagent used to identify N-terminal residues.
E	A reagent that cleaves polypeptides into smaller fragments.
B	A reagent used in sequencing polypeptides from the N-terminus.

14. If the insulin is reduced and alkylated to eliminate disulfide bonding between Cys residues, then endopeptidase V8, which cleaves after Glu residues, will generate six peptides. If the disulfide bonds between the two chains of insulin are intact, the peptidase will generate four fragments.

	Total Protein (mg)	Activity (nkat)	Specific Activity (nkat/mg)	Fold Purification
Crude extract	50,000	10,000,000	200	1
Ammonium sulfate precipitation	5,000	3,750,000	750	3.75
DEAE-cellulose, stepwise KCl gradient	500	500,000	1,000	5
DEAE-cellulose, KCL linear gradient	250	500,000	2,000	10
Gel filtration	25	250,000	10,000	50
Substrate affinity chromatography	1	100,000	100,000	500

15. Since the average mass of an amino acid residue is ~110 D, a peptide with a mass of 4500 D contains ~44 residues. Thus, both peptides are small enough to be directly sequenced. The two peptides are linked by a disulfide bond, so determining the protein's complete sequence requires the following steps: (a) reduce the disulfide bond; (b) alkylate the free thiol groups; (c) separate the peptides; and (d) sequence each peptide. If either peptide contains more than one Cys residue, additional fragmentation and sequencing steps would be necessary to identify the position of the disulfide bond in the intact protein.

16. Yes, there is enough information. The overlap of the relevant fragments is

MLYCRGM

GMNIK

NIKGLM

GLMR

RFM

FMK

Thus, the peptide sequence is **MLYCRGMNIKGLMRFMK**. To confirm the deduced sequence, perform one round of Edman degradation to verify that the N-terminus is methionine.

17. Since there are no cysteines, there is only one N-terminus in this polypeptide. Therefore, treatment (b) reveals that His or Lys must be at the N-terminus (the ε-amino group of Lys can also react with dansyl chloride). Treatment (c) indicates that there is a Lys at the C-terminus, and since trypsin treatment did not cleave the polypeptide, the **YSK** peptide must represent the C-terminal peptide. Since peptide A contains **F** and must end in **F** due to the specificity of chymotrypsin, **T** must precede **F**. We can now deduce the following:

HSE

EGT...

FTS

TSD

DYS

YSK

From the masses of the two peptides produced by chymotrypsin treatment, we know that peptide A is about 6 amino acids long (~110 D per residue) and peptide B is 4 amino acids long. Hence, the sequence of peptide A must be **HSEGTF** and that for peptide B **TSDY**. Since **YSK** is the C-terminal peptide, we can deduce that the sequence of the polypeptide is **HSEGTFTSDYSK**.

18.   The table below shows the percent difference in mass when examining different pairs of peaks. See Sample Calculation 5-1 for the determination of *M*.

m/z pair	*M*	% Difference
1884.7 / 1696.3	16,950.3 D	0.004
1542.3 / 1414.0	16,975.0 D	0.14
893.3 / 848.7	16,959.7 D	0.051

19.   The rate of protein evolution depends on the ability of a protein to accept mutations without significantly altering its function (neutral drift), which varies among proteins. Hence, proteins with slow rates of evolution cannot accept many amino acid substitutions without loss of function, thereby eliminating mutant forms by natural selection.

20.   From inspection of the last row of Table 5-5, it is seen that residue 56 has three different amino acids in the table, 61 has four, 74 has one, 85 has two, and 89 has nine. Hence their order, from least to most conserved is: 89, 61, 56, 85, 74.

21.   The fibrinopeptides are evolving too rapidly to show the more ancient relationships among mammals, since over periods greater than 100 million years, many residues will have mutated two or more times. Histone H4 has evolved too slowly to reveal variations in mammalian species (they all have nearly identical histone H4's). However, cytochrome *c* and hemoglobin evolve at rates that would be useful for assessing phylogenetic relationships among mammalian species, with hemoglobin, the faster-evolving protein of the two, providing a mores sensitive measure.

22.   In the comparison between these two proteins, the follow amino acid pairs are listed as positives:

   H:Y – Histidine and tyrosine
   S:N – Serine and glutamine
   L:V – Leucine and valine
   D:N – Glutamate and glutamine
   I:V – Isoleucine and valine
   L:I – Leucine and isoleucine
   S:A – Serine and alanine
   M:L – Methionine and leucine

   The following pairs are considered positives since they all have nonpolar R groups: L:V, I:V, L:I, and M:L. Similarity based on polarity can be seen in the S:N and D:N pairs being categorized as positives in BLAST. Similarity in size or bulkiness accounts for the H:Y and S:A pairings being scored as positives, with H and Y both being bulky, and S and A having small R groups.

23.   Shown below is an alignment file that can be obtained at the Clustal W site:

```
CAP_Ecoli                 -MVLGKPQTDPTLEWFLSHCH------------------------IHKY 24
CAP_Kpneumoniae           -MVLGKPQTDPTLEWFLSHCH------------------------IHKY 24
CAP_Ypestis               -MVLGKPQTDPTLEWFLSHCH------------------------IHKY 24
Crp_Shewanella            MTIEQNKNRRPAASGCAIHCHDCSMGTLCIPFTLNANELDQLDDIIERKK 50
fnr_Aeromonas_hydrophila  MIPEKKPGRRIQSGGCAIHCQDCSISQLCIPFTLNDNELDQLDSIIERKK 50
                           :             **:            :*

CAP_Ecoli                 PSK--STLIHQGEKAETLYYIVKGSVAVLIKDEEGKEMILSYLNQGDFIG 72
CAP_Kpneumoniae           PSK--STLIHQGEKAETLYYIVKGSVAVLIKDEEGKEMILSYLNQGDFIG 72
CAP_Ypestis               PSK--STLIHQGEKAETLYYIVKGSVAVLIKDEEGKEMILSYLNQGDFIG 72
Crp_Shewanella            PIQKGEQIFKSGDPLKSLFAIRSGTVKSYTITEQGDEQITGFHLAGDVIG 100
fnr_Aeromonas_hydrophila  PIQKGEELFKAGDELKSLYAIRSGTIKSYTITEQGDEQITAFHLAGDLVG 100
                          * :   . ::: *:  ::*: * .*::     *:*.* * .:   **.:*

CAP_Ecoli                 ELGLFEEGQERSAWVRAKTACEVAEISYKKFRQLIQVNPDILMRLSAQMA 122
CAP_Kpneumoniae           ELGLFEEGQERSAWVRAKTACEVAEISYKKFRQLIQVNPDILMRLSSQMA 122
CAP_Ypestis               ELGLFEEGQERSAWVRAKTACEVAEISYKKFRQLIQVNPDILMRLSSQMA 122
Crp_Shewanella            FDGIHAQSHQ--SFAQALETSMVCEIPFNILDELSGSMPKLRQQIMRLMS 148
fnr_Aeromonas_hydrophila  FDAIHKQAHP--SFAQALETAMVCEIPFDVLDDLSGKMPKLRQQIMRLMS 148
                          .:. :.:   ::.:*  :. *.**.:. : :*   *.:  ::   *:

CAP_Ecoli                 RRLQVTSEKVGNLAFLDVTGRIAQTLLNLAKQP-DAMTHPDGMQIKITRQ 171
CAP_Kpneumoniae           RRLQVTSEKVGNLAFLDVTGRIAQTLLNLAKQP-DAMTHPDGMQIKITRQ 171
CAP_Ypestis               NRLQITSEKVGNLAFLDVTGRIAQTLLNLAKQP-DAMTHPDGMQIKITRQ 171
Crp_Shewanella            NEIMSDQEMILLLSKKNAEERLAAFISNLANRFGNRGFSPKEFRLTMTRG 198
fnr_Aeromonas_hydrophila  NEIMGDQEMILLLSKKNAEERLAAFLHNLSVRFSERGFSAKEFRLSMTRG 198
                          ..:   .* : *: :. *:*  : **: :  :  .. :::.:**

CAP_Ecoli                 EIGQIVGCSRETVGRILKMLEDQNLISAHGKTIVVYGTR----------- 210
CAP_Kpneumoniae           EIGQIVGCSRETVGRILKMLEDQNLISAHGKTIVVYGTR----------- 210
CAP_Ypestis               EIGQIVGCSRETVGRILKMLEDQNLISAHGKTIVVYGTR----------- 210
Crp_Shewanella            DIGNYLGLTVETISRLLGRFQKSGLIEVKGKYITILDHHELNLLAGNARI 248
fnr_Aeromonas_hydrophila  DIGNYLGLTVETISRLLGRFQKSGLISVKGKYITVLDHVALGVMAGATRP 248
                          :**: :* : **:.*:*  ::...**..:** *.: .

CAP_Ecoli                 ---
CAP_Kpneumoniae           ---
CAP_Ypestis               ---
Crp_Shewanella            AR- 250
fnr_Aeromonas_hydrophila  PCG 251
```

(a) The N-terminal and C-terminal amino acids show the most divergence across all 5 species; however, note that even in these areas, some species show some similarities between each other. This may be a reflection of the broad range of genes, ranging from sugar catabolism to nitrogen metabolism, which this family of proteins may control.

(b) Indeed, most of this protein seems to be conserved.

(c) Two small domains are identified, the CAP_ED domain and the HTH_CRP domain. The CAP_ED domain is a ligand-binding domain that serves as the sensor for the DNA-binding activity for this protein. Notice that current information suggests a broad range of ligands (cAMP, CoA, iron–sulfur clusters), depending on the polypeptide. The HTH_CRP domain is the DNA-binding domain. The colors indicate degrees of probability that the matches shown are not due to chance. However, since the lengths of these amino acids sequences are similar and not very long, the e-values can be deceptive. Note that in the CAP_ED domain, 55 of 90 amino acids are either identical or similar in chemistry (61.1%), while in the HTH_CRP domain, 50 of 68 amino acids are identical or similar (73.5%), which makes a little more biochemical sense given the putative biological roles of each domain.

6

Proteins: Three-Dimensional Structure

This chapter introduces you to an area of biochemistry that has shown an incredible growth of information in recent years: the three-dimensional structures of proteins. The chapter first discusses in some detail the geometric and physical properties of the planar peptide group, which contains the peptide bond. Here you see that the number of polypeptide conformations, while quite large, is finite due to restrictions imposed by colliding van der Waals spheres of the atoms of the polypeptide. The chapter then examines the major forms of secondary or local three-dimensional structure in polypeptides, in particular, the α helix and the β sheet. Keratin and collagen are the most common and best-characterized fibrous proteins and are largely composed of one type of secondary structure. The structural properties and biology of these proteins are discussed at this point.

The chapter then turns its attention to the methods by which overall three-dimensional structure, or tertiary structure, is determined: X-ray diffraction of protein crystals and nuclear magnetic resonance (NMR) of proteins in solution. You are then introduced to the various kinds of supersecondary structures (motifs) and domains that have been identified in many proteins, including a brief discussion of the symmetries observed in the quaternary structures of proteins that consist of more than one subunit. This chapter also introduces structural bioinformatics, the branch of biochemistry that is involved in the visualization, analysis, and classification of protein three-dimensional structure.

The factors that affect protein stability are then explored, followed by a discussion of how newly synthesized proteins fold. In this presentation, you are introduced to proteins that facilitate protein folding *in vivo:* protein disulfide isomerase and the molecular chaperones (the Hsp70 family and the chaperonins). The section closes with a discussion of diseases related to protein misfolding.

Box 6-1 provides a brief biography of one of the 20th century's greatest chemists and scientists, Linus Pauling, and his major contributions to protein structure. Box 6-2 provides an example of how changes in primary structure affect changes in tertiary and quaternary structures leading to pathology (collagen diseases). Boxes 6-3 and 6-4 discuss the discovery of thermostable proteins and recent progress in predicting the structures of proteins, respectively.

Essential Concepts

1. There are four levels of organization in protein structure:
    (a) Primary structure, which is the amino acid sequence.
    (b) Secondary structure, which is the local three-dimensional structure of a polypeptide without regard to the conformations of its side chains.
    (c) Tertiary structure, which refers to the overall three-dimensional structure of an entire polypeptide.
    (d) Quaternary structure, which refers to the three-dimensional arrangement of polypeptides in a protein composed of multiple polypeptides.

Secondary Structure

2.   Amino acids are joined by the condensation of the amino group of one amino acid with the carboxyl group of another, which results in a linkage called a peptide bond. The peptide group has a resonance structure such that the bond connecting the carboxyl carbon and the amino nitrogen has ~ 40% double-bond character.

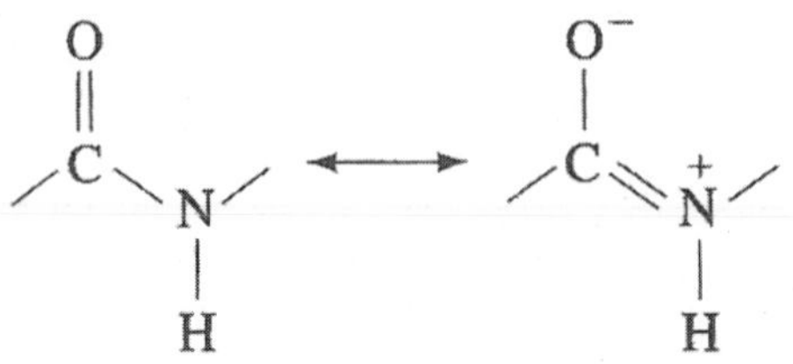

$$
\begin{array}{ccc}
\underset{\underset{|}{\overset{\overset{O}{\|}}{C}}\underset{\underset{H}{|}}{\diagdown N \diagup} & \longleftrightarrow & \underset{\underset{|}{\overset{\overset{O^-}{|}}{C}}\underset{\underset{H}{|}}{= \overset{+}{N} \diagup}
\end{array}
$$

With rare exceptions, the *trans* conformation shown above occurs in proteins.

3.   The backbone of a polypeptide chain can be viewed as a linked series of planar peptide groups that can rotate about the single covalent bonds involving the α carbon. When the peptide groups are fully extended and all lie in a plane, the torsion angles of the C_α—N bond (ϕ) and the C_α—C bond (ψ) are defined as 180° (Figure 6-4). As these planar groups rotate, the van der Waals spheres of their various atoms collide; hence the rotational freedom of the bonds is restricted. The allowed conformations, given as ϕ and ψ values, are summarized in the Ramachandran diagram (Figure 6-6).

4.   In regular secondary structures, successive residues have similar backbone conformations, that is, repeating values of ϕ and ψ (e.g., the α helix and the β sheet).

5.   The α helix is a right-handed helix with 3.6 residues per turn and a pitch of 5.4 Å. Hydrogen bonds connect the peptide $C = O$ group of the *n*th residue with the peptide N—H group of the (*n* + 4)th residue. The core of the helix is tightly packed, and the amino acid side chains project outward and downward from the helix.

6. β sheets result from hydrogen bonding between segments of extended polypeptide chains. Antiparallel β sheets consist of polypeptide strands that extend in alternate directions. Parallel β sheets consist of polypeptide chains that all extend in the same direction. Parallel β sheets are slightly less stable than antiparallel sheets because the hydrogen bonds connecting the polypeptides are distorted. Mixed β sheets, consisting of both parallel and antiparallel polypeptide chains, are common.

7.   Proteins are typically classified as fibrous or globular. A single type of secondary structure dominates fibrous proteins. Two of the best best-characterized fibrous proteins are α keratin and collagen, both of which form higher order structures that are insoluble in water.
      (a) α Keratin consists of two α-helical chains that wrap around each other in a left-handed coiled coil. Each polypeptide chain has a 7-residue pseudorepeat, *a-b-c-d-e-f-g,* such that hydrophobic residues at positions *a* and *d* form the contacts between the helices.

This dimer is assembled further into higher order structures that are less well characterized.

(b) Collagen contains long stretches of the three-residue repeating unit Gly-X-Y, in which X is often Pro, and Y is often 4-hydroxyPro (Hyp; hydroxylation of Pro residues requires ascorbic acid). Each polypeptide strand forms a left-handed helix with around three residues per turn. Three such parallel chains then wrap around each other in a gentle right-handed coil (Figure 6-17). Interchain hydrogen bonds connect all three strands, which are staggered so that a Gly residue occurs at every position along the triple helix (Figure 6-18). Covalent cross-links between modified Lys residues (allysine) and His residues bind together aggregates of collagen fibers.

8. Most proteins contain irregular secondary structures such as β bulges, reverse turns (β bends), and other structures whose residues have nonrepeating backbone conformations.

9. Certain amino acid residues occur more often in α helices or β sheets. Moreover, certain residues tend to disrupt or break secondary structures. These tendencies provide a foundation for predicting protein structure from amino acid sequences and for designing proteins with particular structures.

10. Linus Pauling (1901–1994) laid the foundations for structural biochemistry, culminating with a description of α helices, β sheets, and a precise structural basis for sickle-cell hemoglobin involving a single amino acid substitution.

Tertiary Structure

11. The tertiary structure of a protein can be determined by X-ray diffraction or nuclear magnetic resonance (NMR).

(a) The interaction of a beam of X-rays with the electrons of a crystallized protein yields a diffraction pattern that can be mathematically analyzed to reconstruct a three-dimensional electron density map of the atoms of the protein. Knowledge of the protein's primary structure is necessary to identify the amino acid residues in a three-dimensional model of the protein.

(b) NMR spectroscopy can be used to obtain information about the tertiary structure of a protein in solution. This technique is limited to proteins with ~350 residues but may soon increase to ~900 residues.

12. Protein crystals are typically 40–60% water by volume, giving them a jelly-like consistency. For most crystals, the resolution ranges from 1.5 to 3.0 Å. Several lines of evidence suggest that molecules of crystalline protein have nearly the same structure that they have in solution.

13. In globular proteins, nonpolar residues occur most often in the protein interior. Charged residues most commonly occur on the protein surface. Polar residues that are buried in the protein interior are almost always "neutralized" by forming hydrogen bonds to other internal groups.

14.   Groups of secondary structural elements form motifs, the most common of which are the βαβ motif, β hairpin, αα motif, Greek key motif, and β barrel.

15.   Although nearly 50,000 protein structures are known, about 200 folding patterns account for nearly half of all known protein structures. Hence, it appears that evolution tends to conserve protein structures, motifs and domains in particular, rather than amino acid sequences. This probably reflects biochemical constraints based on the ability of these domains to
      (a)   form stable folding patterns.
      (b)   tolerate amino acid deletions, substitutions, and insertions without significant loss of biological activity.
      (c)   support essential biological functions.

16.   Larger globular domains form in polypeptides containing more than ~200 amino acids. These domains are often associated with a specific biochemical function (e.g., NAD^+ binding). There appears to be a limited number of domains, which allows biochemists to group proteins into families.

17.   The Protein Data Bank archives the atomic coordinates of proteins and nucleic acids. These can be interpreted by a number of computer programs, many of which are freely available on the internet, to generate three-dimensional, manipulatable images. Several internet sites also use this information to compare the structures of proteins, providing investigators powerful tools to analyze potential structure-function relationships and the evolution of protein structure. One should keep in mind that all such bioinformatic tools help scientists *generate hypotheses* that must ultimately be tested in the laboratory.

Quaternary Structure and Symmetry

18.   Most proteins with molecular masses >100 kD consist of more than one polypeptide chain. The arrangement of the chains is the protein's quaternary structure. The contact regions between subunits of a protein resemble the interior of a polypeptide.

19.   The advantages of having multisubunit proteins include the following:
      (a)   A defective region of a protein (e.g., caused by mistranslation or improper folding) can be easily repaired by replacing the defective subunit.
      (b)   The only genetic information necessary to specify a large protein is that specifying its few different self-assembling subunits.
      (c)   It provides a structural basis for the regulation of enzymatic activity.

20.   Proteins with more than one subunit are called oligomers, and their identical subunits (which may contain more than one polypeptide) are called protomers. In most proteins, protomers occupy geometrically equivalent positions, displaying rotational symmetry. The most common are cyclic symmetry (most often involving two protomers) and dihedral symmetry (in which an *n*-fold rotation axis intersects a two-fold rotation axis at right angles).

Protein Stability

21. Native proteins are only marginally stable under physiological conditions. The hydrophobic effect is the principal force that stabilizes protein structures. Electrostatic interactions, especially hydrogen bonds and ion pairs, contribute little to the overall stability of tertiary structures because water interacts similarly with fully unfolded (denatured) proteins. However, these weak interactions play important roles in aligning residues in specific secondary structures and domains.

22. Disulfide bonds cross-link structures in extracellular proteins, which occur in a relatively oxidizing environment. Disulfide bonds are rare in intracellular proteins, presumably because the reducing environment of the cytosol weakens them. In some proteins, metal ions such as Zn^{2+} cross-link small structures that would otherwise be unstable.

23. Theoretical calculations suggest that native protein structure consists of rapidly interconverting conformations (breathing), allowing small molecules to move into and out of the interior of proteins.

24. Most proteins unfold or denature at temperatures well below 100°C, with a sharp transition indicating that denaturation is a cooperative process. Conditions or substances that denature proteins include heat, variations in pH, detergents, and chaotropic agents (e.g., guanidinium ion and urea).

Protein Folding

25. Under proper conditions, most unfolded proteins will renature spontaneously, thereby indicating that tertiary structure is dictated by primary structure.

26. Protein folding is an ordered process rather than a random search for a stable conformation. During folding, secondary structures form first, followed by motifs and domains. A key driving force is a hydrophobic collapse that buries hydrophobic regions in the interior of the protein, out of contact with the aqueous medium (to yield a state referred to as a molten globule).

27. Certain proteins facilitate protein folding *in vivo,* which is often considerably faster than folding *in vitro.* Protein disulfide isomerase mediates the formation of disulfide bonds. Molecular chaperones prevent improper associations between exposed hydrophobic segments of unfolded polypeptides that could lead to non-native folding as well as nonspecific aggregation of the hydrophobic regions of different unfolded polypeptides.

28. The molecular chaperones include several groups of proteins: (1) the Hsp70 family of proteins; (2) chaperonins; and (3) the Hsp90 family of proteins. The Hsp70 chaperones are a family of 70 kD proteins that prevent premature folding of polypeptides as they are being synthesized in the ribosome.

29. The chaperonins, which form barrel-shaped structures, consist of two types of proteins, Hsp60 (GroEL in *E. coli*) and Hsp10 (GroES in *E. coli*). A cycle of conformational

changes in GroEL/ES requiring the hydrolysis of 7 ATP molecules coordinates protein folding in the interior of the GroEL/ES complex. A polypeptide engages in an average of 24 cycles of binding to the GroEL/ES complex before acquiring its native structure.

30. Several neurological diseases are characterized by the intracellular accumulation of proteins in aggregates called amyloid plaques. The most notable of these are Alzheimer's disease, prion diseases, and an array of amyloidoses. The aggregates form fibrils largely comprised of β structures. In prion diseases (e.g., scrapie, "mad cow disease," Creutzfeldt–Jakob disease), an α-helical prion changes its conformation to a mixture of α helices and β sheets and forms insoluble fibers. It has been proposed that prions act as infectious agents by catalyzing the misfolding of other prion proteins. The formation of amyloid fibrils in other diseases is less clear; however, once formed they appear to be kinetically resistant to destruction by cellular proteases.

Exercise

Use a molecular model set to satisfy you that glycine at the center of a tripeptide can assume conformations not shaded in the Ramachandran diagram (i.e., ϕ values between 60° and 180° and all ψ values except those between −45° and 45°).

Questions

Secondary Structure

1. Why does the peptide bond result in a planar configuration for its adjoining functional groups?

2. In the dipeptide below, indicate which bonds are described by ϕ and ψ.

3.   Shown below is a portion of poly-β-alanine$_{10}$. Comment on the secondary structure of this peptide.

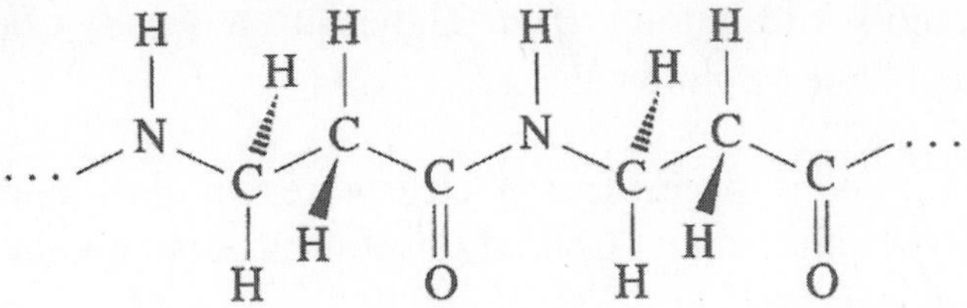

4.   When ϕ = 180° and ψ = 0°, what groups come into closer-than-van-der-Waals contact? (See Figure 6-4.)

5.   How do R groups constrain the potential conformations of a protein? (See Figure 6-4 and Kinemage 3-1.)

6.   Study Figure 6-7 and describe the orientation of the groups that contribute to the dipole moment of the α helix and the overall direction of this dipole moment. *Hint:* The sum of local dipole moments determines the overall dipole moment of the helix.

7.   Why are antiparallel β sheets more stable than parallel β sheets?

8.   How does the 7-residue repeat of α keratin promote formation of a coiled coil?

9.   Wool from sheep is steamed to generate long fibers for knitting. Wool sweaters shrink after subsequent drying with heat. Explain what is happening at the molecular level.

10.   What is the repeating sequence of collagen and how is it essential to the structure of the triple coils of collagen fibers?

11.   Scurvy is caused by a deficiency of _____, which is necessary for the activity of _____.

12.   Organic compounds containing radioactive atoms can be used to follow the biosynthesis of molecules in cells and organisms (see Section 14-4A). In these experiments, the amount of radioisotope in proteins derived from the cells or organisms is monitored. The administration of ^{14}C-labeled 4-hydroxyproline to mice results in no appearance of ^{14}C in the animals' collagen fibers. However, the administration of ^{14}C-labeled proline does result in the appearance of ^{14}C in the collagen fibers of mice. Explain these results.

13.   List the three amino acids that are least likely to occur in a β sheet.

Tertiary Structure

14.   Examine Table 4-1 (pp. 76–77) and determine which amino acids might exhibit identical electron densities at 2.0-Å resolution. *Hint:* The electron density of a hydrogen atom is too small to be apparent at this resolution.

15.   A resolution of ~3.5 Å is necessary to clearly reveal the course of the polypeptide backbone in X-ray crystallography. Can any useful information about protein structure be obtained at 6 Å resolution?

16.   What advantages does NMR provide over X-ray crystallography in characterizing protein structure? What is the major limitation of NMR analysis?

17.   What specific role does knowledge of the primary amino acid sequence play in the determination of tertiary structure by X-ray crystallography?

18.   How does the polarity or charge of an amino acid affect its likely location within a protein?

19.   How does the ratio of hydrophilic to hydrophobic amino acid residues change in a series of globular proteins ranging from 10 to 50 kD?

20.   Go to the National Center for Biotechnology Information (NCBI: http://www.ncbi.nlm.nih.gov/) and click on the word Structure along the menu line (last word on the right). If you choose to download the Cn3D program, you will be able to view the three-dimensional structures directly from the links after you have done a search for a protein structure. While Cn3D is not as powerful as some of the programs discussed in the text, it is a very handy tool. In the search box, type in "cam kinase" for calmodulin-dependent kinase and observe the output. Protein kinases are enzymes (transferases) that transfer a phosphate from ATP to another protein. This covalent modification results in structural changes that either activate or inhibit enzyme activity. Hence, protein kinases are regulatory enzymes in the cell (see Chapter 13).

(a)   Click on 2V7O. This brings up the MMDB Structure Summary page. What kinds of information can you find here?

(b)   This site recognizes two structural domains in calcium-calmodulin-dependent kinase. What structural features seem to dominate each domain? *Hint:* Click on the PDB file name again, and then the folder tab for "Sequence Details."

(c)   Go back to the Structure Summary page, and click on VAST (Vector Alignment Search Tool). Click on each domain and examine the proteins that are retrieved, which show similar structures (this information is based on available 3D structural information). Are any hits for either domain not protein kinases? (See the Structural Genomics Consortium at http://www.sgc.utoronto.ca/SGC-WebPages/StructureDescription/ 2V7O.php for more information about the biochemical functions of each domain.)

Quaternary Structure and Symmetry

21. What are the possible symmetries of (a) a tetrameric protein and (b) a pentameric protein? Assume all the subunits are identical.

22. The subunit composition of an oligomeric protein can be determined by treating the protein with a cross-linking agent (a bifunctional molecule that reacts with and links groups in two different portions of the polypeptide), denaturing the protein, and analyzing the products by SDS-PAGE (Section 5-2D).

  (a) An oligomer analyzed by SDS-PAGE shows a single 40 kD polypeptide. Brief treatment with a cross-linking agent yields the SDS-PAGE banding pattern shown below. What is the protein's probable subunit composition?

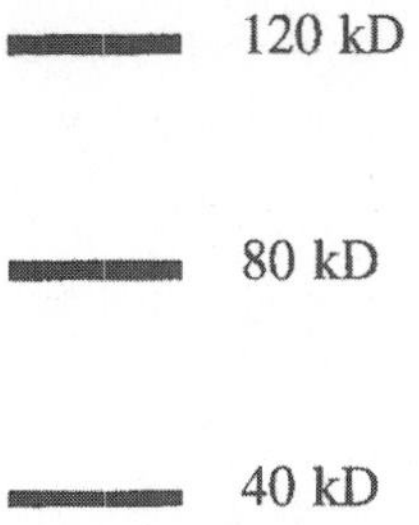

                    ▄▄▄▄▄   120 kD

                    ▄▄▄▄▄   80 kD

                    ▄▄▄▄   40 kD

  (b) Not shown in the electrophoretogram above are faint bands at 240 kD and 360 kD. What do these bands represent?

  (c) Another protein analyzed by SDS-PAGE shows two polypeptides at 20 kD and 50 kD. Chemical cross-linking of the native protein yields the results shown below in SDS-PAGE. Not shown is a very faint band at 540 kD. What is the subunit composition of this protein?

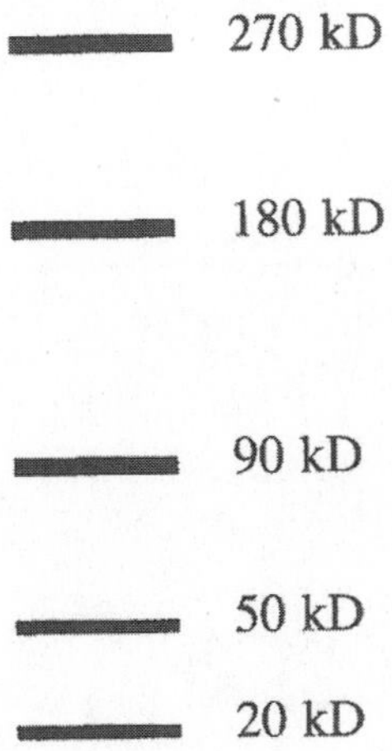

                    ▄▄▄▄▄   270 kD

                    ▄▄▄▄▄   180 kD

                    ▄▄▄▄▄   90 kD

                    ▄▄▄▄   50 kD

                    ▄▄▄▄   20 kD

23.   Consider the equilibrium of an oligomer:

2 monomers $\rightleftharpoons$ dimer

The $\Delta H°$ for this reaction is positive and varies little between 4°C and 45°C. $T\Delta S°$, in contrast, varies widely over the same temperature range. The behaviors of $\Delta H°$ and $T\Delta S°$ are shown in the graph below. How does the position of equilibrium of this oligomer differ at 4°C versus 45°C?

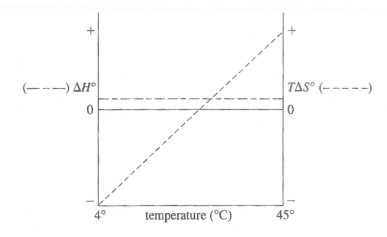

Protein Stability

24.   What thermodynamic consideration prevents ion pairs from contributing significantly to protein stability?

25.   List four denaturants of proteins.

26.   What is the probability of recovering the native conformation (as judged by biological activity) of a dimeric protein, each of whose subunits contains three disulfide bonds? Assume that the disulfide linkages form randomly.

27.   What amino acid differences and molecular forces distinguish proteins found in hyperthermophiles from those found in mesophiles?

Protein Folding

28.   Calculate the maximum length of a polypeptide that could recover its native conformation in about one day if the polypeptide explored all possible conformations (see p. 161).

29.   To appreciate the role of hierarchical assembly in the rapid renaturation of proteins, consider the equation $t = 10^n/10^{13}$ as a reasonable approximation of renaturation time for a globular protein in which n represents the number of nucleating structures (α helices and β sheets). How does this improve the rate of renaturation of a 200-residue globular protein with eight α helices and six β sheets?

30. Assume that the average molecular mass of an amino acid is 112, and that 4 ATP equivalents are expended to form a peptide bond. Using the current model for protein folding in GroEL/ES, compare the cost of protein synthesis with the cost of protein folding for a 70-kD polypeptide in *E. coli*.

31. What statement(s) in the text suggests that native protein structure is highly labile?

Answers to Questions

1. The C—N bond has a partial double-bond character due to resonance interactions. This limits rotation around the C—N bond, constraining the carbonyl and amide groups to lie in the same plane.

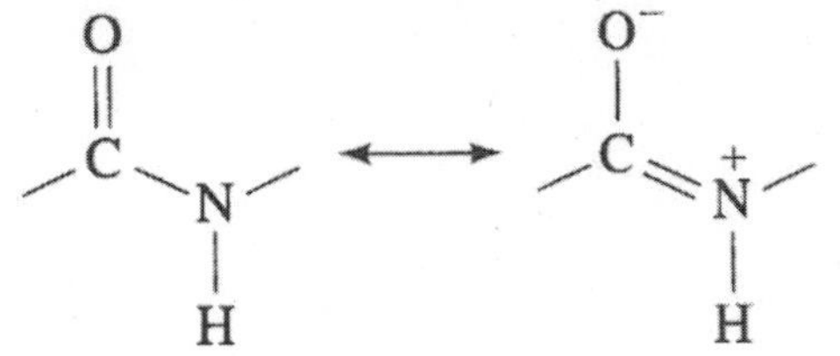

2.

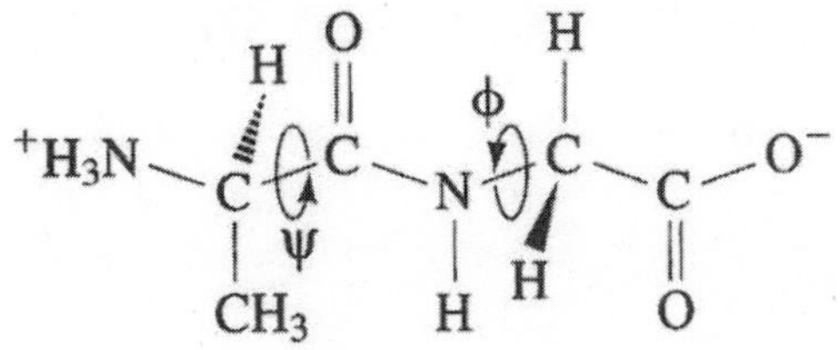

Note that both ϕ and ψ increase clockwise when viewed from C_α.

3. In the structure shown below, the thicker lines represent the peptide bonds. The polypeptide backbone contains three single bonds capable of free rotation between each peptide bond (versus two in a standard polypeptide). This greater flexibility allows for more possible conformations of the peptide chain, so that the polymer is more likely to assume a random, fluctuating conformation rather than a regular secondary structure.

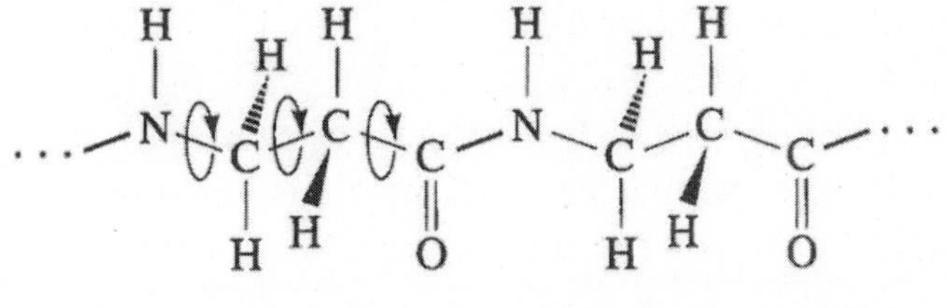

4.   When $\phi = 180°$ and $\psi = 0°$, the van der Waals spheres of the amide hydrogens overlap.

5.   Rotation around ϕ will cause the van der Waals sphere of the carbonyl oxygen of residue n to collide with the van der Waals sphere of the C_β of residue $n + 1$. This can be clearly seen in Kinemage 3-1. Turn on AAs 1–3, PH1, planes, O1, and CB. Then try rotating around ϕ to observe collision of van der Waals spheres.

6.   Shown below is the dipole of a peptide unit. In an α helix, all the hydrogen bonds point in the same direction, so the dipoles of each peptide unit sum up to an overall dipole moment for the entire helix. Since the carbonyl groups all point from the N-terminus toward the C-terminus of the helix, the N-terminus of the helix has a partial positive charge and the C-terminus has a partial negative charge.

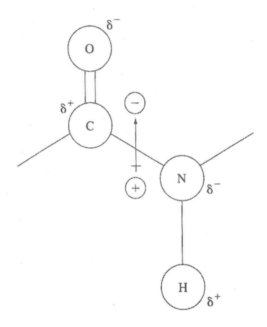

7.   Antiparallel β sheets are more stable because the hydrogen bonds connecting adjacent polypeptide strands are not distorted, as they are in parallel β sheets (see Figure 6-9).

8.   The first and fourth residues of α keratin's seven-residue repeat (residues a and d of the repeating sequence a-b-c-d-e-f-g) are primarily nonpolar residues that form a hydrophobic strip along one side of the α helix (which has $3.6 \approx 7/2$ residues per turn). When the hydrophobic strips of two keratin helices interact, the helices incline slightly, causing them to coil around each other (see Figure 6-15).

9.   The steaming process stretches α keratin into a pleated β sheet conformation. When subsequently heated, the keratin in the wool is converted back to an α-helical conformation, which is more compact.

10. The repeating sequence is Gly-X-Y, where X is often Pro and Y is often Hyp. The glycine residues occur in the interior of the triple helix, where there is no room to accommodate any side chain larger than a hydrogen atom. The highly constrained Pro and Hyp side chains confer rigidity on the helical structure.

11. Scurvy is caused by a deficiency of <u>ascorbic acid (vitamin C)</u>, which is necessary for the activity of <u>prolyl hydroxylase</u>.

12. Hydroxyproline is not one of the 20 standard amino acids used in the synthesis of proteins from mRNA. Specific proline residues in collagen are oxidized by prolyl hydroxylase to form 4-hydroxyproline only after peptide synthesis. Hence, only the administration of ^{14}C-labeled proline can yield radioactively labeled collagen fibers.

13. Glu, Asp, and Pro are the least likely to occur in a β sheet as judged by their low propensity for appearing in the β sheets of known proteins (see Table 6-1 on p. 140).

14. Several pairs of amino acids give indistinguishable electron density maps: Asp and Asn, Glu and Gln, Thr and Val. Other amino acids that may have similar shapes are Ser and Cys, although the S atom of Cys has a much greater electron density than the O atom of Ser.

15. At 6 Å resolution, larger elements of protein structure can be identified, most notably helices, which have a diameter of several Å and are visible as rodlike shapes.

16. The advantages of NMR include (a) structural information from proteins that fail to crystallize and (2) information about protein folding and dynamics since protein movements can be traced over relatively long time scales. The primary disadvantage of NMR is that the protein must be no larger than ~40 D.

17. When interpreting an electron density map, knowledge of the primary structure allows the identification of specific amino acids along the polypeptide chain. This would otherwise be quite difficult if not impossible for R groups with similar shapes (see Question 14)

18. Charged or polar amino acids are commonly on the outside of the molecule, where they are exposed to the aqueous solvent, whereas uncharged or nonpolar amino acids are often buried the interior of the protein, out of contact with the aqueous solvent, due to hydrophobic effects.

19. As protein size increases, the surface-to-volume ratio decreases. Recall that the surface of a sphere increases as the square of the radius (r^2), while the volume of a sphere increases as the cube of the radius (r^3). Hence, the ratio of hydrophilic to hydrophobic residues decreases as the molecular mass increases since the interior (volume) of the protein increases more rapidly than its surface area.

20. (a)   On the MMDB Summary Page, one can obtain
     (i)   The currently accepted name for the protein;
     (ii) Taxonomic information about the organism from which the protein was obtained;

(iii) An overview of structural and sequence information; and

(iv) Links to other data files, particularly the PDB page, which contains a wealth of information about the biochemistry of the protein, and the VAST page, which allows one to examine proteins with similar structural domains.

(b) From the Sequence Details page, one can see that β sheets dominate domain 1 and alpha helices dominate domain 2. Note, however, that the catalytic domain includes both of these structural domains, although domain 2 is where the regulatory function, the binding of Ca^{2+}-calmodulin, occurs (see the Structural Genomics Consortium at http://www.sgc.utoronto.ca/SGC-WebPages/StructureDescription/2V7O.php).

(c) While domain 2 has VAST "hits" mostly to protein kinases, domain 1 has some similarity to different proteins, including a secretion chaperone, interferon regulatory factor (part of a transcription regulatory complex), and an archeabacterial endonuclease.

21. A tetrameric protein can have C_4 or D_2 symmetry. A pentameric protein can have only C_5 symmetry.

22. (a) The protein is probably a trimer of 40 kD subunits. The gel shows cross-linked dimers and trimers in addition to free monomers.

(b) The apparent masses of the faint bands are multiples of 120 kD and therefore probably represent protein trimers that have been chemically cross-linked to other trimers.

(c) The protein is a trimer of two 20 kD subunits and one 50 kD subunit. Some of the 90-kD oligomers have been cross-linked to form larger structures of 180 kD (a dimer), 270 kD (a trimer), and 540 kD (a hexamer).

23. Recall that $\Delta G° = \Delta H° - T\Delta S°$. At 4°C, $T\Delta S°$ is negative and its absolute value is greater than that of ΔH, so $\Delta G°$ is positive, indicating that the equilibrium favors monomer formation (see Table 1-3, p. 15). At 45°C, $T\Delta S°$ is positive and greater than ΔH, so $\Delta G°$ is negative, indicating that the equilibrium now favors the formation of dimers. The cytoskeletal elements called microtubules behave in such a manner.

24. The favorable free energy of formation of an ion pair is nearly equivalent to the loss of solvation free energy that occurs when the charged groups form the ion pair.

25. Heat, pH changes, detergents, and chaotropic agents.

26. The probability for each monomer is $1/5 \times 1/3 \times 1/1 = 1/15 = 0.067$ or 6.7%. The probability that a native dimer can form is $0.067 \times 0.067 = 0.0044$ or 0.44%.

27. Thermostable proteins are very similar in amino acid composition to mesophilic proteins. One distinguishing feature in some is that many such proteins have networks of salt bridges on their surfaces.

28. $t = 10^n/10^{13}$, where t is the time in seconds, and n is the number of amino acid residues. Rearranging terms,

$$n = \log 10^{13} + \log t$$

Here, $t = (60 \text{ s/min})(60 \text{ min/h})(24 \text{ h/day}) = 86{,}400 \text{ s}$, so that

$$n = 13 + \log 86{,}400$$
$$= 13 + 4.9$$
$$= 17.9$$

Therefore, the peptide could have no more than 18 residues.

29.   For a 200-residue protein, $t = 10^n/10^{13} = 10^{200}/10^{13} = 10^{187}$ s or about 3×10^{179} years. If the eight α helices and six β sheets act as nucleating elements for the formation of tertiary structure, $n = 200/(8+6) \approx 14$ so that $t = 10^{14}/10^{13} = 10$ s.

30.   A 70 kD polypeptide would be composed of roughly 625 amino acids (the 70,000 molecular mass of the polypeptide divided by 112, the molecular mass of an amino acid), which means that there are 624 peptide bonds. Therefore, 2496 ATP equivalents (4 ATP equivalents per peptide bond × 624 peptide bonds) are required to synthesize this polypeptide. Folding of this average-sized polypeptide in the GroEL/ES complex would require 168 ATP molecules (7 ATP per cycle × 24 cycles for an average-sized polypeptide). Hence, properly folding the polypeptide costs about 6.7% of the ATP energy needed to synthesize it. Note that this calculation ignores all the energy required to activate the gene and transcribe it as well as the considerable energy required to synthesize amino acids (Section 21-5).

31.   Under physiological conditions, most polypeptides display marginal stability, averaging ~0.4 kJ/mol amino acids (see page 156). Furthermore, some native proteins in *E. coli* appear to repeatedly bind to GroEl/ES complexes long after they have been initially folded into their native structure, which suggests that native proteins frequently denature under physiological conditions (see page 168).

7

Protein Function: Myoglobin and Hemoglobin, Muscle Contraction, and Antibodies

A fundamental question in biochemistry is how protein structure influences the functional properties of proteins. Chapter 7 begins this exploration by closely examining the structure and function of some of the best studied proteins in biochemistry: myoglobin and hemoglobin, actin and myosin, and antibodies. Myoglobin and hemoglobin are globular proteins involved in oxygen storage and transport. Myoglobin, whose primary role is to facilitate oxygen transport in muscle (and, in aquatic mammals, to store oxygen), binds oxygen with a simple equilibrium constant. Hemoglobin transports oxygen, first binding oxygen in the capillaries of the lungs (or gills or skin), where oxygen concentrations are high, and then releasing the oxygen in the capillaries of the tissues, where the oxygen concentrations are lower. The binding of oxygen to hemoglobin is not as simple as that for myoglobin. Several factors influence oxygen binding to hemoglobin, including the partial pressure of oxygen itself, pH, the concentration of CO_2, and 2,3-bisphosphoglycerate (BPG). The study of mutant hemoglobin molecules with amino acid substitutions that affect hemoglobin function provides important insights into the roles of individual amino acids.

Muscle fibers are largely bundles of myofibrils, consisting of interdigitated thick and thin filaments. The thick filaments are made of hundreds of myosin molecules, whose two heavy chains and four light chains form a long rod-like segment with two globular heads. Thin filaments contain three proteins: actin, tropomyosin, and troponin. Interactions between the thick and thin filaments allow the proteins to move past each other in a process that is driven by the hydrolysis of ATP. The myosin molecules undergo large conformational changes that generate easily measurable force and are the canonical examples of force-generating proteins called motor proteins, which include proteins involved in the movement of chromosomes and organelles (kinesins and dyneins), as well as DNA and RNA polymerases.

Antibodies are the major line of defense against extracellular pathogens such as bacteria and viruses. Cellular immunity is mediated by T lymphocytes, which are formed in the thymus. Humoral immunity is mediated by antibodies (immunoglobulins) that are produced by B lymphocytes, which mature in the bone marrow. A B cell makes only one type of antibody. Antigen binding to a specific antibody located on the surface of a B cell triggers an immune response in which the B cells secreting that antibody proliferate. B cells usually live for only a few days; however, memory B cells that recognize specific antigens remain and proliferate when more antigen is present. The specificity of an antibody–antigen interaction arises from variable amino sequences in the immunoglobulin, which create a unique antigen-binding site.

Essential Concepts

Oxygen Binding to Myoglobin and Hemoglobin

1. Myoglobin is a single polypeptide containing a heme group, which consists of a porphyrin ring whose coordinated Fe(II) atom binds molecular oxygen. The protein prevents oxidation of the heme iron to Fe(III), which does not bind oxygen. Oxidized myoglobin is called metmyoglobin.

2. Myoglobin facilitates oxygen diffusion in muscle, which under conditions of high exertion have a particularly high demand for oxygen. It also functions as an oxygen-storage protein in aquatic mammals, where it is in ten-fold higher concentration than in terrestrial animals. A simple equilibrium equation describes O_2 binding to myoglobin (Mb): Mb + O_2 $\rightleftharpoons$ MbO_2. The fractional saturation of myoglobin is defined as $Y_{O_2} = pO_2/(K + pO_2)$, where pO_2 is the partial pressure of oxygen and K is its dissociation constant. A plot of Y_{O_2} versus pO_2 is a hyperbolic curve. It is convenient to define K as p_{50}, the partial pressure of O_2 at which 50% of myoglobin has bound oxygen. The p_{50} for myoglobin is 2.8 torr, which is much lower than the pO_2 in venous blood (30 torr), so that myoglobin is nearly saturated with oxygen under physiological conditions.

3. Hemoglobin binds oxygen in the lungs (pO_2 = 100 torr) and releases it in the capillaries (pO_2 = 30 torr). The efficiency of oxygen transport is greater than expected if oxygen binding were hyperbolic, as in myoglobin. Hemoglobin, which is an $\alpha_2\beta_2$ tetramer, each of whose subunits contains a heme group, binds O_2 cooperatively and thus has a sigmoidal oxygen binding curve. Deoxyhemoglobin is bluish (the color of venous blood), whereas oxyhemoglobin is bright red (the color of arterial blood).

4. The p_{50} of hemoglobin is about 26 torr, which is nearly 10 times greater than that of myoglobin. Because hemoglobin exhibits a sigmoidal oxygen-binding curve, it releases a much greater fraction of its bound O_2 in passing from the lungs to the tissues than if I had a hyperbolic binding curve with the same p_{50} (see Figure 7-6). The Hill equation describes the cooperative nature of oxygen binding, and the Hill constant, n, describes the degree of cooperativity. The Hill constant, which is not necessarily an integer, is obtained experimentally. The binding of O_2 to hemoglobin is said to be cooperative because the binding of O_2 to one subunit increases the O_2 affinity of the other subunits. The fourth oxygen to bind to hemoglobin does so with a 100-fold greater affinity than the first.

5. Hemoglobin and myoglobin are two examples of numerous oxygen-transport proteins that have been identified, among which are leghemoglobins from leguminous plants, chlorocruorins from some annelids, and hemerythrin and hemocyanin from invertebrates. Hemerythrin and hemocyanin are unique as oxygen-transporters since they do not contain heme groups.

6. Hemoglobin has only two stable conformational states, the T state (the conformation of deoxyhemoglobin) and the R state (the conformation of oxyhemoglobin). Oxygen binding causes the T state to shift to the R state, which has greater affinity for oxygen. The T to R shift is triggered by oxygen binding to the heme iron, which pulls the heme iron atom into the heme plane. This movement is transmitted to the F helix through His F8, which ligands the iron atom. Conformational changes in one subunit are transmitted across the α_1–β_2 and α_2–β_1 interfaces. Due to the conformational constraints at these interfaces, the conformational shift of one subunit must be accompanied by the conformational shift of all subunits, thereby increasing the oxygen affinity of the unoccupied subunits.

7. Decreases in pH promote the release of oxygen from hemoglobin. This effect, called the Bohr effect, is driven by dissolved carbon dioxide, which forms bicarbonate ion and a hydrogen ion. The hydrogen ion protonates hemoglobin, thereby stabilizing its T (deoxy) state. In the lungs, the reaction is reversed: Oxygen binding causes a switch to the R (oxy) state. The released hydrogen ions shift the equilibrium between bicarbonate and carbon dioxide, thereby forming carbon dioxide for expulsion from the lungs. Due to the Bohr effect, the low pH in highly active muscles causes the amount of oxygen delivered by hemoglobin to increase by nearly 10%. Carbon dioxide also binds preferentially to the N-terminal amino groups of T-state hemoglobin as carbamates and hence is released by R-state hemoglobin. This accounts for about half of the carbon dioxide released from the blood in the lungs.

8. Hemoglobin stripped of 2,3-bisphosphoglycerate (BPG) binds oxygen more tightly than hemoglobin in the blood. BPG binds to the T state but not the R state, thereby decreasing hemoglobin's oxygen affinity. This allows nearly 40% of the oxygen to be unloaded in venous blood. Fetal hemoglobin does not bind BPG as tightly and therefore has a higher affinity for oxygen than adult hemoglobin.

9. Humans use two mechanisms to adapt to the lower oxygen partial pressure at higher altitudes: (1) erythrocyte BPG concentration increases to promote oxygen delivery to tissues, and (2) the hemoglobin concentration rises. Other mammals that live at high altitude have isoforms of hemoglobin with increased affinity for oxygen.

10. Allosteric proteins are oligomers with multiple ligand-binding sites, in which ligand binding at one site alters the protein's binding affinity for a ligand at another site. Two models for cooperativity have been proposed: (1) the symmetry model and (2) the sequential model.

11. The symmetry model proposes that all the subunits in the protein exist in either the T state or the R state, and ligand binding to one subunit favors the conversion of all subunits to the R state. In contrast, the sequential model proposes that ligand binding induces a conformational change in one subunit (to the R state), which causes progressive changes in conformation in adjacent subunits (compare Fig. 7-15 with Fig. 7-16). Current studies suggest that hemoglobin shows features of both models, neither of which can fully explain the complexity of protein dynamics upon ligand binding.

12. There appear to be nearly 950 allelic variants of hemoglobin in humans, >90% of which are due to single amino acid substitutions. Most do not result in any discernible clinical symptoms. Hemoglobin variants that have demonstrated clinical symptoms are most commonly correlated with (1) decreased ability of globins to bind heme; (2) decreased cooperativity in which the protein is "locked" in either the T- or R-state; (3) increased propensity for oxidation of the ferrous iron to the ferric state; and (4) decreased protein stability resulting in protein degradation.

13. The analysis of mutant hemoglobins with altered functions has conveyed considerable knowledge about protein structure–function relationships. One variant, hemoglobin S, has a substitution of valine for glutamic acid in the β subunit. In the deoxy state, hemoglobin S aggregates, causing erythrocytes to sickle and block the capillaries. Individuals who are homozygous for the gene specifying hemoglobin S have sickle cell anemia, a debilitating and often fatal disease.

Muscle Contraction

14. Striated muscle is made of parallel bundles of myofibrils. As seen in the electron microscope, I bands of lesser electron density alternate with A bands of greater density. The repeating unit, the sarcomere, is bounded by Z disks at the centers of adjacent I bands and includes the A band, which is centered on the M disk. Within the sarcomere, thick filaments are linked to thin filaments by cross-bridges. Muscle contraction occurs when the filaments slide past each other, bringing the Z disks closer together.

15. Thick filaments are made of myosin, which consists of six subunits: two heavy chains and two pairs of light chains. A heavy chain consists of a globular head and a long α-helical tail. Two such tails associate in a 1600-Å-long left-handed coiled coil, yielding a rodlike molecule that has two globular heads. The light chains bind near the globular heads. Thin filaments are composed of actin, tropomyosin, and troponin. Tropomyosin is a coiled coil that winds in the actin polymer's helical grove so as to contact seven successive actin monomers. Troponin, a heterotrimer , interacts with tropomyosin and serves as a sensor molecule, binding Ca^{2+} released from the sarcoplasmic reticulum upon stimulation of muscle contraction. Additional muscle proteins help define sarcomere structure.

16. Dystrophin, a critical protein link between F-actin filaments and a transmembrane glycoprotein, helps prevent large membrane ruptures from the mechanical stress of muscle contraction. Sufferers of Duchenne Muscular Dystrophy (DMD) lack this protein, while patients with milder Becker Muscular Dystrophy (BMD) have reduced levels.

17. The sarcomere contraction cycle involves conformational changes in myosin that drive the movement of actin filaments towards each other. (See Figure 7-32.) This model was born out of the sliding filament model proposed by Hugh Huxley and Andrew Huxley in 1954 (Box 7-4).
    (a) The myosin heads of the thick filaments bind to the thin filaments. When ATP binds to the myosin head, myosin dissociates from the actin filament.
    (b) Subsequent ATP hydrolysis cocks the head of myosin and allows it to rebind weakly to actin.

(c) The release of inorganic phosphate increases the strength of binding and causes the head of myosin to snap back in the power stroke. This pulls the filaments past each other. Each myosin head acts in this manner to cause muscle contraction.

(d) ADP is released, completing the cycle.

18. Ca^{2+}, which binds to troponin C, triggers muscle contraction. Nerve impulses induce the release of Ca^{2+} from intracellular stores. This Ca^{2+} binds to troponin and causes a conformational change that exposes additional myosin binding sites on the thin filament. At lower Ca^{2+} concentrations, the myosin head is blocked from binding actin and the muscle is relaxed.

19. Nonmuscle microfilaments (~70-Å-diameter fibers), which consist of actin, occur in all eukaryotic cells, where their assembly and disassembly drives such cellular processes as ameboid locomotion and cytokinesis.

20. In most eukaryotic cells, actin is the most abundant cytosolic protein. The globular polypeptide has an ATP-binding site on one side of the protein. When actin polymerizes, this binding site marks the (–) end or pointed end of the polymer; the opposite end is referred to as the (+) end or barbed end. Actin also has a divalent cation–binding site that has a greater affinity for Ca^{2+} than Mg^{2+}; however, most G-actin *in vivo* is bound to Mg^{2+} due to its greater concentration in cells.

    (a) Polymerization of G-actin into F-actin differs at each end of the growing filament: The rate of subunit addition at the (+) is 5- to 10-fold greater than that at the (–) end, so new growth tends to occur at the (+) end (hence its name).

    (b) The rate of addition of G-actin monomers may be matched by the rate of G-actin monomer dissociation. This steady state process, called treadmilling, plays an important role in the amoeboid crawling by cells.

Antibodies

21. Antibodies (immunoglobulins) are proteins produced by the immune system of higher organisms to protect them against pathogens such as viruses and bacteria. Antibodies are produced by B lymphocytes, which recognize foreign macromolecules (antigens). The primary response to an antigen requires several days for B cells to generate the required antibodies. If the organism subsequently encounters the same antigen, a secondary response results, in which large amounts of the antibody are produced more rapidly.

22. Antibodies contain at least four subunits, two identical light chains and two identical heavy chains, which are held together in part by interchain disulfide bonds to form a Y-shaped molecule. Of the five classes of antibodies, IgG is the most abundant. The classes are distinguished by the type of heavy chain (α, δ, ε, γ, and μ). There are two types of light chain (κ and λ). IgG heavy chains each consist of three domains of constant sequence, designated C_H1, C_H2, and C_H3, and one variable domain, designated V_H. The light chains each consist of one constant domain, C_L, and one variable domain, V_L. The V_H and V_L are located at the two split ends of the Y-shaped protein.

23. The variable regions contain the antigen-binding sites in which hypervariable sequences determine the exquisite specificity of antibody–antigen interactions.

24. The binding of an antibody to its antigen is highly specific and has a dissociation constant ranging form 10^{-4} to 10^{-10} M. Because antibodies each have two identical antigen-binding sites, a population of antibodies can form large antigen–antibody aggregates that hasten the removal of the antigen and induce B cell proliferation.

25. A B cell produces one kind of antibody. Antigens stimulate the proliferation of a population of pre-existing B cells whose diversity arises from genetic changes that occur in B cell development, including somatic recombination and somatic hypermutation. Sequence variation allows the synthesis of potentially billions of immunoglobulins with different antigen-binding specificities.

26. In autoimmune diseases, the organism loses its self-tolerance and produces antibodies against its own tissues. This process is sometimes triggered by trauma or infection and may result from the resemblance of a self-antigen to some foreign antigen. Autoimmune diseases have variable symptoms ranging from mild to lethal.

30. Monoclonal antibodies, developed by César Milstein and Georges Köhler, are derived from clones of fused cells that make antibodies specific for one antigen (Box 7-5). Monoclonal antibodies can be used as the stationary phase of affinity chromatography columns for the purification of proteins, they can be used to identify specific infectious agents and test for the presence of drugs and other diagnostic molecules in body tissues, and they can be used in chemotherapy against specific cancers.

Key Equations

$$Y_{O_2} = \frac{pO_2}{K + pO_2}$$

$$Y_{O_2} = \frac{(pO_2)^n}{(p_{50})^n + (pO_2)^n}$$

Questions

Oxygen Binding to Myoglobin and Hemoglobin

1. Match the descriptions on the left with the terms on the right.

    _____ A component of cytochromes   A. methemoglobin
    _____ Binds O_2 B. myoglobin
    _____ Contains iron in the Fe(III) state   C. hemoglobin
    _____ Found in muscle only   D. hemoglobin S
    _____ Forms filaments in the deoxy state   E. heme

2. The heme moiety by itself can bind oxygen. What physiological function does the globin serve?

3. How do tissues with high metabolic activity facilitate oxygen delivery?

4. How would a lower p_{50} affect hemoglobin's oxygen acquisition in the lungs and oxygen delivery to the peripheral tissues?

5. Describe, on the molecular level, the role of myoglobin in O_2 transport in rapidly respiring muscle tissue.

6. What function(s) does carbamate formation in hemoglobin serve?

7. Which of the following modulators of O_2 binding to hemoglobin counteract each other? CO_2, H^+, BPG.

8. Explain why individuals with severe carbon monoxide poisoning are often given transfusions instead of oxygen-rich gas.

9. Match each of the structural elements of myoglobin or hemoglobin with its function below.

A.	His F8	F.	α_1–β_1 interface/α_2–β_2 interface
B.	His E7	G.	oxymyoglobin/deoxymyoglobin
C.	E and F helices	H.	oxyhemoglobin/deoxyhemoglobin
D.	O_2–Fe(II)	I.	α_1–β_2 interface/α_2–β_1 interface
E.	Val E11	J.	C-terminal salt bridges

____ Forms a coordination bond with Fe(II)
____ Involved in the binding of heme
____ Partially occludes the O_2-binding site
____ Conformations are nearly superimposable
____ Forms a hydrogen bond with O_2
____ Is associated with a rotational shift of the hemoglobin $\alpha_1\beta_1$ dimer with respect to the $\alpha_2\beta_2$ dimer.
____ Disrupts the N- and C-terminal salt bridges in hemoglobin
____ Stabilizes the T state
____ Involved in the interactions of hemoglobin subunits

10. Oxygen binding to hemoglobin decreases the pK of the imidazole group of His 146β from 8.0 to 7.1. How does this contribute to the Bohr effect?

11.   Describe how BPG decreases the O_2-binding affinity of hemoglobin in terms of the T $\leftrightarrow$ R equilibrium.

12.   What kind of allosteric effect is inconsistent with the symmetry model?

13.   How many conformational states are possible for a trimeric binding protein, whose three subunits each contain a ligand-binding site, when binding follows the (a) symmetry model or (b) the sequential model of allosterism?

14.   Examine the hemoglobin variants in Table 7-1 and provide plausible answers to the following questions.
      (a) Which variant might be advantageous for an individual living in the Andes?
      (b) Which variants are likely to be the least stable, resulting in hemolytic anemia?
      (c) Which variants are likely to result in increased hematocrit (elevated erythrocyte content)?
      (d) Which variants are most like to affect the Hill constant?

15.   What physiological condition leads to hemoglobin S fiber formation in the capillaries?

16.   Some hemoglobin S individuals have significant levels of fetal hemoglobin in their erythrocytes. Why is this an advantage?

Muscle Contraction

17.   Describe how ATP hydrolysis is involved in muscle contraction.

18.   How does calcium regulate muscle contraction?

19.   The graph below, the time course of microfilament polymerization *in vitro*, shows three phases of growth. (a) Describe the biochemical events taking place during each phase. Assume that ATP is in excess. (b) How might this graph look if ATP concentration becomes limiting as ATP is hydrolyzed by actin?

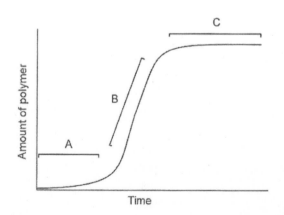

Antibodies

20. In a typical protocol for preparing antibodies for laboratory use, an animal is injected with an antigen several times over a period of weeks or months. Why are multiple injections useful?

21. Indicate which class of immunoglobulin (IgA, IgD, IgE, IgG, or IgM) best corresponds to each characteristic listed below (more than one may be implicated by each description).
    (a) First to be secreted in response to an antigen
    (b) Implicated in allergic reactions
    (c) Occurs in the intestinal tract
    (d) Contains J chains
    (e) The most abundant antibody

22. The following questions refer to the production of monoclonal antibodies.
    (a) Why are myeloma cells used?
    (b) Do all cells growing in the selective medium produce antibodies to antigen X?
    (c) Why do only hybrid cells grow in the selective medium?

23. You wish to isolate a large amount of Fab fragments in order to examine their binding to protein X by X-ray crystallography. At first, you inject a large rabbit with protein X to obtain antibody. On further reflection, you decide to inject the antigen into a mouse in order to produce monoclonal antibodies.
    (a) How can you purify protein X–specific antibodies from rabbit serum or the hybridoma medium?
    (b) How do the rabbit and mouse antibody preparations differ?
    (c) Would the rabbit or mouse antibodies be more suitable for X-ray crystallography?

24. Why are the hypervariable sequences of immunoglobulins located in loops?

Answers to Questions

1.
E	A component of cytochromes
B, C, D	Binds O_2
A	Contains iron in the Fe(III) state
B	Found in muscle only
D	Forms filaments in the deoxy state

2. The globin prevents the oxidation of its bound heme to the Fe(III) state, and in the case of hemoglobin, permits cooperative O_2 binding, which is responsible for the efficient transport of O_2 from the lungs to the tissues.

3. High metabolic activity generates CO_2, which reacts with water to form H_2CO_3, which in turn ionizes to yield $HCO_3^- + H^+$. The protons preferentially bind to T-state hemoglobin (the Bohr effect) and thereby cause O_2 to be released.

4. A lower p_{50}, or higher affinity of hemoglobin for O_2, would have almost no effect on O_2 uptake in the lungs, since hemoglobin is nearly saturated at arterial pO_2. However, the lower p_{50} would result in less O_2 released at the peripheral tissues, where the pO_2 is ~30 torr. The p_{50} of normal hemoglobin (26 torr) is an evolutionary compromise that allows hemoglobin to become saturated with O_2 in the lungs but to be deoxygenated in the oxygen-poor peripheral tissues.

5. The O_2 that is released in the capillaries diffuses to the mitochondria, where it is reduced to water. However, the solubility of O_2 in aqueous solution is too low to support its required rate of diffusion in rapidly respiring muscle. The presence of myoglobin, in effect, increases the solubility of O_2 in muscle tissue. Thus, the O_2 released by hemoglobin is passed from myoglobin molecule to myoglobin molecule in a kind of molecular bucket brigade until it is taken up by the mitochondria. In this way, myoglobin increases the rate that O_2 can diffuse through the tissues.

6. CO_2 reacts preferentially with amino groups in deoxyhemoglobin to form carbamates. This helps transport CO_2 and facilitates the release of O_2 by stabilizing the deoxy state.

7. None; CO_2, H^+, and BPG all bind preferentially to deoxyhemoglobin to reduce the affinity of hemoglobin for O_2.

8. CO binds to hemoglobin with so much higher affinity than O_2 that CO cannot be displaced by high concentrations of oxygen. For practical purposes, CO binding to hemoglobin is therefore irreversible (at least in the short term). Therefore, a fresh blood transfusion is required to counteract the effects of this poison.

9.

A	Forms a coordination bond with Fe(II)
A, C	Involved in the binding of heme
E	Partially occludes the O_2-binding site
G	Conformations are nearly superimposable
B	Forms a hydrogen bond with O_2
I	Is associated with a rotational shift of the hemoglobin $\alpha_1\beta_1$ dimer with respect to the $\alpha_2\beta_2$ dimer.
D	Disrupts the N- and C-terminal salt bridges in hemoglobin
J	Stabilizes the T state
F, I	Involved in the interactions of hemoglobin subunits

10. The decrease in pK promotes deprotonation of the imidazole group at pH 7.4. The release of H^+ contributes to the Bohr effect.

11. BPG binds preferentially to the T state and stabilizes it through the formation of salt bridges between BPG and the β subunits. This makes the shift to the R state less energetically favorable, thereby decreasing hemoglobin's ability to bind O_2.

12. The symmetry model cannot account for negative cooperativity, in which the binding of a ligand to one subunit decreases the binding affinity of the other subunits.

13. (a) Only two conformational states are possible, either R or T, since all subunits change conformation simultaneously.

(b) Four conformations are possible, corresponding to a trimer in which 0, 1, 2, or 3 subunits have bound ligand.

14. (a) Hb Kansas. Individuals with this variant might be relatively symptom-free in the Andes, since this mutation, which destabilizes oxyhemoglobin, would allow more oxygen to be given up to tissues.

(b) All the variants listed, except for Hb Kansas and Hb Yakima, are likely to be unstable, since all will affect key electrostatic interactions within the globin polypeptides or the quaternary structure of the variant hemoglobin.

(c) Hb Yakima. The mutation in this variant destabilizes the T state, which leads to a decrease in oxygen dissociation in the peripheral tissues. The body thereby compensates by increasing the amount of hemoglobin.

(d) All the variants listed may show changes in the Hill constant, since cooperativity will be affected by all the changes described in Table 7-1.

15. The low pO_2 of the capillaries, which leads to an increase in the concentration of deoxyhemoglobin, promotes polymerization of hemoglobin S.

16. Fetal hemoglobin (which contains γ globin chains rather than β chains) dilutes the hemoglobin S in erythrocytes, so that the deoxyhemoglobin S in the venous blood is less likely to achieve the critical concentration for fiber formation.

17. Myosin moves along the actin filament via repeated cycles of conformational changes in the head region of the myosin molecule. This cycle of conformational changes is unidirectional because it is coupled to the irreversible hydrolysis of ATP. When the myosin head binds ATP, it releases the actin filament to which it is bound. ATP hydrolysis follows, resulting in a change in the conformation of the myosin head to the "cocked" or high-energy state. The myosin head then binds weakly to the actin filament at a new position. The strength of this binding interaction increases upon the release of P_i. The myosin head then undergoes a second major conformational change, producing the power stroke that causes the translocation of the actin filament relative to the myosin filament. Upon completion of the power stroke, the myosin head releases ADP but remains bound to the actin filament. It can then bind another ATP molecule to begin the cycle anew. The thick filament's multiple myosin heads, each undergoing multiple ATP-driven reaction cycles, cause the thick filament to "walk" along the thin filament. As the thick and thin filaments slide past each other, the sarcomere decreases in length and the muscle thereby contracts.

18. The thin filaments of striated muscle contain actin in complex with tropomyosin and troponin. At resting Ca^{2+} concentrations, the muscle is relaxed because tropomyosin blocks the myosin binding sites on the actin filament. When the intracellular Ca^{2+} concentration increases in response to a nerve impulse, troponin C (a subunit of troponin) binds Ca^{2+}, causing a conformational change in troponin. This results in the movement of tropomyosin, which uncovers the myosin binding sites, and thus permits the myosin heads to interact with the thin filament.

19. (a) During phase A, most of the actin exists in the form of small oligomers. Only a few oligomers nucleate microfilament growth. During phase B, rapid growth occurs at the ends of growing filaments. During phase C, the rate of filament growth is equal to the rate of dissociation of ADP–G-actin from the filaments, so there is no net increase in F-actin.

    (b) If the ATP concentration becomes limiting as it is hydrolyzed in actin subunits, the entire population of actin subunits will eventually contain bound ADP. ADP–G-actin can still form filaments, but since ADP–F-actin is more likely to disassemble than ATP–F-actin, the F-actin concentration will decrease until a new equilibrium is established, as shown below.

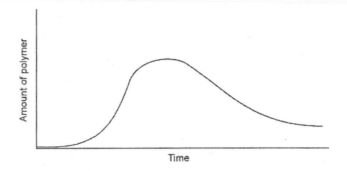

20. A secondary immune response is greater than the first (see Figure 7-37), so more antibody can be recovered.

21. (a) IgM   (b) IgE   (c) IgA   (d) IgA, IgM   (e) IgG

22. (a) Antibody-producing lymphocytes have a limited proliferative capacity and are therefore unsuitable for growing in large numbers in culture to produce large amounts of antibody. Myeloma cells, like all cancer cells, have an unlimited proliferative capacity and impart their immortal phenotype to the hybrid cells.

    (b) All the lymphocytes harvested from the immunized animal can potentially fuse with the myeloma cells. The resulting hybridomas will therefore secrete a variety of different antibodies, only a small fraction of which are specific for protein X.

    (c) Unfused lymphocytes have no metabolic deficiency but grow for only a limited time, and then die out. In the selective medium, unfused myeloma cells cannot synthesize purines, which are necessary for DNA replication and cell division. Only the fused cells can both synthesize purines and proliferate without limit.

23. (a) Affinity chromatography using immobilized protein X would be the best procedure.
    (b) It is likely that protein X contains several antigenic features that are recognized by B cells and elicit antibody production. The anti-protein X antibodies isolated from the rabbit are therefore a mixture of different IgG molecules that recognize different antigenic features on protein X. Even antibodies that recognize the same portion of the antigen may have different amino acid sequences and different antigen-binding affinities. In contrast, each mouse monoclonal antibody preparation is homogeneous. Because monoclonal antibodies are all the products of identical B cells, they all have the same unique sequence and antigen-binding specificity.
    (c) One of the mouse monoclonal antibodies would be most suitable, since it would yield identical Fab fragments that would be likely to form regular crystals. In contrast, Fab fragments isolated from a heterogeneous population of rabbit antibodies could not form such crystals.

24. The loops can accommodate a wide variety of amino acid sequences because they are on the surface of the protein. Such sequence variation in the β sheet core of the protein would likely disrupt its structure.

8
Carbohydrates

This chapter is concerned with the structures and properties of carbohydrates. These molecules contain just three elements, namely carbon, hydrogen and oxygen (although their derivatives may also contain nitrogen and sulfur). Carbohydrates are not only important metabolic energy sources, as detailed in Chapters 15 and 16, but they also have key functions in molecular and cellular recognition events. You will first learn the structures and chemical characteristics of monosaccharides and some of their derivatives, and then of oligosaccharides and polysaccharides. This is followed by a presentation of the composition, structure, and function of molecules in which carbohydrate are covalently linked to polypeptides, including proteoglycans, bacterial cell wall components, and glycoproteins. It is important to realize that a large proportion of proteins contain covalently attached carbohydrates and that the structures of these carbohydrate chains can vary enormously.

Essential Concepts

Monosaccharides

1. Monosaccharides can be defined as aldehyde or ketone derivatives of straight-chain polyhydroxy alcohols containing a minimum of three carbon atoms. Sugars that have aldehyde groups are called aldoses, whereas those with ketone moieties are termed ketoses. Depending on the number of carbon atoms, monosaccharides are referred to as trioses, tetroses, pentoses, hexoses, etc.

2. D-Glucose has four chiral centers and is therefore one of 16 possible aldohexose stereoisomers. D Sugars have the same absolute configuration at the asymmetric center most distant from the carbonyl group as does D-glyceraldehyde. Epimers are sugars in which the configuration around one carbon atom differs. Ketoses, which have a ketone function at C2, have one less chiral center than aldoses with the same number of carbons. Therefore, a ketohexose has only 8 possible stereoisomers.

3. Sugars can be represented in their cyclic hemiacetal and hemiketal forms as planar Haworth projections. Sugars that form six-membered rings are known as pyranoses, whereas those with five-membered rings are known as furanoses. A cyclic monosaccharide exists as either an α or a β anomer. Anomers freely interconvert in aqueous solution via the linear (open chain) form.

4. Five- and six-membered sugar rings are most the abundant, probably because of their stability. The tetrahedral bonding angles of carbon prevent the rings from being truly planar. The pyranose ring prefers the chair conformation and exists predominantly in the form that minimizes steric interactions among bulky ring substituents (i.e., bulky groups tend to occupy equatorial rather than axial positions).

5. Sugars can undergo reactions characteristic of aldehydes and ketones:
    (a) Oxidation of the aldehyde group of an aldose yields an aldonic acid. Thus, D-glucose oxidation results in D-gluconic acid.
    (b) Oxidation of the primary hydroxyl group produces a uronic acid, such as D-glucuronic acid from D-glucose.
    (c) Reduction of aldoses and ketoses yields polyhydroxy alcohols called alditols. Thus, D-glucose reduction gives glucitol (also known as sorbitol).

6. Monosaccharides in which an OH group is replaced by an H are called deoxy sugars. The most important of these is β-D-2-deoxyribose, a component of DNA.

7. In amino sugars, an OH group is replaced by an amino group that is usually acetylated. A common amino sugar is *N*-acetylglucosamine. *N*-acetylneuraminic acid, an important constituent of glycoproteins (see below), is composed of an acetylated amino sugar, *N*-acetylmannosamine, covalently linked to pyruvic acid.

8. The anomeric carbon of a sugar can form a covalent bond with an alcohol to form an α or β glycoside. The bond is called a glycosidic bond. An *N*-glycosidic bond links an anomeric carbon and a nitrogen atom, as in the covalent bond between ribose and a purine or pyrimidine.

9. Reducing sugars are saccharides with anomeric carbons that have not formed glycosides. The free aldehyde of these sugars can reduce mild oxidizing agents. Hence, the identification of a nonreducing sugar represents evidence that the saccharide is a glycoside.

Polysaccharides

10. Polysaccharides are made up of monosaccharides covalently linked by glycosidic bonds. A homopolysaccharide contains one type of monosaccharide, whereas a heteropolysaccharide can contain diverse monosaccharides. Polysaccharides may be linear or branched, because glycosidic bonds can form between an anomeric carbon and any of the hydroxyl groups of another monosaccharide. Naturally occurring polysaccharides incorporate only a few types of monosaccharides and glycosidic linkages.

11. Disaccharides consist of two sugars linked by a glycosidic bond. One example is lactose, in which C1 of galactose is linked to C4 of glucose by an α-glycosidic bond. Lactose is a reducing sugar because the free anomeric carbon on the glucose residue can reduce a mild oxidizing agent. Another disaccharide is sucrose, common table sugar, in which the C1 anomeric carbons of glucose and fructose are joined by an α-glycosidic bond. Sucrose is therefore a nonreducing sugar.

12. Cellulose, the most abundant polysaccharide, is a large, linear polymer in which glucose units are linked by β(1→4) glycosidic bonds. Cellulose forms a highly hydrogen-bonded structure of enormous strength that contributes to the rigidity of plant cell walls. Cows and other herbivores can utilize cellulose as a nutrient because they harbor microbes that produce cellulases which cleave the glycosidic bonds. Another widely distributed

polysaccharide is chitin, which comprises the exoskeletons of many invertebrates. It is a homopolymer of *N*-acetylglucosamine residues linked in β(1→4) fashion. Another key component of plant cell walls are pectins, heterogeneous polysaccharides (similar to glycosaminoglycans described below).

13. Starch is the principal food reserve in plants. It has two components: α-amylose, a linear polymer of α(1→4) linked glucose units, and amylopectin, an α(1→4) linked glucose polymer bearing periodic branches linked by α(1→6) bonds. Digestion of starch by animals begins with the action of salivary amylase, an enzyme that cleaves α(1→4) glycosidic bonds, and is continued in the small intestine by pancreatic amylase, α-glucosidase, and a debranching enzyme that can cleave α(1→6) links. The resulting oligosaccharides are eventually converted to glucose, which can be absorbed by the intestine.

14. Glycogen is a polysaccharide that is synthesized and stored by animals, primarily in skeletal muscle and liver. The structure of glycogen resembles that of amylopectin but is more highly branched. When needed for metabolic energy, glycogen is broken down through the combined action of glycogen phosphorylase, which cleaves α(1→4) bonds, and glycogen debranching enzyme.

15. Glycosaminoglycans are major constituents of the extracellular matrix in eukaryotes and in bacterial biofilms. Most of these rigid, linear polysaccharides are composed of alternating uronic acid and hexosamine residues. For example, hyaluronic acid is composed of D-glucuronate linked β(1→3) to *N*-acetylglucosamine which in turn is linked β(1→4) to the next glucuronate residue. Because of its polyanionic nature, hyaluronic acid forms viscoelastic solutions, a property that makes it an effective biological shock absorber and lubricant.

16. Other types of glycosaminoglycans, all of which are composed of sulfated disaccharide units, include chondroitin sulfates, dermatan sulfate, keratin sulfate, and heparin. The last-named substance is found in mast cells and inhibits blood clotting.

Glycoproteins

17. A large proportion of all proteins have covalently bound carbohydrates and are therefore glycoproteins. The polypeptide chains of glycoproteins are encoded by nucleic acids, whereas the attached oligosaccharide chains are products of enzymatic reactions. This is the source of microheterogeneity, the variability in composition of the carbohydrate component in a population of glycoprotein molecules that all have the same polypeptide chain.

18. Proteoglycans occur mainly in the extracellular matrix and are combinations of proteins and glycosaminoglycans that associate by both covalent and noncovalent bonds. These molecules have a bottlebrush-like structure (e.g., Figure 8-15) in which up to 100 core proteins with attached glycosaminoglycans and both *N*-linked and *O*-linked oligosaccharides are linked to hyaluronate. This assembly of protein and carbohydrate is a

huge, space-filling macromolecule. Proteoglycans are highly hydrated, so that, in combination with collagen, they account for the high resilience of cartilage.

19. Bacteria possess rigid cell walls that are responsible in part for their virulence. In gram-positive bacteria, the cell wall consists of polysaccharide and polypeptide chains that are covalently attached to form a baglike structure called peptidoglycan that completely envelops the cell. Gram-negative bacteria have a relatively thin peptidoglycan cell wall surrounded by a complex outer membrane.

20. The polysaccharide of some bacterial cell walls consists of alternating residues of $\beta(1\rightarrow4)$-linked N-acetylmuramic acid and N-acetylglucosamine in which the N-acetylmuramic acid residues are linked via an amide bond to a tetrapeptide containing D-amino acids. A continuous meshlike framework is formed by cross-linking adjacent peptidoglycan chains through their tetrapeptide side chains (Figure 8-17). The enzyme lysozyme can degrade peptidoglycan by cleaving the glycosidic bond between N-acetylmuramic acid and N-acetylglucosamine. The antibiotic action of penicillin rests on its ability to inhibit the formation of cross-links in peptidoglycan.

21. Glycoproteins include nearly all membrane-bound and secreted eukaryotic proteins. The oligosaccharide chains are attached to the proteins by either N-glycosidic or O-glycosidic bonds. In an N-glycosidic bond, the amide group of an asparagine in the sequence Asn-X-Ser/Thr is linked to N-acetylglucosamine. N-glycosylation occurs in stages:
    (a) An oligosaccharide rich in mannose and containing glucose and N-acetylglucosamine is attached cotranslationally, that is, while the polypeptide is being synthesized on ribosomes bound to the endoplasmic reticulum.
    (b) The oligosaccharide undergoes trimming, the enzymatic removal of some sugars, as the glycoprotein moves from the endoplasmic reticulum to the Golgi apparatus.
    (c) Further processing occurs in the Golgi apparatus, where monosaccharides such as N-acetylglucosamine, galactose, L-fucose, and N-acetylneuraminic acid are enzymatically added to the trimmed chain by glycosyltransferases. N-linked glycoproteins exhibit great diversity in their oligosaccharide chains due to differences in the extent of processing.

22. O-glycosidically linked oligosaccharide chains are covalently linked to a Ser or Thr side chain in a protein and vary considerably in structure. O-linked oligosaccharides are added to completed polypeptides in the Golgi apparatus and are built up by stepwise addition of monosaccharides in reactions catalyzed by glycosyltransferases.

23. The number and structure of N- and O-linked oligosaccharides attached to a given polypeptide chain can vary, giving rise to glycoprotein variants called glycoforms.

24. Particular functions can sometimes be attributed to the presence of oligosaccharides. Oligosaccharides attached to proteins may modulate the conformational freedom of the polypeptide. The hydrophilic oligosaccharide chains may take up considerable volume and thereby tend to protect the protein from enzymatic attack or modify its activity.

25.   The enormous number of oligosaccharide structures suggests that they contain biological information and are important in molecular recognition. All cells have a coat of glycoconjugates (a mixture of glycoproteins and glycolipids). Proteins called lectins specifically recognize and bind to individual monosaccharides or small oligosaccharides in discrete glycosidic linkages at the cell surface. For example, the leukocyte selectins bind to cell-surface carbohydrates on endothelial cells, an interaction that helps direct leukocytes to a site of blood vessel injury.

26.   Different oligosaccharide components of certain glycolipids distinguish the ABO blood group antigens.

27.   Oligosaccharide chains mediate a variety of biological functions that depend on molecular recognition between proteins and carbohydrates. Among these are delivery of proteins to appropriate destinations within cells and the regulation of cell growth.

Questions

Monosaccharides

1.   Indicate which of the following is an aldose, a ketose, a pentose, a hexose, a uronic acid, an alditol, a deoxy sugar, or a reducing sugar.

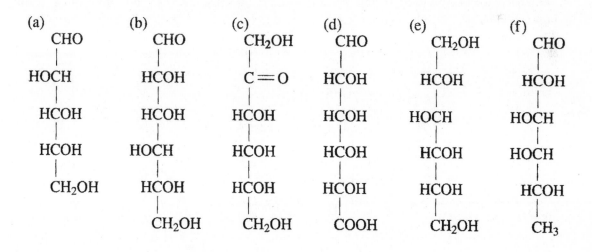

2.   Draw the following monosaccharides as Haworth projections:
   (a)   α anomer of D-ribose
   (b)   β anomer of D-glucose
   (c)   β anomer of D-fructose
   (d)   methyl-β-D-galactose.

Which of these compounds contains a glycosidic bond?

3.   An equilibrium mixture of D-glucose contains approximately 63% β-D-glucopyranose and 36% α-D-glucopyranose. There are trace amounts of three other forms. What are they?

4. Although β-D-glucopyranose is the predominant form of glucose in solution, crystalline glucose consists almost exclusively of α-D-glucopyranose. What accounts for this difference?

5. Draw the most stable chair conformation of α-D-galactose.

Polysaccharides

6. Match each term at the top with its definition below.
   A. Alditol
   B. Epimer
   C. Glycan
   D. Anomer
   E. Glycoside
   _____ Differs in configuration at the anomeric carbon
   _____ Polyhydroxy alcohol
   _____ Product of condensation of anomeric carbon with an alcohol
   _____ Differs in configuration at one carbon atom.
   _____ Polymer of monosaccharides

7. Which of the following di- and trisaccharides contains fructose, contains an α anomeric bond, and/or is a reducing sugar?

(a)   sucrose

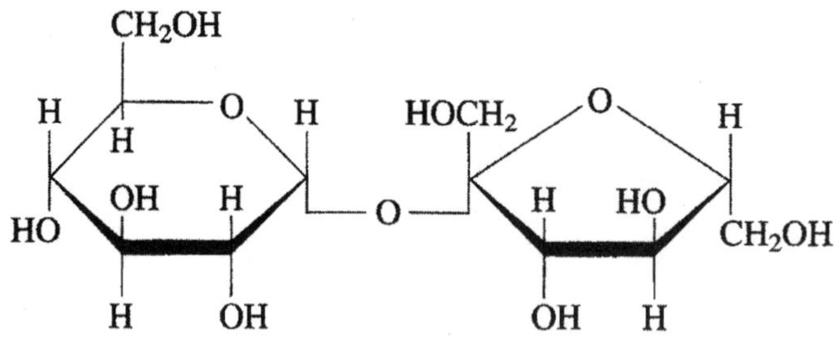

(b)   cellobiose

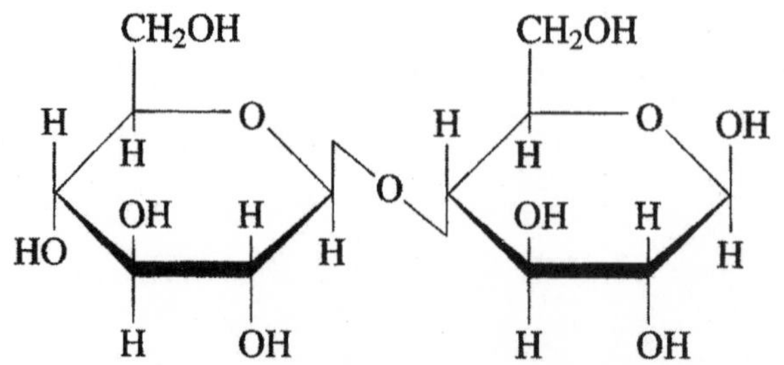

(c)   lactose

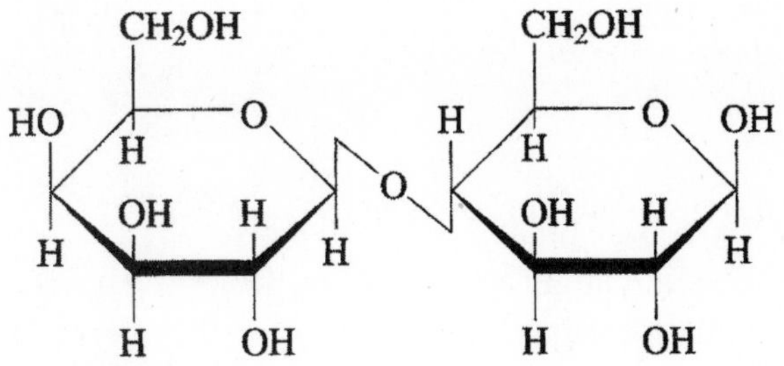

(d)   melibiose

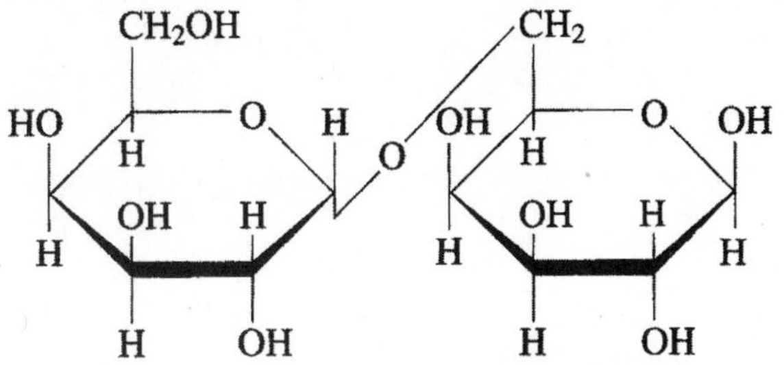

8.   An unknown trisaccharide was treated with methanol in HCl (to methylate its free OH groups) then subjected to acid hydrolysis (to break glycosidic bonds). The products were 2,3,4,6-tetra-*O*-methylgalactose, 2,3,4-tri-*O*-methylglucose, and 2,3,6-tri-*O*-methylglucose. Treatment of the intact trisaccharide with β(1→6)-galactosidase yielded D-galactose and a disaccharide. Treatment of this disaccharide with α(1→4)-glucosidase yielded D-glucose. Draw the structure of this trisaccharide and give its systematic name.

9.   How do the chemical differences between starch and cellulose result in their very different polymeric structures?

10.   Glycogen and starch are extensively branched high-molecular-weight polymers. Give two reasons why such a structure is advantageous for a fuel-storage molecule.

Glycoproteins

11.   How could lectins be used to purify polysaccharides and glycoproteins?

12.   Figure 8-19 presents structures of typical *N*-linked oligosaccharides. (a) Which sugars make up the core oligosaccharide? (b) What other sugars are typically found at the termini of branched oligosaccharides?

13.   A major protein of saliva contains several hundred identical covalently attached disaccharides containing *N*-acetylneuraminic acid in α(2→6) linkage to *N*-acetylgalactosamine that is linked to a Ser residue. Solutions of the intact protein are extremely viscous. However, when the protein is treated with sialidase, the viscosity decreases markedly. What features of the glycoprotein structure give rise to the high viscosity and why does sialidase treatment bring about the observed change?

14.   Explain why a type AB individual can receive a transfusion of type A or type B blood, but a type A or type B individual cannot receive a transfusion of type AB blood.

Answers to Questions

1.   (a) aldose, pentose, reducing sugar
      (b) aldose, hexose, reducing sugar
      (c) ketose, hexose, reducing sugar
      (d) aldose, hexose, uronic acid, reducing sugar
      (e) hexose, alditol
      (f) aldose, hexose, deoxy sugar, reducing sugar

2.

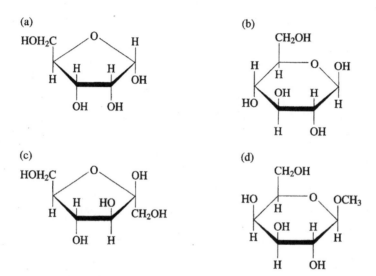

Only methyl-β-D-galactose (d) contains a glycosidic bond.

3.   The three other forms are the open chain form of glucose, α-D-glucofuranose and β-D-glucofuranose.

4.   The α anomer of glucose is less soluble than the β anomer and therefore comes out of solution more readily. As it crystallizes, the β-glucose remaining in solution interconverts with α-glucose thereby maintaining the 36% α–63% β equilibrium ratio. Thus, the α anomer is continually generated and deposited in the crystal.

5.

The most stable conformation has its bulky CH_2OH group and two OH groups in equatorial positions, which minimizes their steric interference.

6.

_D_   Differs in configuration at the anomeric carbon
_A_   Polyhydroxy alcohol
_E_   Product of condensation of anomeric carbon with an alcohol
_B_   Differs in configuration at one carbon atom.
_C_   Polymer of monosaccharides

7.   (a)   Contains fructose, contains an α anomeric bond.
     (b)   Is a reducing sugar.
     (c)   Is a reducing sugar.
     (d)   Contains an α anomeric bond, is a reducing sugar.
     (e)   Contains fructose, contains an α anomeric bond.

8.

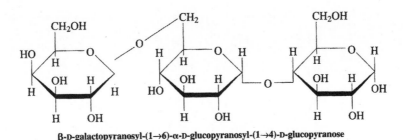

β-D-galactopyranosyl-(1→6)-α-D-glucopyranosyl-(1→4)-D-glucopyranose

β-D-galactopyranosyl-(1→6)-α-D-glucopyranosyl-(1→4)-D-glucopyranose

9.   The β(1→4) glycosidic bonds that join glucose groups in cellulose give rise to extended polymers that line up in parallel and are stabilized by extensive intrachain and interchain hydrogen bonds. This gives cellulose fibers unusual strength and renders them water insoluble. In contrast, the α(1→4) glycosidic bonds linking glucose residues in starch give rise to a totally different structure, namely a linear chain that assumes a relatively open helical conformation.

10.   Two advantages are (1) the reduction in osmotic pressure in a cell when many residues are combined into a single polymeric molecule, and (2) the rapid mobilization of glucose when it is removed from the ends of many branches simultaneously.

11. Lectins are proteins that bind specific sugars with high affinity. Thus, a lectin attached to an immobile support can be used to selectively adsorb polysaccharides or glycoproteins via affinity chromatography. The adsorbed molecules can then be eluted by applying a solution containing an excess of the free sugar to which the lectin preferentially binds.

12. (a)   The core oligosaccharide is $(mannose)_3(GlcNAc)_2$.
     (b) Typical terminal sugars are galactose, *N*-acetylneuraminic acid (sialic acid), and fucose.

13. Each disaccharide unit bears a negative charge (from the ionized COOH group of *N*-acetylneuraminic acid). Because of the strong repulsion between charged groups, the protein assumes a rigid elongated shape, thereby accounting for the high viscosity of the solution. Removal of the *N*-acetylneuraminic acid residues by sialidase abolishes the large net negative charge and permits the protein to assume a more compact shape. As a result, the viscosity of the solution decreases.

14. Type AB individuals have both type A and type B carbohydrate structures on their cell surfaces, so they do not recognize either type A or type B blood cells as foreign. Consequently, they can receive either type A or type B blood. Type A individuals synthesize antibodies to type B carbohydrate antigens, so transfused blood cells bearing the type B carbohydrate antigen will agglutinate in the blood vessels. Similarly, type B individuals synthesize antibodies to the type A carbohydrate antigen and therefore cannot receive type A blood.

9

Lipids and Biological Membranes

Lipids are vital components of all cells. Unlike nucleic acids, proteins, and carbohydrates, lipids do not have unifying structural features. They are defined operationally as a diverse group of nonpolar substances. They are therefore poorly or not at all water-soluble but dissolve readily in many organic solvents. Lipids are essential constituents of biological membranes. In addition, the acyl chains of lipids serve as energy sources. Finally, a number of cell signaling processes involve lipids. This chapter begins by focusing on the structures and physical characteristics of lipids and on the properties of the lipid bilayer. The chapter then turns to biological membranes, which are composed of a great variety of proteins as well as lipids. Membrane proteins act as enzymes, facilitate transmembrane transport, and use information concerning the extracellular milieu to prompt the correct intracellular response. This chapter describes the features of membrane proteins that enable them to interact with lipids, and describes the overall arrangement of lipids and proteins in membranes. The chapter concludes with discussions of the synthesis of membrane and secreted proteins (a process in which a polypeptide is translocated through the endoplasmic reticulum membrane), intracellular trafficking of vesicles during exocytosis and endocytosis, and membrane fusion.

Essential Concepts

Lipid Classification

1. Fatty acids are common components of other lipids, where they occur in ester or amide linkages. They are straight-chain carboxylic acids, usually having between 14 and 22 carbons. Eukaryotic fatty acids usually possess an even number of carbon atoms and may contain one or more double bonds. Although the systematic names of fatty acids reveal their structures, many fatty acids also have common names.

2. Saturated fatty acids, of which the most common are palmitic acid (C16; hexadecanoic acid) and stearic acid (C18; octadecanoic acid), contain no double bonds and are highly flexible molecules that tend to assume a fully extended conformation. In the pure compound, neighboring saturated fatty acyl chains pack together tightly. The resulting van der Waals interactions cause fatty acid melting points to increase as chain length increases.

3. The double bonds of unsaturated fatty acids nearly always have the cis configuration, which introduces a 30° bend in the acyl chain. This prevents unsaturated fatty acids from packing together as closely as saturated fatty acids. As a result, the melting point of an unsaturated fatty acid is always lower than the melting point for a saturated fatty acid with the same number of carbons.

4. Most fatty acid groups in eukaryotic lipids are unsaturated. The most common fatty acid with one double bond is oleic acid (C18; 9-octadecenoic acid). Fatty acids with two or more double bonds are termed polyunsaturated.

5. Triacylglycerols, which contain three fatty acids esterified to the hydroxyl groups of d-glycerol, serve as energy reserves in plants and animals. A triacylglycerol generally contains more than one type of fatty acyl group. Mixtures of triacylglycerols may be fats (which are solid at room temperature) or oils (which are liquid at room temperature), depending on the properties of their component fatty acyl groups.

6. The highly reduced nature of triacylglycerols makes them an efficient metabolic energy store. Adipocytes are cells specialized for the biosynthesis, storage, and breakdown of triacylglycerols. Adipose tissue also provides thermal insulation.

7. The major lipid constituents of biological membranes are glycerophospholipids. These substances are composed of d-glycerol with fatty acids esterified to C1 and C2 and a phosphate group esterified to C3. The phosphate is almost always also esterified to a hydrophilic moiety. Thus, glycerophospholipids are amphipathic molecules with a polar head group and a nonpolar tail. Glycerophospholipids frequently contain saturated or monounsaturated acyl groups at C1 of glycerol and more highly unsaturated acyl moieties at C2.

8. Phospholipases are enzymes that hydrolyze glycerophospholipids. For example, phospholipase A2, which occurs in bee and snake venoms, specifically cleaves the C2 ester linkage to yield a free fatty acid and a lysophospholipid. Several phospholipase-catalyzed hydrolysis products of glycerolipids play roles in cell signaling processes. These include 1,2-diacylglycerol and lysophosphatidic acid.

9. Plasmalogens are glycerophospholipids in which a fatty acyl group is linked to C1 via an α,β-unsaturated ether bond.

10. Sphingolipids all contain a long-chain nitrogen-containing alcohol named sphingosine. Ceramides have a fatty acyl group linked to the primary amino group of sphingosine and can be regarded as the basic building block of more complex sphingolipids. Among these are (a) sphingomyelin, in which ceramide is esterified to a phosphocholine or phosphoethanolamine head group; (b) cerebrosides, in which ceramide forms a glycosidic bond with either glucose or galactose; and (c) gangliosides, in which oligosaccharides containing one or more N-acetylneuraminic acid groups form the ceramide head group. Cerebrosides and gangliosides are therefore glycosphingolipids. On hydrolysis, sphingolipids, like glycerophospholipids, give rise to products that have signaling activity.

11. Steroids are derivatives of the cyclopentaneperhydrophenanthrene fused-ring system. Cholesterol, the most abundant animal steroid, is weakly amphiphilic because it has a hydroxyl group at C3 (it is therefore classified as a sterol). It is a major component of biological membranes in animals. In a cholesteryl ester, the hydroxyl group of cholesterol is esterified to a fatty acid.

12. Cholesterol is the precursor of steroid hormones, which include (a) glucocorticoids (e.g., cortisol), which modulate metabolic processes, inflammatory reactions, and stress responses; (b) mineralocorticoids (e.g., aldosterone), which regulate salt and water excretion; and (c) androgens and estrogens, which influence sexual development and reproductive functions.

13. Vitamins D2 and D3 are produced from steroid precursors through photolysis by ultraviolet light. These two vitamins are then converted to active forms by enzymatic hydroxylation. Active vitamin D promotes absorption of dietary $Ca2+$ and enhances the deposition of $Ca2+$ from the blood into bone.

14. Isoprenoids are a large class of lipids that serve multiple functions in animals and plants. This group includes ubiquinone and plastoquinone (essential components of the electron transport chains of mitochondria and chloroplasts, respectively) and vitamins A, K, and E.

15. Eicosanoids are derived from the highly unsaturated C20 lipid arachidonic acid after its release from membrane glycerophospholipids by phospholipase A2. The eicosanoids include prostaglandins, thromboxanes, leukotrienes, and lipoxins, all of which are biologically active at extremely low concentrations. They are produced in a tissue-specific manner and play roles in inflammatory reactions, the regulation of the cardiovascular system, and reproduction.

Lipid Bilayers

16. The physical properties of lipids in aqueous solution cause them to aggregate. Water tends to exclude the hydrophobic portions of amphiphilic lipids, whereas the polar head groups remain in contact with the aqueous environment. Amphiphiles with a single nonpolar tail, such as soaps, many detergents, and lysophospholipids, form globular micelles. Amphiphiles with two hydrocarbon tails, such as glycerophospholipids, instead form disk-like micelles that are actually lipid bilayers. A bilayer of phospholipids in an aqueous milieu may form a vesicle with a solvent-filled interior, called a liposome. Liposomes are useful as models of biological membranes and as water-soluble drug delivery vehicles.

17. The movement of an amphiphilic lipid across a lipid bilayer, a process called transverse diffusion or a flip-flop, occurs infrequently, because the passage of a polar head group through the hydrophobic interior of the bilayer is thermodynamically unfavorable. In contrast, lipids readily diffuse laterally in the plane of the bilayer, indicating that the hydrocarbon core of the bilayer is highly fluid. The motion of the acyl chains is greatest in the center of the bilayer and decreases markedly near the polar head groups.

18. Bilayer fluidity is a function of temperature. At high temperatures, the lipids form a highly mobile liquid-crystal state, and at low temperatures, they form a gel-like solid as the hydrocarbon tails tightly associate. The transition temperature, the temperature at which the phase changes, depends on bilayer composition and increases with increasing acyl chain length and decreasing unsaturation of the component lipids. The presence of cholesterol also reduces a bilayer's transition temperature. Living organisms alter their membrane composition in order to maintain constant fluidity as the ambient temperature varies.

19. Cholesterol stabilizes the bilayer over a range of temperatures: At high temperatures, its rigid ring system interferes with fatty acyl chain mobility and thereby decreases membrane fluidity. At low temperatures, cholesterol prevents tight packing of adjacent hydrocarbon tails and thereby promotes membrane fluidity.

Membrane Proteins

20. The types and relative amounts of membrane proteins and lipids vary among biological membranes. Membrane proteins, although extremely diverse, can be classified as integral, lipid-linked, or peripheral, depending on how they are associated with the membrane.

21. Integral proteins (also known as an intrinsic proteins) are amphiphilic in that some regions of each molecule strongly associate with nonpolar membrane components by means of hydrophobic effects, whereas hydrophilic portions extend into the aqueous surroundings. Integral proteins can be solubilized and extracted from membranes only by using relatively harsh reagents, such as detergents and organic solvents, to disrupt the membrane structure.

22. Membrane proteins contribute to the asymmetry of a biological membrane because they either are present on one side of a membrane or, if they extend through it, are oriented in only one direction. The transmembrane domain of an integral protein may consist of one α helix (as in glycophorin A), a bundle of helices (as in bacteriorhodopsin), or a β barrel (as in porin). The backbones of both α helices and β barrels can form all their possible hydrogen bonds and are therefore stable in the interior of the lipid bilayer. In all cases, hydrophobic residues contact the hydrocarbon chains in the membrane core, whereas charged polar groups tend to predominate at the membrane surface.

23. Certain membrane proteins are covalently linked to a lipid moiety, which provides the protein with a hydrophobic anchor in the membrane. There are three main types of lipid-linked proteins:
    (a) Prenylated proteins have an attached isoprenoid moiety, a polymer built of C_5 isoprene units. The most abundant of these are farnesyl (C_{15}) and geranylgeranyl (C_{20}) groups. As a rule, prenylation occurs at the C-terminus of a protein.
    (b) Fatty acylated proteins have a covalently attached myristoyl or palmitoyl group. Myristoylation is a stable modification that occurs through an amide bond to the α amino group of an N-terminal glycine. Palmitoylation occurs via a thioester linkage to a cysteine and is reversible.
    (c) Glycosylphosphatidylinositol (GPI)-linked proteins are located at the external surface of the cell and are anchored to the membrane by GPI, a glycolipid that is covalently attached to the C-terminus of the protein via an amide bond.

24. Peripheral membrane proteins (also known as extrinsic proteins) associate with membrane surfaces via electrostatic and hydrogen bonding interactions. As a result, these proteins can be removed from membranes by relatively gentle methods, such as extraction with salt solutions or variations in pH, that do not greatly disturb native membrane structure. Once dissociated, they exhibit properties typical of water-soluble proteins.

Membrane Structure and Assembly

25. The widely accepted concept of biological membrane structure, for which there is much evidence, is the fluid mosaic model. The model envisions integral proteins that float in a sea of lipid and can diffuse laterally unless restrained by other cell constituents. The rate of membrane protein diffusion can be assessed by fluorescent recovery after photobleaching, which measures the rate at which a fluorescent-labeled membrane component diffuses into an area of membrane that has been previously bleached by a laser beam.

26. The structure and function of the erythrocyte membrane have been extensively investigated, largely through the use of erythrocyte ghosts, which are cells from which hemoglobin and other soluble cell constituents have been removed by osmotic lysis. The erythrocyte's biconcave disklike shape, which is essential for its O_2-carrying properties, is maintained by the membrane skeleton, a network of proteins located just beneath the membrane. The chief element in the skeleton is spectrin, a fibrous heterotetrameric protein that is cross-linked to other skeletal and membrane proteins, including actin and ankyrin. Proteins similar to those of the erythrocyte membrane skeleton occur in many other cell types.

27. The lipid and protein components of biological membranes are unevenly distributed. The cytoskeleton apparently restricts the movements of many membrane proteins and hence contributes to this heterogeneity. Localized membrane domains that have different lipid compositions may result from the association of certain lipids with membrane proteins. Lipid rafts are domains of closely packed lipids that have a more crystalline consistency, associate with certain proteins, and may have distinct functions.

28. Membrane constituents are asymmetrically distributed with respect to the outer and inner leaflets of the bilayer in natural membranes. For example, the carbohydrate components of plasma membrane glycoproteins and glycolipids are always located on the extracellular membrane surface. The inner and outer leaflets also contain lipids in different proportions.

29. Lipid asymmetry, which can be assessed through the use of specific phospholipases, results in part from the asymmetric synthesis of lipids on the cytoplasmic face of the plasma membrane (in prokaryotes) or the endoplasmic reticulum (in eukaryotes). Lipids are subsequently redistributed by two mechanisms: (a) Enzymes called flipases facilitate the flip-flop of specific phospholipids from one leaflet to the other; and (b) ATP-dependent translocases establish a nonequilibrium distribution of lipids. In eukaryotes, lipids are transported as components of vesicles that bud off the endoplasmic reticulum and ultimately fuse with other cellular membranes.

30. Protein synthesis starts at the N-terminus and proceeds to the C-terminus of a polypeptide chain. Synthesis takes place either on free ribosomes (for soluble proteins) or on ribosomes bound to the endoplasmic reticulum (for integral membrane and secretory proteins). The signal hypothesis explains how polypeptides traverse the endoplasmic reticulum membrane:

(a) Proteins synthesized on ER-bound ribosomes have an N-terminal amino acid sequence known as a signal peptide.

(b) After an ~40-residue polypeptide chain has been synthesized, the signal peptide binds to the signal recognition particle (SRP), concomitantly with an exchange of GTP for GDP in the SRP. This event stops further polypeptide synthesis until the SRP–ribosome complex docks with the ER membrane.

(c) The SRP–ribosome complex, bearing the nascent polypeptide chain, binds to a docking protein (the SRP receptor) on the ER surface. The SRP receptor is associated with a pore structure, the translocon, which allows the growing polypeptide to pass through the ER membrane into the lumen of the ER. Protein synthesis resumes, and the N-terminal sequence is translocated to the lumen of the ER via a transmembrane channel. At this point, the SRP dissociates from the complex.

(d) Once in the ER lumen, the signal peptide is removed from the polypeptide by hydrolysis catalyzed by a signal peptidase.

(e) The growing polypeptide undergoes folding, facilitated by Hsp70, and posttranslational modification, such as addition of core *N*-linked oligosaccharides.

(f) When synthesis is complete, secretory proteins have passed entirely through the ER membrane into the lumen, whereas transmembrane proteins remain embedded in the membrane with their C-terminus on the cytoplasmic face.

31. The translocon is a multifunctional transmembrane pore that either allows soluble polypeptides to traverse the ER membrane or mediates the insertion into the bilayer of membrane-spanning polypeptide segments.

32. Secreted proteins, lysosomal (and other vesicular) proteins, and membrane proteins are transported between cellular compartments in coated vesicles. The "coat" of these vesicles is composed of one of three classes of proteins:

(a) A complex of tripartite clathrin proteins that form a polyhedral network. These vesicles are usually bound for the plasma membrane or participate in endocytosis;

(b) COPI proteins, which mediate transport within the Golgi apparatus;

(c) COPII proteins, which mediate transport of vesicles from the ER to the Golgi apparatus.

33. Proteins bound for the ER or lysosomes have specific signals that target their final localization. Mannose-6-phosphate residues covalently attached to lysosomal hydrolases target these proteins to lysosomes. Similarly, most proteins localized to the ER contain the amino acid sequence KDEL at the C-terminus. It is currently hypothesized that other carbohydrate moieties or specific peptide sequences are essential for targeting proteins to other cellular compartments.

34. Membrane fusion is critical for the trafficking of membrane components, secretion (exocytosis), and viral entry into cells. Membrane fusion is mediated by complementary pairs of proteins called SNAREs. Both vesicular SNAREs (R-SNAREs) and target membrane SNAREs (Q-SNAREs) comprise a large family of proteins, wherein they provide specificity for vesicular fusion with target membranes.

35. Bacterial toxins from anaerobic bacteria (eg., Clostridium tetani, Clostridium botulinum) cleave SNAREs, thereby interfering with neuromuscular and neuronal synaptic exocytosis (Box 9-3). Animal viruses such as influenza virus contain membrane fusion proteins (e.g., hemagglutinin) that stimulate receptor-mediated endocytosis, followed by membrane fusion, which allows the virus access to the host cell cytoplasm. In the low pH environment of endosomes, hemagglutinin undergoes dramatic conformational changes that turn it from a membrane receptor into a membrane fusion protein.

Questions

Lipid Classification

1. Draw structures for the following fatty acids:
   (a) 18:0 octadecanoic acid
   (b) 20:4 5,8,11,14-eicosatetraenoic acid
   (c) 22:6 4,7,10,13,16,19-docosahexaenoic acid
   (d) 13-(2-cyclopentenyl)-tridecanoic acid (a cyclic fatty acid)

2. How do the following alterations in the structures of fatty acids affect their physical properties?
   (a) Increasing the chain lengths of saturated fatty acids
   (b) Increasing the number of double bonds in unsaturated fatty acids
   (c) Changing a cis double bond in a fatty acid to a trans double bond

3. Draw and name a typical triacylglycerol.

4. List the two major functions of triacylglycerols.

5. Why are glycerophospholipids considered amphiphilic molecules?

6. Draw the structures of the following lipids:
   (a) 1-palmitoyl-2-oleoyl-3-glycerophosphatidylethanolamine
   (b) 1-stearoyl-2-arachidonoyl-3-glycerophosphatidylinositol

7. What distinguishes a plasmalogen from other glycerophospholipids?

8. What chemical moieties do sphingomyelin and gangliosides have in common and which are unique to each?

9. The compound shown below has been advertised as a dietary supplement that purportedly prevents obesity, heart disease, and the ill effects of aging. Based on its structure, what physiological function is it most likely to actually perform?

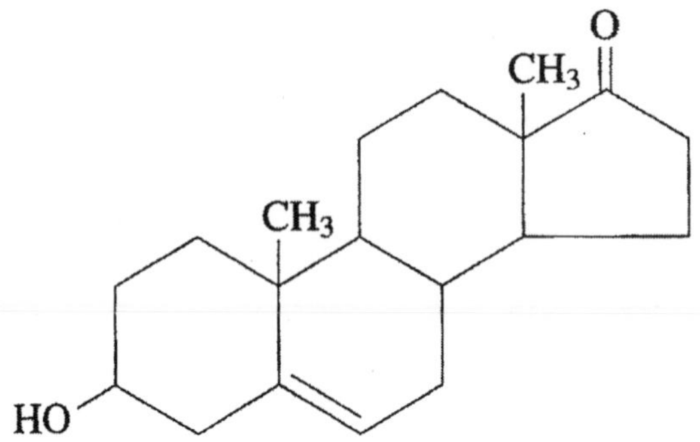

10. Why are eicosanoids called local mediators rather than hormones?

11. Arachidonic acid can arise from phospholipid precursors by the action of:
    (a) Phospholipase A_1
    (b) Phospholipase A_2
    (c) Phospholipase C
    (d) Phospholipase D

Lipid Bilayers

12. Explain why single-tailed amphiphiles tend to form micelles whereas two-tailed amphiphiles tend to form bilayers.

13. Why is a polar solute unlikely to penetrate a lipid bilayer?

14. Describe the structural changes that occur when a pure phospholipid bilayer is warmed and passes through its transition temperature. Explain what would happen if the bilayer contained a significant amount of cholesterol.

Membrane Proteins

15. Summarize the physical properties of a typical integral protein, including the forces involved in its interaction with the membrane, its orientation, and its movement within the membrane.

16. Lipids linked covalently to proteins often _____ the proteins in the membrane.

17. The three main types of lipid moieties that can be covalently attached to proteins are ____, _____ _____ , and _____ .

18. Mutated forms of the protein p21^{c-ras} are associated with the development of a substantial proportion of tumors. However, a second mutation that changes the Cys residue in the C-terminal Cys-X-X-Y sequence of p21^{c-ras} to some other residue inhibits its cancer-causing potential. What does this suggest about the molecular process leading to tumor formation?

Membrane Structure and Assembly

19. In verifying the fluid mosaic model of biological membranes, Michael Edidin observed that fusion of mouse and human cell membranes was temperature dependent, such that fusion did not occur below 15°C but proceeded readily at 37°C. Explain this difference in terms of membrane properties (see Figure 9-26).

20. Explain why an erythrocyte membrane that synthesizes too little spectrin is a sphere rather than a biconcave disk.

21. How did Kennedy and Rothman demonstrate that newly synthesized bacterial membrane phospholipids rapidly cross from the cytoplasmic side to the external side of the membrane (see Fig. 9-34)?

22. Explain why the flip-flop rate of phospholipids in biological membranes is far greater than in artificial lipid membranes.

23. For the signal hypothesis, match each of the following terms with its description below.

    _____ free ribosome
    _____ signal peptide
    _____ signal recognition particle (SRP)
    _____ core glycosylation
    _____ membrane anchor
    _____ translocon
    _____ signal peptidase
    _____ membrane-bound ribosomes
    _____ SRP receptor

    A. A hydrophobic sequence in a transmembrane protein that arrests movement of the protein through the membrane.
    B. A posttranslational modification that occurs while peptide bond formation continues.
    C. Cleaves the signal peptide.
    D. An N-terminal hydrophobic sequence flanked by hydrophilic sequences.
    E. The site of membrane and secretory protein synthesis.
    F. Binds the SRP and reinitiates ribosomal polypeptide synthesis.
    G. The site of soluble protein synthesis.
    H. Binds to a ribosome bearing an extruded signal sequence and stops polypeptide synthesis.
    I. Pore that provides conduit between the cytosol and the lumen of the ER.

24. A membrane glycoprotein–synthesizing system was devised that contained endoplasmic reticulum membranes and all the necessary ingredients for the *in vitro* synthesis of a protein. The complete synthesis of the protein molecule required 40 minutes. When the detergent Triton X-100 was added within a few minutes after the initiation of polypeptide synthesis, the resulting protein was devoid of glycosyl groups. (a) Explain the action of the detergent. (b) Would the protein have been glycosylated if the detergent had been added 35 minutes after initiation of protein synthesis?

25. The antibiotic monensin inhibits some steps of posttranslational protein modification in the Golgi apparatus. What effect would monensin have on protein targeting?

26. Besides providing specificity to membrane fusion, SNAREs (and probably hemagglutinnin, too) are thought to play a role in the energetics of membrane fusion. Indeed, these proteins, beyond simple recognition, induce large conformation changes in the fusing membranes. Can you suggest an energetic barrier that these proteins may help traverse?

Answers to Questions

1.

(a)

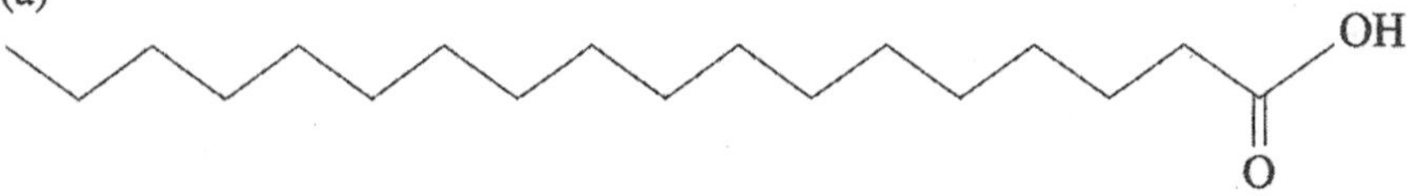

(b)

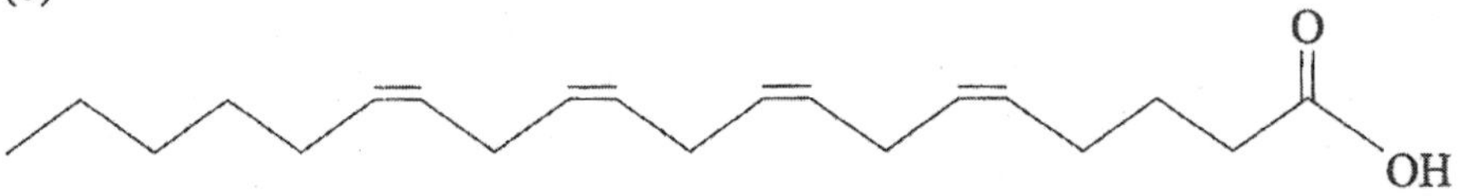

(c)

(d)

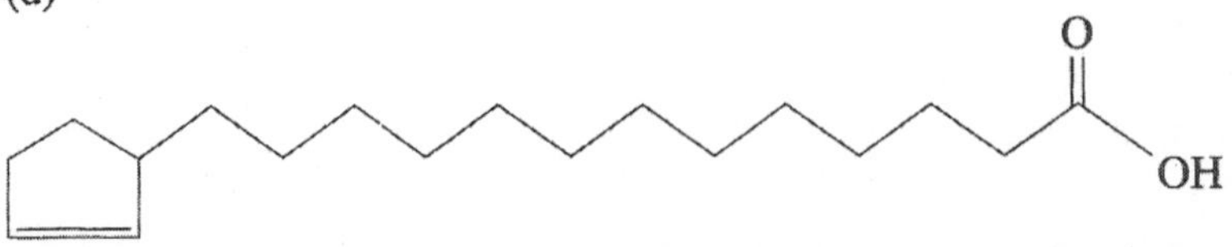

2. (a) The melting point increases with increasing hydrocarbon chain length. Fully saturated hydrocarbon chains have the least amount of steric hindrance and thus can pack together closely.
   (b) The melting point decreases with an increasing number of double bonds.

(c) Fatty acid residues that have a cis double bond exhibit a 30° bend in their hydrocarbon chain that interferes with close packing. In contrast, trans double bonds distort the acyl chain conformation much less, allowing closer packing of molecules. Hence a fatty acid with trans double bonds has a higher melting point than the corresponding cis fatty acid of the same chain length and with the same number of double bonds.

3.   For example, 1-palmitoyl-2-palmitoleoyl-3-oleoyl-glycerol

$$
\begin{array}{ccc}
CH_2 & CH & CH_2 \\
| & | & | \\
O & O & O \\
| & | & | \\
C{=}O & C{=}O & C{=}O \\
| & | & | \\
CH_2 & CH_2 & CH_2 \\
| & | & | \\
CH_2 & CH_2 & CH_2 \\
| & | & | \\
CH_2 & CH_2 & CH_2 \\
| & | & | \\
CH_2 & CH_2 & CH_2 \\
| & | & | \\
CH_2 & CH_2 & CH_2 \\
| & | & | \\
CH_2 & CH_2 & CH_2 \\
| & | & | \\
CH_2 & CH_2 & CH_2 \\
| & | & | \\
CH_2 & CH & CH \\
| & \| & \| \\
CH_2 & CH & CH \\
| & | & | \\
CH_2 & CH_2 & CH_2 \\
| & | & | \\
CH_2 & CH_2 & CH_2 \\
| & | & | \\
CH_2 & CH_2 & CH_2 \\
| & | & | \\
CH_2 & CH_2 & CH_2 \\
| & | & | \\
CH_3 & CH_3 & CH_2 \\
& & | \\
& & CH_2 \\
& & | \\
& & CH_3 \\
\end{array}
$$

4.   Triacylglycerols serve as metabolic energy sources and thermal insulators.

5.   Glycerophospholipids are amphiphilic because they are composed of a hydrophobic diacylglycerol "tail" attached to a hydrophilic phosphoryl derivative "head."

6.

(a)

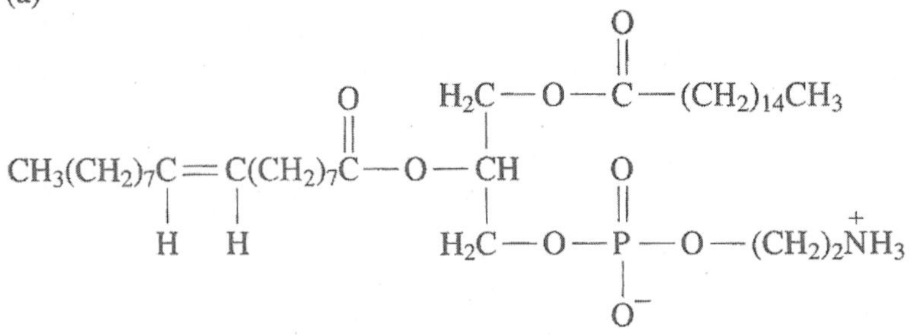

$$CH_3(CH_2)_7C=C(CH_2)_7\overset{\overset{\displaystyle O}{\|}}{C}-O-CH$$

with H, H below the double bond, and the CH connected to:

$$H_2C-O-\overset{\overset{\displaystyle O}{\|}}{C}-(CH_2)_{14}CH_3$$

$$H_2C-O-\overset{\overset{\displaystyle O}{}}{\underset{\underset{\displaystyle O^-}{|}}{P}}-O-(CH_2)_2\overset{+}{N}H_3$$

(b)

$$CH_3(CH_2)_4(CH=CHCH_2)_4(CH_2)_2-\overset{\overset{\displaystyle O}{\|}}{C}-O-CH$$

$$H_2C-O-\overset{\overset{\displaystyle O}{\|}}{C}-(CH_2)_{16}CH_3$$

$$H_2C-O-\overset{\overset{\displaystyle O}{\|}}{\underset{\underset{\displaystyle O_-}{|}}{P}}-O$$

with a sugar ring bearing HO, H, OH, OH, H, OH, H, H, HO substituents.

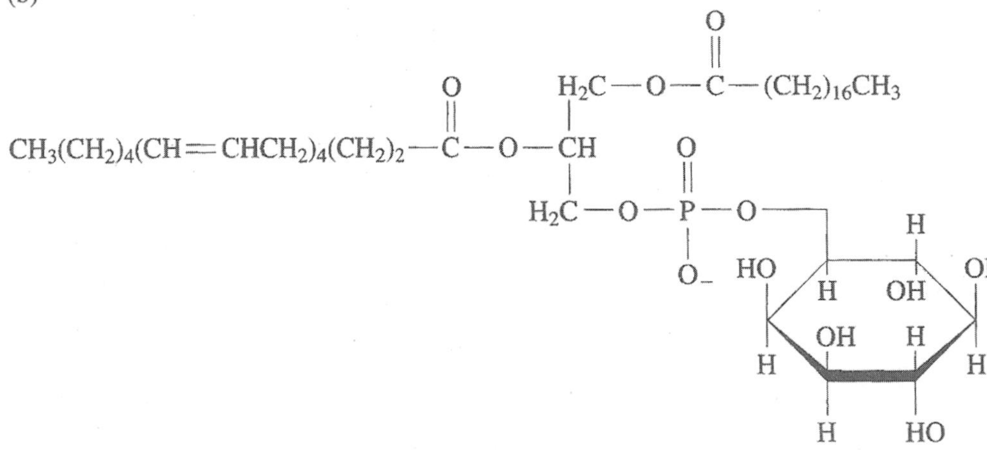

7. A plasmalogen differs from a diacylglycerophospholipid in that a hydrocarbon chain is attached to the C1 position of glycerol via an α,β-unsaturated ether bond in the cis configuration instead of an ester bond.

8. Sphingomyelin and gangliosides both contain a sphingosine moiety to which is attached a fatty acyl group in an amide linkage. They differ in that sphingomyelin has a phosphocholine (or less commonly a phosphoethanolamine) head group, whereas the ganglioside head group is an oligosaccharide that includes one or more sialic acid residues.

9. This compound is a sterol that closely resembles testosterone and estradiol (shown in Figure 9-11). It is most likely to exert its effects by influencing physiological processes that depend on the sex hormones. In fact, it is dehydroepiandrosterone (DHEA), a metabolic precursor of androgens and estrogens.

10. Although eicosanoids, like hormones, are synthesized by a variety of cells and exert their effects on other cells, they are considered local mediators rather than hormones because they act near their sites of synthesis rather than being carried throughout the body via the bloodstream, and because they decompose quickly.

11. (b)

12. The difference arises from geometrical considerations. Single-tailed amphiphiles, such as fatty acid anions, form micelles because of their tapered shape. Their hydrated head groups are wider than their tails, enabling them to pack efficiently into a spheroidal micelle. The cylindrical shape of two-tailed amphiphiles, such as glycerophospholipids, prevents their packing into a spheroidal micelle. Instead they pack together to form disklike micelles, which are really extended bilayers.

13. In order for a polar solute to enter a lipid bilayer, its interactions with surrounding water molecules must first be disrupted (i.e., it must lose its hydration shell), and new interactions with the hydrophobic bilayer constituents would have to form. This would require an increase in free energy for the system; that is, the process is unfavorable.

14. When a pure phospholipid bilayer in an orderly gel-like state is warmed, it is converted (at its transition temperature) to a liquid crystal form that is more fluid. The presence of cholesterol in a phospholipid bilayer both decreases its fluidity and broadens the transition temperature range. These effects occur because the sterically rigid cholesterol molecules insert between the fatty acyl chains of the phospholipids and thus restrict their motion.

15. Integral proteins are strongly associated with membranes through hydrophobic effects and can typically be extracted only by agents such as detergents, organic solvents, or chaotropic agents, which disrupt membrane structure. Some integral proteins bind lipids so tightly that they are dissociated from them only under denaturing conditions. Integral proteins are amphiphilic molecules that are oriented in the membrane such that the regions buried in the interior of the membrane have a surface composed mainly of hydrophobic residues, whereas the portions exposed to the aqueous environment have mostly polar residues. These polar regions may occur on one side of the membrane or the other, or may asymmetrically extend from both sides (transmembrane proteins). Integral proteins can diffuse laterally in the plane of the membrane, but their flip-flop rate is nil.

16. Lipids linked covalently to proteins often <u>anchor</u> the proteins in the membrane.

17. The three main types of lipid moieties that can be covalently attached to proteins are <u>isoprenoid groups</u> , <u>fatty acyl groups</u> , and <u>glycosylphosphatidylinositol (GPI) groups</u> .

18. The cysteine residue in the C-terminal sequence is the site where p21$^{c\text{-}ras}$ is prenylated. Since this event permits the protein to insert into membranes, it appears that p21$^{c\text{-}ras}$ must be membrane-bound to promote tumor formation.

19. Fusion of the two membranes could not occur at low temperature because the fluidity of the membranes was too low to allow the free diffusion of lipids and proteins (presumably, at the lower temperature they became gel-like solids). At the higher temperature, the membranes were more fluid, making the lateral movement of lipids and proteins possible.

20. In normal erythrocytes, the cross-linking of spectrin and other membrane proteins creates a submembrane network of proteins, somewhat akin to a geodesic dome, that gives the cells a biconcave disklike shape. A cell that contains too little spectrin to assemble a complete

membrane skeleton is not constrained in shape and becomes a sphere, the simplest possible shape for a membrane enclosing the cell contents.

21. Growing bacteria were exposed briefly to radioactive phosphate so that newly synthesized phospholipids (mostly phosphatidylethanolamine in *E. coli*) would become labeled. Trinitrobenzenesulfonic acid (TNBS), a membrane-impermeable reagent that reacts with phosphatidylethanolamine (PE), was then added to the cells. At various times, samples of cells were tested for the presence of ^{32}P-labeled PE that also contained the TNB group. The appearance of doubly labeled PE after three minutes indicated that PE initially synthesized on the cytoplasmic side of the membrane (where it was labeled with ^{32}P) traversed the membrane and reacted with TNBS in the extracellular medium.

22. Biological membranes differ from artificial lipid membranes in having protein constituents (e.g., flipases and phospholipid translocases) that greatly accelerate the rate at which phospholipids move from one side of the bilayer to the other.

23.

G	free ribosome
D	signal peptide
H	signal recognition particle (SRP)
B	core glycosylation
A	membrane anchor
I	translocon
C	signal peptidase
E	membrane-bound ribosomes
F	SRP receptor

24. (a) The membrane protein was synthesized on ribosomes bound to the endoplasmic reticulum and passed partially through the membrane into the ER lumen. Treatment with Triton X-100 disrupted the membrane structure, which prevented the nascent protein from being glycosylated by the glycosyltransferases that are normally present in the ER lumen. Thus, posttranslational glycosylation requires an intact ER membrane.

    (b) Because glycosylation is initiated as the protein is being synthesized, the protein will almost certainly have been glycosylated by the ER enzymes before the membrane was disrupted by the detergent.

25. Disruption of posttranslational protein modification reactions would likely also disrupt protein-targeting mechanisms, many of which depend on modifications, such as glycosylation, that occur in the Golgi apparatus.

26. Most likely, these proteins bring the two membranes sufficiently close to overcome the repulsion energy of the charged phospholipids of each membrane.

10

Membrane Transport

Lipid bilayers are impermeable or weakly permeable to a variety of polar and nonpolar compounds that are essential to cellular metabolism. Hence, these substances must enter (and exit) cells by way of specialized membrane proteins that facilitate their transport across lipid bilayers. In order to function normally, cells must not only facilitate diffusion of ions and small molecules across their membranes (passive-mediated transport), they must also concentrate some substances on one side of the membrane (active transport). This chapter introduces the thermodynamics of transport and then discusses passive-mediated transport and active transport. Examples of passive-mediated transporters presented in this chapter include ionophores, porins, ion channels, aquaporins, and transport proteins such as the glucose and oxalate transporters. Many ion channels are gated so that they allow the movement of ions only in response to certain stimuli. Examples of active transporters discussed in this chapter include the ABC transporters, $(Na^+–K^+)$–ATPase, Ca^{2+}–ATPase, and ion gradient–driven active transport systems.

Essential Concepts

Thermodynamics of Transport

1. The thermodynamics of diffusion across a membrane can be expressed in terms of a chemical equilibrium. The difference in concentration of substance A on two sides of a membrane generates a chemical potential difference, $\Delta \bar{G}_A$. When A is ionic, an electrical potential difference ($\Delta \Psi$) may also develop. Thus, the equation for the electrochemical potential difference contains terms for the concentration and the charge of substance A:

$$\Delta \bar{G}_A = RT \ln ([A]_{in}/[A]_{out}) + Z_A \mathcal{F} \Delta \Psi$$

where Z_A is the ionic charge of A and $\mathcal{F}$ is the Faraday constant ($96{,}485 \ J{\cdot}V^{-1}{\cdot}mol^{-1}$). A negative value for $\Delta \bar{G}_A$ indicates spontaneous transport of substance A from the outside to the inside.

2. Transport may be classified as either mediated or nonmediated. Diffusion accounts for nonmediated transport. The chemical potential difference determines the direction of nonmediated transport, and the substance moves in the direction that will eliminate the concentration difference and at a rate proportional to the size of the gradient. Mediated transport occurs in two forms, based on the thermodynamics of the solutes traversing the membrane:
    (a) Passive-mediated transport or facilitated diffusion, in which solutes travel down a concentration gradient.
    (b) Active transport, in which solutes are transported against their concentration gradient via coupling to a sufficiently exergonic process.

3.   Mediated transport makes use of carrier molecules, called permeases, transporters, or translocases. In passive-mediated transport (also called facilitated diffusion), a molecule is transported from a high to a low concentration. In active transport, an energy-yielding process must be coupled to the movement of a substance from a lower to a higher concentration.

Passive-Mediated Transport

4.   Ionophores are carrier molecules that greatly increase the permeability of a membrane to certain ions. Some of these organic compounds (such as valinomycin) bind the ion on one side of the membrane and carry it to the other side; others form transmembrane channels through which the ion can pass.

5.   Porins are β-barrel transmembrane proteins that display varying levels of solute selectivity. Some porins are weakly selective for cations or anions, while others, such as maltoporin, show a high degree of solute selectivity.

6.   All cells contain ion-specific channels, the best characterized of which is the K^+ channel from *Streptomyces lividans* (KcsA).

7.   The KscA K^+ channel is a homotetrameric transmembrane protein. The extracellular surface has a selectivity filter that preferentially allows K^+ to be dehydrated and enter the central cavity wherein K^+ is rehydrated. The geometry of the selectivity filter does not allow the entry of smaller Na^+ ions; hence, the channel is specific for K^+.

8.   Most ion channels are gated, meaning that they open or close in response to a variety of stimuli including (a) local deformations in the membrane (mechanosensitive channels); (b) binding of an extracellular signal such as a neurotransmitter (ligand-gated channels); (c) binding of an intracellular signal (signal-gated channels); and (d) changes in voltage across the membrane (voltage-gated channels).

9.   Voltage-gated channels propagate action potentials in nerve cells. Voltage gating in voltage-gated K^+ (K_V) channels is mediated by α-helical "paddles" that respond to critical voltage changes across the plasma membrane. The K_V channel has two voltage-sensitive gates: one to open the channels and another to close it. Closure is mediated by the N-terminal domain, which swings into place 2–3 milliseconds after channel opening.

10. Aquaporins are ubiquitous transmembrane channels that allow selective movement of H_2O molecules across otherwise impermeable lipid bilayers.

11. Integral membrane proteins that mediate passive transport cycle between conformational states in which binding sites for the molecule to be transported are alternately accessible on one side of the membrane and then on the other. For example, the glucose transporter binds glucose at the external cell surface then changes conformation so as to release glucose at the cytoplasmic surface. It then reverts to its previous conformation.

12. Gap junctions are pairs of juxtaposed channels between two cells that allow non-selective movement of a variety of solutes such as ions, amino acids, sugars, and nucleotides. Gap junctions provide electrical coupling to cells, which is critical to heart function in vertebrates. The core protein is connexin, which forms a hexagonal ring in each membrane (Box 10-1).

13. Mediated transport exhibits four properties common to enzymes: (a) speed and specificity, (b) saturability, (c) competition from structurally similar compounds, and (d) inactivation by compounds that chemically modify proteins. Indeed, mediated transporters can be viewed as "diffusion enzymes" (Box 10-2).

14. Proteins that mediate passive transport can operate in either direction, depending on the concentrations of the transported substance on both sides of the membrane. A uniport mechanism transfers a single molecule across the membrane; symport involves two substances moving in the same direction; and antiport is the movement of two substances in opposite directions.

Active Transport

15. Active transport is coupled to the hydrolysis of ATP. A well-studied example is the plasma membrane $(Na^+–K^+)$–ATPase, which pumps 3 Na^+ out and 2 K^+ into the cell, thereby moving both ions against their concentration gradients, with every ATP hydrolyzed. The resulting electrochemical gradient of Na^+ and K^+ is the basis for the electrical excitability of nerve cells.

16. The mechanism of the $(Na^+–K^+)$–ATPase involves a series of reaction steps that operate in only one direction because ATP breakdown and ion movement are coupled vectorial processes. The Ca^{2+}–ATPase, which pumps cytosolic Ca^{2+} to the cell exterior against a large concentration gradient, functions in a similar manner to the $(Na^+–K^+)$ pump. Cardiac glycosides (sugar components covalently linked to a steroid component) modulate cardiac function by blocking ATP hydrolysis in the $(Na^+–K^+)$–ATPase reaction sequence (Box 10-3).

17. ABC (ATP-binding cassette) transporters are a large class of transporters that pump ions and a wide variety of polar and nonpolar molecules. These proteins include two transmembrane domains and two ATP-binding domains.
    (a) In prokaryotes, a so-called ligand-binding domain is sometimes present, which concentrates the solute near the transporting transmembrane domains. Also, the three components (ligand-binding domain, transmembrane permease domain, ATP-binding domain) often exist as separate polypeptides in prokaryotes.
    (b) In eukaryotes, these domains are found as a single polypeptide. The multidrug resistance (MDR) transporter or P-glycoprotein transports drugs and other substances out of eukaryotic cells.

(c) CFTR, the cystic fibrosis transmembrane regulator (Box 3-1), is an unusual ABC transporter in that ATP hydrolysis regulates the opening of a gate for Cl^- to diffuse through, instead of actively transporting Cl^- up a concentration gradient. CFTR is not a highly selective transporter of Cl^-, ions suggesting that it probably lacks a selectivity filter.

18. In secondary active transport, the energy generated by an electrochemical gradient is utilized to drive another, endergonic transport process. One example is the uptake and concentration of glucose by the intestinal epithelial cells. This task is accomplished by using the Na^+ gradient produced by the (Na^+–K^+) pump. Another example is the bacterial lactose permease, which utilizes a proton gradient across the bacterial membrane to power the transport of lactose.

Key Equations

$$\Delta \bar{G}_A = RT \ln ([A]_{in}/[A]_{out}) + Z_A \mathcal{F} \Delta \Psi$$

Questions

Thermodynamics of Transport

1. Suppose the concentration differences across a membrane for glucose and sodium are both 10^4 (that is, the concentrations are 10^4 greater on one side of the membrane). Is the free energy change for the transmembrane movement of these solutes the same? Explain.

2. What is the electrochemical potential difference when the intracellular $[Ca^{2+}] = 1$ μM and the extracellular $[Ca^{2+}] = 1$ mM? Assume $\Delta \Psi = -100$ mV (inside negative) and $T = 25°C$.

3. Rank the molecules below from lowest to highest according to their ability to diffuse across a lipid bilayer. Explain your rationale.

4. The following data (using arbitrary units) were obtained for the transmembrane movements of compounds A and B from outside to inside a cell:

A		B	
Extracellular concentration	*Flux into cell*	*Extracellular concentration*	*Flux into cell*
2	1.2	2	4.5
5	3.4	5	6.2
10	6.5	10	7.5

Which compound enters the cell by mediated transport? Explain.

5. Which graph below shows the expected relationship between temperature and flux (J, rate of flow) of an ion transported across a biological membrane via a carrier ionophore? Explain.

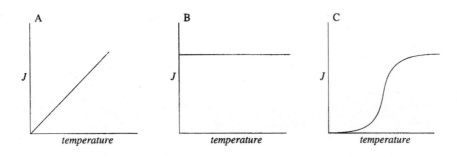

Passive-Mediated Transport

6. K^+ channels have openings that are much wider than Na^+ ions, yet sodium ions cannot pass through efficiently. Explain.

7. The eggs of freshwater invertebrates and vertebrates are hypertonic (higher solute concentration) to their surrounding medium (pond water). What protein is likely to be absent from the egg plasma membranes?

8. The model for glucose transport shown in Figure 10-13 shows two conformational states of the transporter when it is not bound to glucose. Is one of these conformational states likely to be preferred?

Active Transport

9. Does the activity of the (Na^+-K^+)–ATPase tend to make the cell interior electrically more negative or more positive with respect to the outside?

10. Determine whether each of the following transport systems is uniport, symport, or antiport. Which systems are active transport systems?

(a) glucose transporter in erythrocytes
(b) valinomycin
(c) plasma membrane (Na^+-K^+)–ATPase
(d) Na^+–glucose transporter of intestinal epithelium
(e) *E. coli* lactose permease
(f) (H^+-K^+)–ATPase of gastric mucosa

11. Fill in the diagram below showing glucose transport across a brush border cell from the intestinal lumen to the capillary bed. Also, draw arrows to show the flow of solutes and to indicate which transporter requires ATP hydrolysis. *Note: not all blanks and circles may be relevant.*

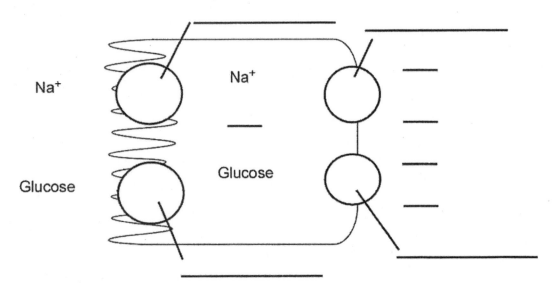

Answers to Questions

1. The free energy change is not the same. For glucose Z_A is zero, since glucose has no charge. Hence, the free energy change for glucose movement relative to Na^+ movement is reduced by $Z_A\mathscr{F}$

2.
$$\Delta\bar{G} = RT\ln([Ca^{2+}]_{in}/[Ca^{2+}]_{out}) + Z\mathscr{F}\Delta\Psi$$
$$= RT\ln(0.001/1) + (2)\mathscr{F}\Delta\Psi$$
$$= (8.3145\ J\cdot K^{-1}\cdot mol^{-1})(298\ K)(-6.91) + (2)(96{,}485\ J\cdot V^{-1}\cdot mol^{-1})(-0.1V)$$
$$= -17{,}121\ J\cdot mol^{-1} - 19297\ J\cdot mol^{-1}$$
$$= -36{,}418\ J\cdot mol^{-1} = -36.4\ kJ\cdot mol^{-1}$$

3.   **1, 2, 3.** The charged molecule is least able to penetrate the hydrophobic core of the bilayer. Although compounds 2 and 3 bear the same polar groups, the larger hydrocarbon chain of compound 3 lets it diffuse more easily through the bilayer.

4.   Compound B enters the cell by mediated transport since its rate of entry falls off as [B] increases. This is consistent with a carrier molecule that becomes saturated with B. Compound A enters by a nonmediated process, since its flux is directly related to its concentration at all values of [A].

5.   C. A carrier ionophore must diffuse through the membrane. The gel-like state of the lipids at low temperatures slows ionophore movement. The steep portion of the curve corresponds to the transition of the membrane to a more fluid state that allows free ionophore diffusion.

6.   The hydration shells around the K^+ and Na^+ ions must be stripped in order for the ions to enter the channel. In K^+ channels, the selectivity filter can accommodate dehydrated K^+ ions but is not flexible enough to coordinate dehydrated Na^+ ions, which are smaller. Consequently, the Na^+ ions remain hydrated and do not enter the K^+ channel.

7.   Aquaporin is lacking. The presence of aquaporin would allow water to enter the eggs, which would cause the eggs to swell and burst.

8.   Probably not, since this transporter allows equilibration of glucose concentration across the membrane in either direction.

9.   The cell interior becomes more negative since 3 Na^+ exit the cell for every 2 K^+ that enter.

10. (a) uniport
     (b) uniport
     (c) antiport
     (d) symport
     (e) symport
     (f) antiport

     The active transport systems are c, d, e, and f.

11.

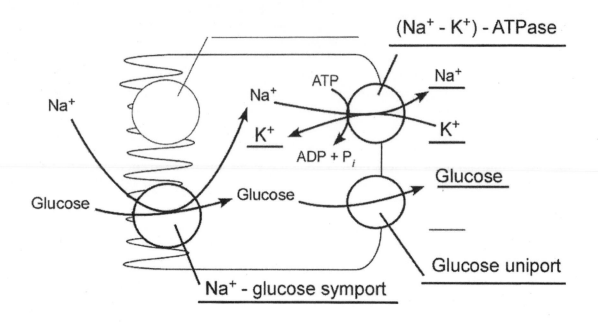

11
Enzymatic Catalysis

Enzymes are biological catalysts that increase the rates of biochemical reactions. In some cases, the enzyme-catalyzed reaction is nearly 10^{15} times faster than the uncatalyzed reaction. Enzymes are proteins (or in some cases, RNA) with very specific functions and are active under mild conditions. Enzymes function by lowering the free energy of the reaction's transition state. This chapter first discusses the various types of enzymes, their substrate specificity, the roles of coenzymes and cofactors, and the reaction coordinate. Next, the five modes of catalysis are described in detail. The chapter ends with full descriptions of the catalytic activity of lysozyme and serine proteases.

Essential Concepts

General Properties of Enzymes

1. Enzymes differ from ordinary catalysts in their higher reaction rates, their action under milder reaction conditions, their greater reaction specificities, and their capacity for regulation.

2. The IUBMB enzyme classification system divides enzymes into six groups (each with subgroups and sub-subgroups) based on the type of reaction catalyzed. A set of four numbers identifies each enzyme.

3. Enzymes are highly specific for their substrates and reaction products. Hence, the enzyme and its substrate(s) must have geometric, electronic, and stereospecific complementarity. Enzymes, for example, aconitase, can distinguish prochiral groups.

4. Many enzymes require cofactors for activity. Cofactors may be metal ions or organic molecules known as coenzymes. Many vitamins are coenzyme precursors. Coenzymes may be cosubstrates, which must be regenerated in a separate reaction, or prosthetic groups, which are permanently associated with the enzyme. An enzyme without its cofactor(s) is called an apoenzyme and is inactive, and the enzyme with its cofactor(s) is a called a holoenzyme and is active.

Activation Energy and the Reaction Coordinate

5. According to transition state theory, the reactants of a reaction pass through a short-lived high-energy state that is structurally intermediate to the reactants and products. This so-called transition state is the point of highest free energy in the reaction coordinate diagram. The free energy difference between the reactants and the transition state is the free energy

of activation, $\Delta G^{\ddagger}$. The reaction rate decreases exponentially with the value of $\Delta G^{\ddagger}$; that is, the greater the value of $\Delta G^{\ddagger}$, the slower the reaction. A catalyst provides a reaction pathway whose $\Delta G^{\ddagger}$ is less than that of the uncatalyzed reaction and hence increases the rate that the reaction achieves equilibrium.

Catalytic Mechanisms

6. Enzymes use several types of catalytic mechanisms, including acid–base catalysis, covalent catalysis, metal ion catalysis, catalysis by proximity and orientation effects, and catalysis by preferential binding of the transition state.

7. Acid–base catalysis occurs when partial proton transfer from an acid and/or partial proton abstraction by a base lowers the free energy of a reaction's transition state. The catalytic rates of enzymes that use acid–base catalysis are pH-dependent. RNase A has two catalytic His residues, which act as general acid and general base catalysts.

8. In covalent catalysis, the reversible formation of a covalent bond permits the stabilization of the transition state of the reaction through electron delocalization. Nucleophilic attack on the substrate by the enzyme to form a Schiff base intermediate capable of stabilizing (lowering the free energy of) a developing negative charge is an example of covalent catalysis. Common nucleophiles, which are negatively charged or contain unshared electron pairs, include imidazole and sulfhydryl groups.

9. Metalloenzymes tightly bind catalytically essential transition metal ions. Metal-activated enzymes loosely bind metal ions such as Na^+, K^+, or Ca^{2+} that play a structural role. Metal ions may orient substrates for reaction, mediate oxidation–reduction reactions, and electrostatically stabilize or shield negative charges. For example, the Zn^{2+} ion of carbonic anhydrase makes a bound water molecule more acidic, thereby increasing the concentration of the nucleophile OH^-.

10. Enzymes lower the activation energies of the reactions they catalyze by bringing their reactants into proximity, by properly orienting them for reaction, by using charged groups to stabilize the transition state (electrostatic catalysis), and, most importantly, by freezing out the relative motions of the reactants and the enzyme's catalytic groups.

11. An enzyme's preferential binding of the transition state lowers $\Delta G^{\ddagger}$ and thereby increases the rate of the reaction. For this reason, an unreactive compound that mimics the transition state may be an effective enzyme inhibitor.

Lysozyme

12. Hen egg white lysozyme, an enzyme that cleaves the glycosidic linkage between NAG and NAM residues in bacterial cell walls, has a substrate-binding cleft that accommodates six sugar residues such that the cleavage occurs between the residues in subsites D and E.

13. In the mechanism for lysozyme, the NAM residue that binds in the D site is distorted toward the half-chair conformation. Glu 35 (a general acid catalyst) then transfers a proton to O1 to cleave the C1—O1 bond of the substrate, generating an oxonium ion transition state. The nucleophilic carboxylate group of Asp 52 simultaneously attacks C1 to form a covalent intermediate (covalent catalysis). Hydrolysis of this bond completes the catalytic cycle. Because the substrate is distorted toward the transition-state conformation on binding, transition state stabilization is an important catalytic mechanism.

Serine Proteases

14. Serine proteases are a widespread family of enzymes that have a common mechanism. The active-site Ser (identified through its inactivation by diisopropylphosphofluoridate), His (identified through affinity labeling with a chloromethylketone substrate analog), and Asp (identified by X-ray crystallography) form a hydrogen-bonded catalytic triad. Nonhomologous serine proteases have developed the same catalytic triad through convergent evolution.

15. The differing substrate specificities of trypsin, chymotrypsin, and elastase depend in part on the shapes and charge distribution of the substrate-binding pocket near the active site.

16. Catalysis by serine proteases is a multistep process in which nucleophilic attack on the scissile bond by Ser 195 (using the chymotrypsinogen numbering system) results in a tetrahedral intermediate that decomposes to an acyl–enzyme intermediate. The replacement of the amine product with water is necessary for the formation of a second tetrahedral intermediate, which yields the carboxyl product and regenerated enzyme.

17. Serine proteases use acid–base catalysis (involving the Ser-His-Asp triad), covalent catalysis (formation of the tetrahedral intermediates), and catalysis through binding of the transition state in the oxyanion hole. The tight binding of bovine pancreatic trypsin inhibitor to trypsin inhibits the enzyme by preventing full attainment of the tetrahedral intermediate, as well as the entry of water into the active site.

18. Low-barrier hydrogen bonds (LBHBs) have been proposed to play important roles in the stabilization of transition states. LBHB's are strong hydrogen bonds, isolated from other water molecules, which have predicted bond energies –4 times those of normal hydrogen bonds in aqueous environments.

Questions

General Properties of Enzymes

1. What is an enzyme's EC number?

2. Explain why enzymes are stereospecific.

3. What is an apoenzyme and how does it differ from a holoenzyme? Which form is active?

4. What is the relationship between vitamins and coenzymes?

5. Proteins can be chemically modified by a variety of reagents that react with specific amino acid residues. How can such reagents be used to identify residues involved in an enzyme's activity? What are the shortcomings of this method?

Activation Energy and the Reaction Coordinate

6. What is the rate-determining step of an enzyme-catalyzed reaction?

7. Answer *yes* or *no* to the following questions and explain your *answer*.
   (a) Can the absolute value of ΔG for a reaction be larger than $\Delta G^{\ddagger}$?
   (b) Can $\Delta G^{\ddagger}$ for an enzyme-catalyzed reaction be greater than $\Delta G^{\ddagger}$ for the nonenzymatic reaction?
   (c) In a two-step reaction, such as the one diagrammed in Figure 11-6, must the intermediate (I) have less free energy than the reactant (A)?
   (d) In a multistep reaction, does the transition state with the highest free energy always correspond to the rate-determining step?

8. An increase in temperature increases the rate of a reaction. How does the temperature affect $\Delta G^{\ddagger}$?

9. $\Delta\Delta G^{\ddagger}$ for an enzymatic reaction at 25°C is 13 kJ·mole^{-1}. (a) Calculate the rate enhancement. (b) What is $\Delta\Delta G^{\ddagger}$ when the rate enhancement is 10^5?

Catalytic Mechanisms

10. List the amino acid residues that are most likely to participate in general acid–base catalysis. Why isn't glycine among those listed?

11. If a protonated His residue acts as the proton donor in an acid-catalyzed enzymatic reaction, what happens to the enzyme's activity as the pH increases to a value that exceeds the pK_R of that residue?

12. The pK values of two essential catalytic residues in RNase A are 5.4 and 6.4. (a) Which corresponds to His 12 and which to His 119 (see Figure 11-10)? (b) Draw a titration curve for these two residues.

13. What is the difference between nucleophilic catalysis and general base catalysis?

14. A good covalent catalyst is highly nucleophilic and can form a good leaving group. What structural properties support these seemingly opposite characteristics?

15. Classify each of the following groups as an electrophile or nucleophile:
     (a) amine
     (b) carbonyl
     (c) cationic imine
     (d) hydroxyl
     (e) imidazole

16. For each of the following reactions, indicate the nucleophilic center and the electrophilic center. Draw curved arrows to indicate the movement of electrons and draw the reaction products.

     (a)   carbonyl phosphate + ammonia $\rightleftharpoons$ carbamic acid + P$_i$

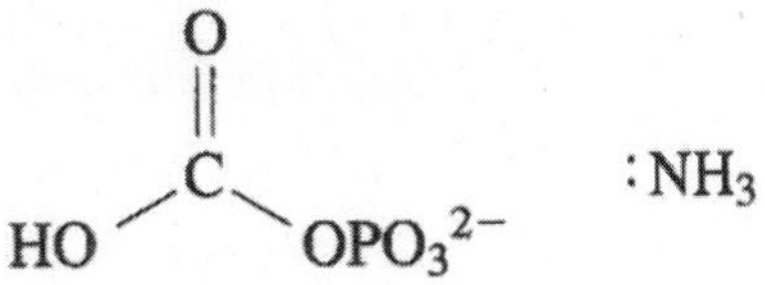

     (b)   NADH + acetaldehyde $\rightleftharpoons$ ethanol + NAD$^+$

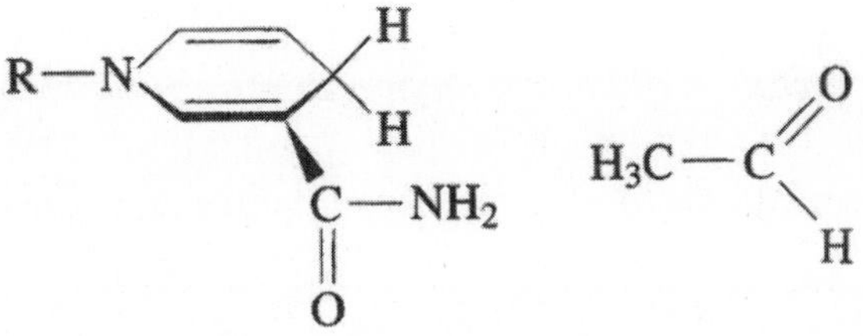

     (c)   2 amino acid $\rightleftharpoons$ dipeptide + H$_2$O

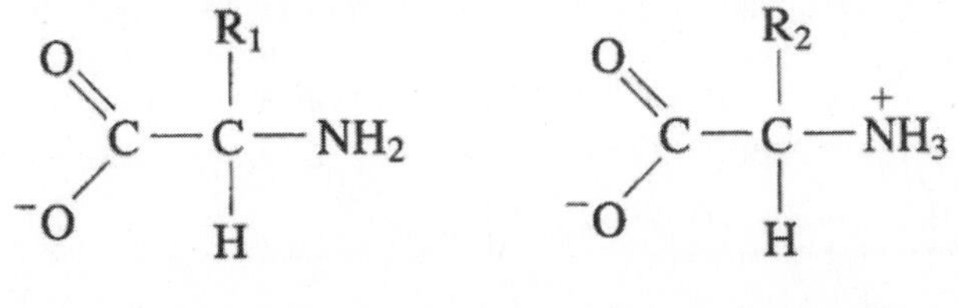

17. What is the role of Zn^{2+} in carbonic anhydrase? *Hint*: See Interactive Exercise 7.

Lysozyme

✦ ***See Interactive Exercise 8 before answering the following questions.***

18. What are the roles of Glu 35 and and Asp 52? What is the role of water in catalysis?

19. Why does lysozyme appear to bind only NAG residues at subsites C and E?

20. Would $(NAG)_6$ or $(NAM)_6$ be a better substrate for lysozyme and why?

21. Hydrogen bonding of substrates to enzymes often involves the polypeptide backbone rather than amino acid side chains. What backbone–substrate hydrogen bond helps distort NAM in the D subsite of lysozyme? Can this hydrogen bond form when N-acetylxylosamine is in the active site?

N-Acetylxylosamine

22. Draw the resonance forms of the half-chair conformation of the oxonium ion transition state of the lysozyme reaction.

Serine Proteases

✦ *See Guided Exploration 10 before answering the following questions.*

23. What is the role of the following amino acids in the active site of chymotrypsin?
    (a) Asp 102
    (b) His 57
    (c) Ser 195
    (d) Gly 193
    (e) Which three constitute the "catalytic triad?"

24. What two catalytic residues in chymotrypsin were identified by chemical modification? Could the same reagents used to identify these residues be used to label the catalytic residues of other serine proteases?

25. The different substrate specificities of chymotrypsin and trypsin have been attributed to the presence of different amino acid residues in the binding pocket. What problems can arise when site-directed mutagenesis is used to test predictions about the roles of such residues in substrate specificity?

26. The cleavage of the ester *p*-nitrophenylacetate (shown below) by chymotrypsin occurs in two stages. In the first stage, the product *p*-nitrophenolate is released in a burst, in amounts equivalent to the amount of active enzyme present. In the second stage, *p*-nitrophenolate is generated at a steady but much reduced rate. Explain this phenomenon in terms of the catalytic mechanism presented in Figure 11-29.

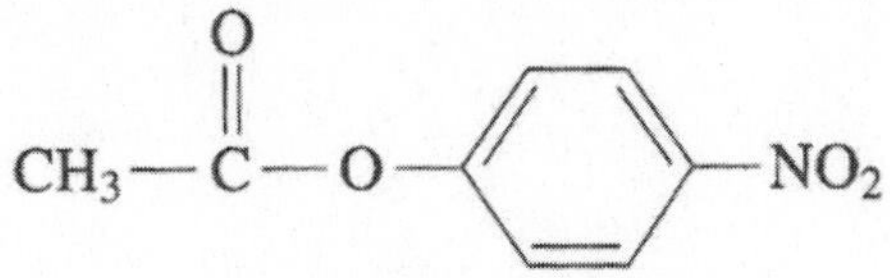

p-Nitrophenylacetate

27. What catalytic mechanism contributes the most to chymotrypsin's rate enhancement?

28. Lys 15 of bovine pancreatic trypsin inhibitor binds to the active site of trypsin but is not cleaved. Explain why the proteolytic reaction cannot proceed.

29. Since trypsin activation is autocatalytic, what is the role of enteropeptidase in activating trypsin?

Answers to Questions

1. The enzyme commission (EC) number is unique to each enzyme. An enzyme is assigned an EC number according to the type of reaction it catalyzes.

2. The protein (or RNA) enzyme is a chiral molecule whose binding clefts and catalytic residues are arranged in a specific three-dimensional asymmetric array. Hence, only substrates with the appropriate stereochemistry can bind to the enzyme, and the enzyme transforms the substrate to product according to the spatial arrangement of interacting functional groups.

3. An apoenzyme is the protein portion of an enzyme that has lost its cofactor (a metal ion or coenzyme). A holoenzyme is an active enzyme containing both the protein and the cofactor.

4. Many coenzymes are synthesized from precursors that are vitamins (substances that an animal cannot synthesize and must obtain from its diet). However, not all coenzymes have vitamin precursors and not all vitamins are precursors of coenzymes.

5. Protein-modifying reagents can provide clues to the identities of catalytic residues. If chemical modification of a residue does not result in loss of activity, that residue can be ruled out as essential for catalysis. Loss of enzymatic activity on modification of a residue may indicate that the modified residue plays an essential role. However, chemical modification of residues at sites other than the active site may interfere with catalysis nonspecifically by disrupting protein structure.

6. The rate-determining step is the slowest of the steps in the reaction mechanism, the step with the greatest free energy of activation.

7.

(a) Yes. ΔG is the difference in free energy between reactants and products, whereas $\Delta G^{\ddagger}$ is the difference in free energy between the reactants and the transition state.

(b) No. By definition, a catalyst decreases $\Delta G^{\ddagger}$ of a reaction.

(c) No. The free energy of the intermediate may be greater than that of the reactant. The reaction will proceed as long as the ΔG of the overall reaction A→P is negative.

(d) No. The rate-determining step is the one whose $\Delta G^{\ddagger}$ is greatest. This does not always correspond to the step with the highest free energy, since $\Delta G^{\ddagger}$ depends on the difference in free energies between a reactant or an intermediate and the following transition state, not just on the free energy of this transition state.

8. $\Delta G^{\ddagger}$ is largely independent of temperature. An increase in temperature increases the rate of a reaction by increasing the number of reacting molecules that reach the transition state.

9. [

(a) The rate enhancement is

$$e^{\Delta \Delta G^{\ddagger}/RT}$$
$$= e^{(13 \text{ kJ·mol}^{-1})/(8.3145 \text{ J·K}^{-1}\cdot\text{mol}^{-1})(298 \text{ K})}$$
$$= e^{5.25}$$
$$= 190$$

The enzyme-catalyzed reaction therefore proceeds 190 times faster than the uncatalyzed reaction.

(b) When the rate enhancement is 10^5,

$$100{,}000 = e^{\Delta \Delta G^{\ddagger}/RT}$$
$$\Delta \Delta G^{\ddagger} = RT \ln 100{,}000$$
$$= (8.3145 \text{ J·K}^{-1}\cdot\text{mol}^{-1})(298\text{K})(11.5)$$
$$= 28.5 \text{ kJ·mol}^{-1}$$

10. Aspartate, cysteine, glutamate, histidine, lysine, and tyrosine are likely to participate in general acid–base catalysis. Glycine does not have an ionizable side chain.

11. As the pH approaches the pK_R, the His become progressively deprotonated and less effective as an acid catalyst. When pH = pK_R, 50% of the His residues are deprotonated and the enzyme is half-active. When the pH significantly exceeds pK_R, the His is almost completely deprotonated and the enzyme is inactive.

12. (a) His 12 has a pK of 5.4, since it is active when unprotonated. It would lose activity when the pH decreases.

   His 119 has a pK of 6.4 since it is active when protonated. It would lose activity when the pH increases.

   (b)

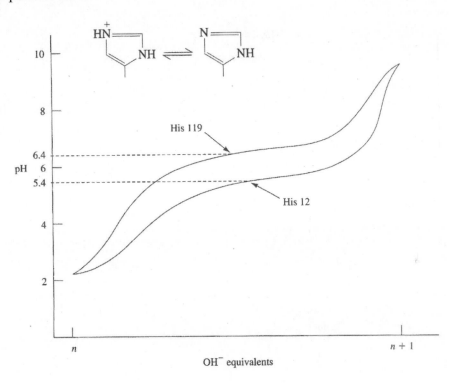

13. In general base catalysis, a proton is abstracted from the substrate, whereas in nucleophilic catalysis, a covalent bond forms.

14. Amino acid side chains that have highly delocalized electrons or high polarizabilities are good covalent catalysts. The hydroxyl group of serine, the carboxyl group of aspartate, and the thiol group of cysteine are highly polarizable groups, and the electrons are highly delocalized in the imidazole group of histidine.

15.
   (a) nucleophile
   (b) electrophile
   (c) electrophile
   (d) nucleophile
   (e) nucleophile

16. (a)

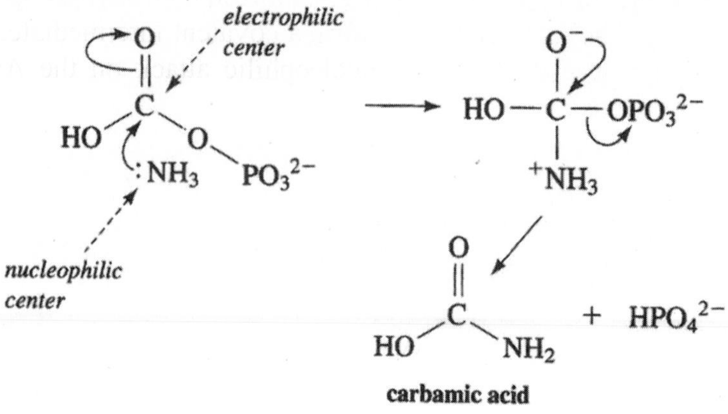

carbamic acid

(b)

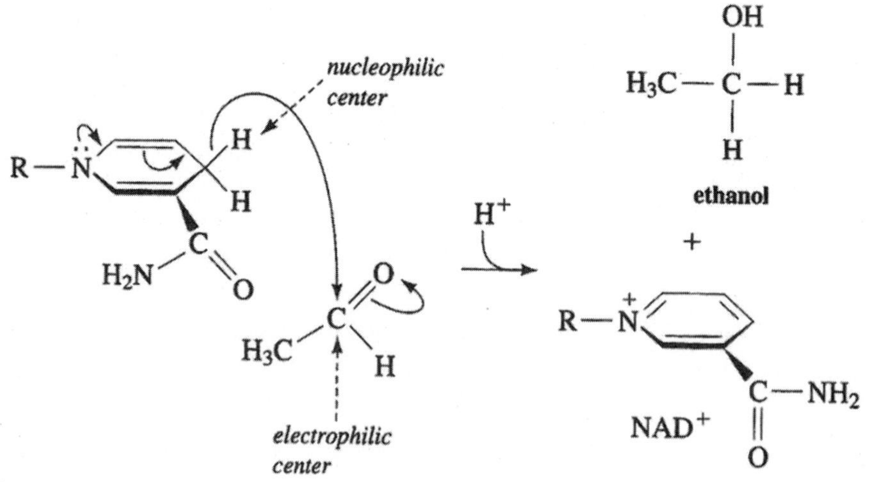

ethanol

(c)

$+ H_2O$

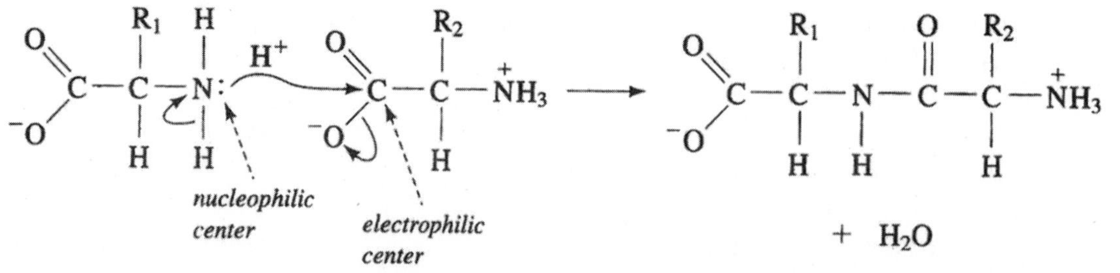

17. The Zn^{2+} in carbonic anhydrase polarizes a water molecule and thereby causes it to ionize. The resulting Zn^{2+}—OH^- is the nucleophile that attacks carbon dioxide, converting it to HCO_3^-. The metal ion stabilizes the negative charge of the OH^-, which would otherwise not form at neutral pH.

18. Glu 35 gives up its proton to O1 of the D ring of the substrate (see Figure 11-21), which leads to the creation of the oxonium ion transition state. The ionized hydroxyl group of Asp 52 attacks the electron-poor C1 of the oxonium ion to form a covalent intermediate. Water regenerates the original form of the enzyme by its nucleophilic attack on the Asp 52– substrate covalent intermediate, releasing the substrate.

19. The lactyl side chain of NAM residues sterically prevents NAM binding to subsites C and E. Hence only NAG residues bind to these subsites.

20. NAG_6 would be better because the C and E residues of NAM_6 would not bind to their subsites.

21. The backbone NH group of Val 109 forms a hydrogen bond with O6 of the D-site residue, helping stabilize its half-chair conformation. *N*-acetylxylosamine lacks O6 and therefore cannot form this hydrogen bond.

22.

23.
   (a) Asp 102 polarizes the imidazole ring of His 57.
   (b) His 57 extracts a proton from the nearby hydroxyl group of Ser 195.
   (c) Ser 195 provides the ionized hydroxyl group that will act as a nucleophile to break the nearby amide bond of the substrate.
   (d) Gly 193 stabilizes the oxyanion tetrahedral intermediate of the reaction.
   (e) Asp 102, His 57, and Ser 195 form the "catalytic triad."

24. Ser 195 was identified through the use of diisopropylphosphofluoridate (DIPF), which reacts with the Ser —OH group at the active site of chymotrypsin and irreversibly inactivates the enzyme. His 57 was identified through affinity labeling using the substrate analog tosyl-L-phenylalanine chloromethylketone. DIPF also reacts with the active-site Ser in other serine proteases and therefore would label this residue. However, these other enzymes, not all of which share chymotrypsin's specificity for Phe-containing substrates, would require chloromethylketone analogs that incorporated residues corresponding to their substrate specificities, in order to react with their active-site His residues.

25. Site-directed mutations often change (or fail to change) enzymes in unexpected ways. For example, site-directed mutation of trypsin's Asp 189 to Ser did not change trypsin's specificity to that of chymotrypsin. Several other changes involving amino acids found on the surface loops surrounding the binding pocket were required for trypsin to emulate chymotrypsin.

26. Chymotrypsin acts as an esterase, attacking the carbonyl C of the substrate's ester bond to form a tetrahedral intermediate. This is followed by decomposition to an acyl–enzyme intermediate with release of the first product, *p*-nitrophenolate.

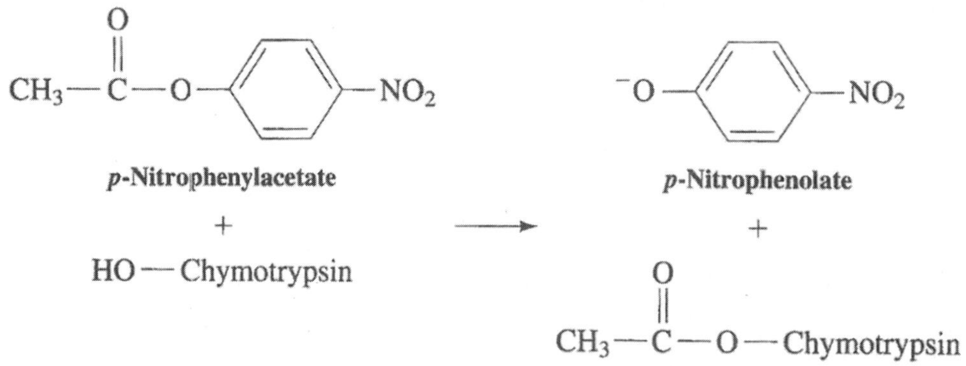

p-Nitrophenylacetate

p-Nitrophenolate

+

HO — Chymotrypsin

+

CH_3-C-O-Chymotrypsin

Release of the second product, acetate, via hydrolysis of the acyl–enzyme intermediate, is much slower. Consequently, every active site attacks the substrate and releases the first product in a stoichiometric fashion, accounting for the initial burst of product formation. Because the second phase of the reaction is slow, the enzyme is only slowly regenerated and made available to catalyze additional rounds of the esterolytic reaction.

27. Stabilization of the transition state via the oxyanion hole is responsible for the largest portion of chymotrypsin's rate enhancement.

28. Bovine pancreatic trypsin inhibitor binds trypsin so tightly that the complex is too rigid to allow formation of the tetrahedral intermediate, to allow release of the first product, or to allow water to enter the active site of trypsin.

29. Trypsinogen's catalytic activity is too low to activate other trypsinogen molecules at a biologically significant rate. Enteropeptidase cleaves trypsinogen, thereby generating a small amount of trypsin that then commences autocatalytic activation. The role of enteropeptidase is to initiate this process in a controlled manner.

12

Enzyme Kinetics, Inhibition, and Control

This chapter introduces chemical kinetics—the study of reaction rates—followed by the kinetics of enzymatic reactions. An enzyme-catalyzed reaction can be described by the Michaelis–Menten equation, which expresses the reaction velocity in terms of its Michaelis constant, K_M, and its maximum velocity, V_{max}. Detailed knowledge of the kinetics of a reaction can contribute to the understanding of its step-by-step reaction mechanism. The effects of different substrates, inhibitors, and other factors may also reveal an enzyme's physiological function. This knowledge can be exploited to develop drugs that are enzyme inhibitors. In this chapter, the three types of reversible enzyme inhibition and the equations that describe them are presented. The chapter also describes the control of enzymes, using aspartate transcarbamoylase as an example of allosteric control, and glycogen phosphorylase as an example of control by covalent modification. The chapter concludes with a discussion of how drugs are developed and tested, and the basis of adverse drug reactions.

Essential Concepts

Reaction Kinetics

1. A chemical reaction may proceed through several simple steps, called elementary reactions. The overall reaction pathway may therefore involve several short-lived intermediates.

2. The rate, or velocity (v), at which a reactant is consumed or a reaction product appears can be mathematically described. Thus, for the conversion of reactant A to product P,

$$v = \frac{d[P]}{dt} = -\frac{d[A]}{dt} = k[A]$$

3. The rate of an elementary reaction varies with the concentration(s) of the reacting molecule(s). For example, for a single-reactant reaction (a unimolecular or first-order reaction), the rate is directly proportional to the concentration of the reactant. For a two-reactant reaction (a bimolecular or second-order reaction), the rate is directly proportional to the product of the concentrations of the reactants or to the square of the concentration of a reactant that reacts with itself.

4. The proportionality constant in the equation above, which is known as the rate constant, k, can be determined graphically. The rate equation for a first-order reaction is

$$\ln[A] = \ln[A]_o - kt$$

and that for a second-order reaction is

$$\frac{1}{[A]} = \frac{1}{[A]_o} + kt$$

where $[A]_o$ is the initial concentration of the reactant and t is time. Consequently, if a plot of ln $[A]$ versus t yields a straight line, the reaction is first-order, and if a plot of $1/[A]$ versus t yields a straight line, the reaction is second-order. The slope of the line reveals the value of the corresponding rate constant, k.

5. The kinetics of enzyme-catalyzed reactions are more complicated because the enzyme and substrate (reactant) combine to form a complex that then decomposes to product and free enzyme. For reactions involving a single substrate, S, the reaction velocity is typically measured under conditions where $[S] > [E]$. At very high substrate concentrations, the velocity is independent of $[S]$ and the enzyme is said to be saturated with substrate.

6. The Michaelis–Menten equation, which describes an enzymatic reaction, is based on the assumption that the enzyme–substrate complex maintains a steady state; that is, its concentration does not change. This assumption is valid over most of the course of a typical enzymatic reaction.

7. The Michaelis–Menten equation is

$$v_o = \frac{V_{max} [S]}{K_M + [S]}$$

where v_o is the initial velocity of the reaction (before more than ~10% of the substrate has been consumed), V_{max} is the maximum rate of the reaction, and K_M is the Michaelis constant. This equation describes a rectangular hyperbola (the shape of the curve generated by a plot of v_o versus $[S]$) whose asymptote is V_{max}.

8. The Michaelis constant, K_M, is unique to each enzyme–substrate pair. Its value is the substrate concentration at which the reaction velocity is half-maximal. It is therefore a measure of the affinity of the enzyme for its substrate.

9. The catalytic constant (k_{cat}), or turnover number, of an enzyme can be derived from V_{max}:

$$k_{cat} = \frac{V_{max}}{[E]_T}$$

where $[E]_T$ is the total enzyme concentration. The overall catalytic efficiency of an enzyme can be expressed as k_{cat}/K_M, which is an apparent second-order rate constant for the enzymatic reaction.

10. The kinetic parameters for an enzymatic reaction can be determined by taking the reciprocal of the Michaelis–Menten equation:

$$\frac{1}{v_o} = \left(\frac{K_M}{V_{max}}\right)\frac{1}{[S]} + \frac{1}{V_{max}}$$

A plot of $1/v_o$ versus $1/[S]$, a so-called Lineweaver–Burk or double-reciprocal plot, yields a straight line whose slope and intercepts yield the values of K_M and V_{max}.

11. Steady state kinetics cannot unambiguously establish a reaction mechanism because there are an infinite number of mechanisms that are consistent with a given set of kinetic data. However, mechanisms that are not consistent with the kinetic data can be ruled out.

12. Many enzymes have multiple (usually two) substrates and products. For example, transferase reactions are bisubstrate reactions. In a sequential reaction, all the substrates bind to the enzyme before products are formed. In sequential reactions, a particular order of substrate addition may be obligatory (an Ordered mechanism) or not (a Random mechanism).

13. In a Ping Pong reaction, one or more products of a transferase reaction are released before all substrates bind.

Enzyme Inhibition

14. Many substances, often pharmaceutically active compounds, alter the activities of specific enzymes. Detailed information about the mechanism of an enzyme can aid in the design of drugs with the desired properties. Studies of the kinetics of enzymes in the presence of specific inhibitors help reveal how the inhibitors act and provide a way to quantitatively compare the effects of different inhibitors. An inhibitor may reversibly interact with an enzyme to interfere with its substrate binding, its catalytic activity, or both.

15. Three modes of reversible enzyme inhibition can be distinguished by their effects on the kinetic behavior of enzymes: competitive, uncompetitive, and mixed (noncompetitive) inhibition. Double-reciprocal plots of data collected in the presence of different concentrations of an inhibitor reveal the value of K_I, the dissociation constant of the inhibitor from the enzyme.

16. A competitive inhibitor competes with a normal substrate for binding to the enzyme. It therefore reduces the apparent affinity of the enzyme for its substrate (increases K_M). A large excess of substrate can overcome the effect of the inhibitor, so V_{max} is not affected.

17. An uncompetitive inhibitor binds only to the enzyme–substrate complex and apparently distorts the active site. It decreases the apparent K_M and V_{max}.

18. A mixed inhibitor binds to both free and substrate-bound enzyme and may interfere with both substrate binding and catalysis. As a result, the apparent V_{max} decreases, and the apparent K_M may increase or decrease. When only V_{max} is affected, the inhibition is said to be noncompetitive.

Control of Enzyme Activity

19. Enzyme activity can be controlled either by altering the amount of enzyme available for reaction or by modifying its catalytic activity through allosteric effects or by covalent modification. For example, in feedback regulation, the end product of a metabolic pathway inhibits the first committed step in the pathway. This ensures that the pathway is active when product concentrations are low but is inactive when product concentrations exceed the levels needed by the cell.

20. *E. coli.*aspartate transcarbamoylase (ATCase) provides an example of allosteric regulation that includes feedback inhibition by CTP (the ultimate product of the pyrimidine synthesis pathway, which begins with the ATCase reaction) and activation by ATP (which ensures that the concentrations of pyrimidine nucleotides keep pace with those of purines). The binding of either effector molecule (ATP or CTP) to the regulatory subunits of ATCase induces changes in the enzyme's quaternary structure that alter the activity of the catalytic subunits. ATP stabilizes the R (high activity) state, whereas CTP stabilizes the T (low activity) state of this allosteric enzyme.

21. Enzymes controlled by covalent modification usually undergo phosphorylation and dephosphorylation of a Ser, Thr, or Tyr residue, such that the enzyme shifts to a more or less active state. The modifications are catalyzed by kinases and phosphatases that are often part of signaling cascades.

Drug Design

22. The majority of drugs act by modifying the activity of a receptor protein. The second largest class of drugs is enzyme inhibitors.

23. Methods of screening drugs to develop a potential candidate drug, called a lead compound, have evolved dramatically in the past thirty years, but in most cases, initial screening begins with an assessment of K_I. Rational or structure-based drug design takes advantage of knowledge of the three-dimensional structure of the target protein. Combinatorial approaches coupled with robotic systems have also quickened the pace of drug development.

24. Once a drug candidate has been developed, scientists must determine whether it can be delivered in sufficiently high concentrations to the target organ and protein. All drugs must ultimately be tested in humans, which involves expensive, federally regulated clinical trials. Currently, the Food and Drug Administration approves only 1 drug candidate in 5000.

25. Differences in patients' reactions to drugs arise from differences in genetic makeup as well disease states. The cytochrome P450 superfamily of heme-containing proteins oxidizes most drugs (and other xenobiotics). Humans express 57 isozymes of cytochrome P450, many of which are highly polymorphic. The resulting variation among individuals in the rates that specific drugs are metabolized are, in part, responsible for the variation in patients' responses to the same drug. Similarly, drug–drug interactions may result from cytochrome P450 activity on simultaneously administered drugs.

Key Equations

$$\ln [A] = \ln [A]_o - kt$$

$$\frac{1}{[A]} = \frac{1}{[A]_o} + kt$$

$$v_o = \frac{V_{max} [S]}{K_M + [S]}$$

$$k_{cat} = \frac{V_{max}}{[E]_T}$$

$$\frac{1}{v_o} = \left(\frac{K_M}{V_{max}}\right)\frac{1}{[S]} + \frac{1}{V_{max}}$$

Questions

Reaction Kinetics

1. For each of the following reactions, write a rate equation and determine the reaction order.
   (a) $A \rightarrow P$
   (b) $A + B \rightarrow P + Q$
   (c) $2A \rightarrow P$

2. List two different ways to measure the progress of a chemical reaction.

3. A first-order reaction has a $t_{1/2}$ of 20 minutes.
   (a) What is the rate constant k?
   (b) What time is required to form 20% of the product?
   (c) What time is required to form 80% of the product?
   (d) How much starting material remains after 15 min?
   (e) Compare the rate constant for this reaction to that of the decay of ^{32}P, which has a half-life of 14 days.

4. The energy of binding a transition state complex ($X^{\ddagger}$) can be determined by writing an equilibrium expression for the formation of the complex.
    (a) For the reaction $A \leftrightarrows X^{\ddagger}$, what is the equilibrium expression?
    (b) What is the expression for the free energy of binding to the transition state ($\Delta G^{\ddagger}$)?
    (c) What factors are needed to equate the rate constant k to the free energy of activation?

5. For the following reaction:

$$E + S \underset{k_{-1}}{\overset{k_1}{\rightleftharpoons}} ES \overset{k_2}{\rightleftharpoons} P + E$$

    (a) What is meant by the term "enzyme–substrate complex"?
    (b) Write a rate equation for the production of ES.
    (c) What is the rate of product formation from ES?
    (d) If all the enzyme is bound to substrate, what is the effect of adding more substrate on the forward rate of the reaction?

6. What is meant by (a) the steady state assumption, (b) K_M, (c) k_{cat}, (d) turnover number, (e) catalytic efficiency, and (f) diffusion-controlled limit?

7. The following data were obtained for the reaction $A \leftrightarrows B$, catalyzed by the enzyme Aase. The reaction volume was 1 mL and the stock concentration of A was 5.0 mM. Seven separate reactions were examined, each containing a different amount of A. The reactions were initiated by adding 2.0 µL of a 10 µM solution of Aase. After 5 minutes, the amount of B was measured.

Reaction	Volume of A added (µL)	Amount of B present at 5 minutes (nmoles)
1	8	26
2	10	29
3	15	39
4	20	43
5	40	56
6	60	62
7	100	71

    (a) Calculate the initial velocity of each reaction (in units of $\mu M \cdot min^{-1}$)
    (b) Determine the K_M and V_{max} of Aase from a Lineweaver–Burk plot.
    (c) Calculate k_{cat}.

8. Can you use kinetic data to prove that a particular model for an enzymatic reaction mechanism is correct? Explain.

9. Why is it possible for sequential bisubstrate reactions to be Ordered or Random, whereas a Ping Pong reaction always has an invariant order of substrate addition and product release?

10. Assume that $k_2 = k_{cat}$ for a highly purified enzyme for which you seek to determine $\Delta G^{\ddagger}$. Using the Arrhenius equation, $k = Ae^{-\Delta G^{\ddagger}/RT}$, where k is a reaction rate constant and A is a constant, you can determine the activation energy to a good approximation.
    (a) Derive a mathematical relationship that will allow you to calculate $\Delta G^{\ddagger}$ and show a graph that can used with experimental data. *Hint*: Convert the Arrhenius equation into a linear form, in which the constant A need not be known beforehand.
    (b) What are the easily measurable parameters that you can obtain?
    (c) What assumptions do you need to make in order to obtain a good approximation of the activation energy?

Enzyme Inhibition

11. There are three general mechanisms for the reversible inhibition of enzymes that follow the Michaelis–Menten model. How does the mode of inhibitor–enzyme binding differ among the three mechanisms?

12. The catalytic behavior of an enzyme may depend on ionizable amino acids. Therefore, a change in pH may influence an enzyme's catalytic behavior. How can you tell whether pH affects substrate binding or catalytic activity?

13. The movement of glucose across the erythrocyte membrane is "catalyzed" by a transport protein (Section 10-2E and Box 10-2).
    (a) What is the kinetic behavior of this process?
    (b) Can glucose transport be subject to competitive, uncompetitive, or mixed inhibition? Explain.

14. In hen egg white lysozyme (Section 11-4), the substitution of Ala for Asn at position 37 or for Trp at position 62 may alter the enzyme's kinetics. What changes would you predict and why?

Control of Enzyme Activity

15. Draw velocity versus [Asp] curves for the reaction catalyzed by the ATCase catalytic trimer and by the intact enzyme. Explain why the curves differ.

16. How do carbamoyl phosphate, aspartate, ATP, and CTP affect the T $\leftrightarrows$ R equilibrium of ATCase?

Drug Design

17. What biochemical parameters are likely to vary among different allelic isozymes of cytochrome P450?

18. Draw a flow diagram that summarizes the steps involved in developing a drug from the generation of a lead compound through FDA approval. Which is the most time-consuming step?

Answers to Questions

1.

(a) $v = -d[A]/dt = k[A]$ This is a first-order reaction.

(b) $v = -d[A]/dt = -d[B]/dt = k[A][B]$ This is a second-order reaction (A and B must collide to form product).

(c) $v = -d[A]/dt = k[A]^2$ This is also a second-order reaction (A must collide with another molecule of A for the reaction to proceed).

2. The progress of a reaction can be followed by measuring the rates of the appearance of the product(s) or the disappearance of the reactant(s). In practice, any physical property, such as light absorbance, pH, or an NMR signal, can be followed, provided that it changes in proportion to the concentration(s) of the reactant(s) or product(s).

3.

(a) For a first-order reaction, $t_{1/2} = 0.693/k$ (Equation 12-9). Therefore, $k = 0.693/20$ min $= 0.035$ min^{-1}.

(b) Since $\ln([A]/[A]_o) = -kt$ (from Equation 12-6), $t = \ln([A]/[A]_o)/-k$. When 20% of A has been converted to product, $[A]/[A]_o = 0.8$ and $t = (\ln 0.8)/(-0.035$ min$^{-1}) = (-0.22)/(-0.035$ min$^{-1}) = 6.4$ min.

(c) When 80% of A has been converted to product, $[A]/[A]_o = 0.2$ and $t = (\ln 0.2)/(-0.035$ min$^{-1}) = (-1.61)/(-0.035$ min$^{-1}) = 46$ min.

(d) When $t = 15$ min, $\ln([A]/[A]_o) = (-0.035$ min$^{-1})(15$ min$) = -0.525$. Since $e^{-0.525} = 0.59$, 59% of A remains at 15 minutes.

(e) For the decomposition of ^{32}P, $k = (0.693/14$ days$)$ $(1$ day$/1440$ min$) = 3.4 \times 10^{-5}$ min^{-1}. This is ~1000 times slower than the reaction described above.

4.

(a) $K^{\ddagger} = [X^{\ddagger}]/[A]$.

(b) $\Delta G^{\ddagger} = -RT \ln K^{\ddagger}$.

(c) As described in Box 12-3, $d[P]/dt = k[A] = k'[X^{\ddagger}]$ where k' is the rate constant for the decomposition of $X^{\ddagger}$ to from products. k' can be expressed in terms of the Boltzman constant (k_B) and Planck's constant (h): $k' = k_B T/h$. Thus, since $[X^{\ddagger}] = K^{\ddagger}[A]$

$$e^{-\Delta G\ddagger/RT},$$

$$\frac{d[P]}{dt} = \frac{k_B T}{h} e^{-\Delta G\ddagger/RT} [A]$$

and

$$k = \left(\frac{k_B T}{h}\right) e^{-\Delta G\ddagger/RT}$$

5.

(a) The enzyme–substrate (ES) complex is the species formed by the interaction between an enzyme and its substrate.

(b) The rate equation for the net formation of ES is the rate of formation of ES minus the rate of ES degradation: $d[ES]/dt = k_1[E][S] - k_{-1}[ES] - k_2[ES]$.

(c) The rate of product formation, $d[P]/dt = v = k_2[ES]$.

(d) Increasing substrate concentration when all the enzyme is in the ES complex does not increase the rate of the reaction since then $v = V_{max} = k_2[E]_T$ where $[E]_T$ is the total enzyme concentration; that is, the enzyme is working at its maximal rate; it can work no faster.

6.

(a) The steady state assumption assumes that during the course of an enzyme-catalyzed reaction, the concentration of the ES complex does not change.

(b) K_M is the Michaelis constant: $K_M = (k_{-1} + k_2)/k_1$. K_M is the substrate concentration at which the reaction velocity is half-maximal.

(c) The k_{cat} is the maximum velocity divided by the total enzyme concentration: $k_{cat} = V_{max}/[E]_T$. It is the number of reaction processes (turnovers) that each active site catalyzes per unit time. For the simple kinetic scheme used to derive the Michaelis–Menten equation, $k_{cat} = k_2$.

(d) The turnover number is the same as k_{cat}.

(e) Catalytic efficiency, calculated as k_{cat}/K_M, is the apparent second-order rate constant for the reaction of E + S and indicates how often the enzyme catalyzes a reaction upon encountering its substrate.

(f) If an enzyme catalyzes a reaction every time it collides with its substrate, it has reached catalytic perfection and the rate is controlled by how often the molecules collide, that is, by their rate of diffusion. At this point, the rate is said to have reached its diffusion-controlled limit.

7.

(a) To calculate v_o for Reaction 1, for example,

v_o = (26 nmol/5 min)/(1.0 mL) × (10^3 mL/1 L) × (0.001 μmol/nmol) = 5.2 μM·min^{-1}

Reaction	v_o (μM·min^{-1})
1	5.2
2	5.8
3	7.8
4	8.6
5	11.2
6	12.4
7	14.2

(b) First, calculate [S] for each reaction. In Reaction 1, for example, [A] = (0.008 mL)(5 mM)/(1 mL) × (1000 μM/1 mM) = 40 μM. Next, convert the data to values of 1/[S] and 1/v.

Reaction	[S] (μM)	1/[S] ($μM^{-1}$)	v (μM·min^{-1})	1/v (min·$μM^{-1}$)
1	40	0.025	5.2	0.192
2	50	0.02	5.8	0.172
3	75	0.0133	7.8	0.128
4	100	0.010	8.6	0.0116
5	200	0.005	11.2	0.089
6	300	0.0033	12.4	0.081
7	500	0.002	14.2	0.070

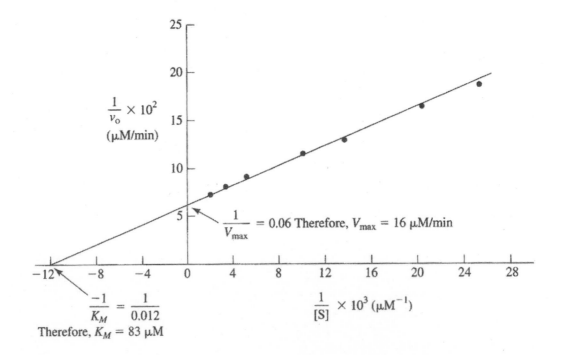

(c) First calculate $[E]_T = (0.002\ mL)(10\ \mu M\ Aase)/(1\ mL) = 0.02\ \mu M$.
Using the value of V_{max} determined above and Equation 12-27, $k_{cat} = V_{max}/[E]_T = (16$
$\mu M\cdot min^{-1})/(0.02\ \mu M) = 800\ min^{-1}$.

8. Kinetics can support but cannot prove a particular reaction mechanism. A single kinetic model may explain several mechanisms, so additional experiment must be performed to establish a particular mechanism. However, kinetic data can rule out mechanisms that are inconsistent with the observed behavior.

9. In a Sequential reaction, both substrates must bind before any product is released. Hence it is possible for the two substrates to bind in an Ordered or Random fashion. In a Ping Pong reaction, a group is transferred from the first substrate to the enzyme to form the first product; then a second substrate binds and is converted to a second product. Consequently, the second product cannot bind first or yield the second product until the first substrate has bound and been converted to the first product.

10.

(a) We can approximate $\Delta G^{\ddagger}$ by assuming we have conditions where $k_2 = k_{cat}$ and $k_{cat} = k$ in the Arrhenius equation, $k = Ae^{-\Delta G^{\ddagger}/RT}$. We can substitute and convert this equation to a logarithmic form in the form $y = mx + b$.

$$k_{cat} = Ae^{-\Delta G^{\ddagger}/RT}$$

$$\ln k_{cat} = \ln A - \Delta G^{\ddagger}/RT$$

$$\ln k_{cat} = (-\Delta G^{\ddagger}/R)(1/T) + \ln A$$

Hence, now we can plot data in graphical form as shown below.

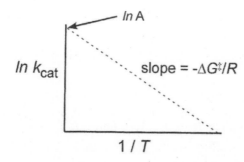

(b) Since our enzyme is in highly purified form, we can calculate our total enzyme concentration in an experiment as a molar quantity. k_{cat} can be calculated after we experimentally obtain V_{max}. We then do kinetic experiments at several temperatures at which the enzyme is stable so that we can determine k_{cat} at each temperature. These data are then plotted as shown above so that the slope provides $\Delta G^{\ddagger}$.

(c) A variety of assumptions need to be made:
  (i) Small errors in $[E]_T$ can lead to substantive differences in the $\Delta G^\ddagger$;
  (ii) We assume that $[E]_T = [X^\ddagger]$; which may only be an approximation for enzymes with low values for k_{cat} and one transition-state complex (see Box 12-3).
  (iii) The enzyme's structure, and hence its catalytic activity, is not significantly altered at the temperatures tested.

11. The three general mechanisms of inhibition are competitive, uncompetitive, and mixed inhibition. In competitive inhibition, the inhibitor competes with the substrate for binding to the active site, increasing the apparent K_M. The inhibitor binds only to the free enzyme and not to the ES complex. In uncompetitive inhibition, the inhibitor binds only to the ES complex to inhibit the enzyme. Although the inhibitor does not directly affect substrate binding, it decreases the apparent K_M. In mixed inhibition, the inhibitor can bind to both the free enzyme and the ES complex. Binding of a mixed inhibitor may alter the binding of the substrate and therefore may alter the apparent K_M. If the inhibitor does not alter the apparent K_M, inhibition is said to be noncompetitive.

12. Aside from extremely low or high pH's, which can denature an enzyme, changes in pH may affect the protonation/deprotonation of residues involved in substrate binding and/or catalysis. Constructing a Lineweaver–Burk plot at several different pH values should yield a set of lines that indicate whether K_M (substrate binding) and/or V_{max} (catalytic activity) is affected (just as Lineweaver–Burk plots can reveal the different types of enzyme inhibition).

13.
  (a) The process obeys Michaelis–Menten (saturation) kinetics:

  $$\text{Glucose}_{out} + T_{out} \rightarrow \text{glucose}_{out}{\cdot}T_{out} \rightarrow \text{glucose}_{in}{\cdot}T_{in} \rightarrow \text{glucose}_{in} + T_{in}$$

  (b) Yes. All three inhibitory modes are possible. An inhibitor could compete with glucose for binding to the transport protein (competitive inhibition); it could bind to the transporter–glucose complex and interfere with the conformational change that exposes glucose to the other side of the membrane (uncompetitive inhibition); or it could interfere with the transporter's function with glucose bound or not bound (mixed inhibition).

14. These substitutions would weaken substrate binding to lysozyme by removing hydrogen bonds that hold the substrate in place. The K_M would increase, but V_{max} would remain nearly the same, since the catalytic residues Asp 52 and Glu 35 are not changed.

15. The catalytic trimer does not display cooperativity in the absence of the regulatory subunits. Hence the rate profile is hyperbolic, much like the binding of O_2 to myoglobin. Intact ATCase exhibits a sigmoidal curve characteristic of cooperative substrate binding, which is possible when the regulatory subunits mediate conformational changes between catalytic subunits.

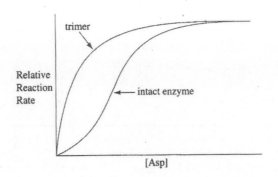

16. Carbamoyl phosphate, aspartate, and ATP have higher affinity for and stabilize the R (more active) state, whereas CTP has higher affinity for and stabilizes the T (less active) state.

17. The intrinsic biochemical parameters that are likely to vary among different allelic isozymes of cytochrome P450 include:
    (a) The range of specificities for different substrates;
    (b) The K_M values for different substrates;
    (c) The V_{max} values for different substrates;
    (d) The concentration of the isozyme.

    A key biological parameter that may vary among allelic isozymes includes the steady-state concentration of the isozyme, which is a function of the rate of synthesis of the protein (as determined by the steady-state concentration of its mRNA), and its intracellular stability (as determined by its amino acid sequence, especially at the N-terminus).

18. A lead compound for a drug that interacts with a target protein can come from several sources:
    (a) Natural products, most often from plants, including potential candidates from so-called folk remedies;
    (b) Rational or structure-based design using the three-dimensional structure of the protein to explore chemical structures that may fit in an active or allosteric site;
    (c) Combinatorial libraries of peptides and other compounds (see Fig. 12-18, for example), which can generate a variety of chemical structures that may bind effectively to the target protein. This has been recently facilitated by automated high-throughput screening technologies; and
    (d) A combination of combinatorial libraries, high-throughput screening, and structure-based design.

The most time-consuming (and expensive) step is the identification of adverse side reactions during clinical trials.

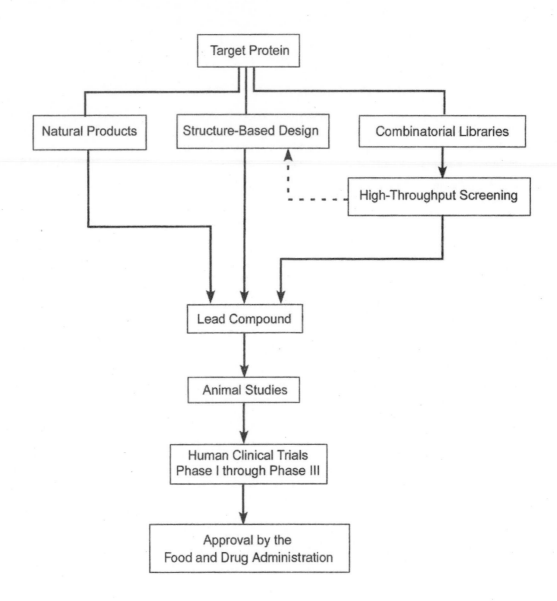

13
Biochemical Signaling

This chapter introduces the fundamental biochemical signaling pathways that allow the cell to respond to chemical and physical signals in its environment. The focus of this chapter is on biochemical signaling pathways that respond to intercellular signals called endocrine hormones. The binding of a hormone to its receptor induces a conformational change that transduces the signal into new biochemical activity inside the cell. Three signal transduction pathways are discussed in detail: signaling pathways activated by receptor tyrosine kinases; the adenylate cyclase signaling system activated by G protein–coupled receptors; and the phosphoinositide pathway. Alterations of signaling pathways, which often result from mutations in genes (oncogenes) that encode signaling proteins, are important to the development of cancers, as indicated in Box 13-3. Signal transduction pathways are the target of a variety of drugs and toxins, as discussed in Box 13-4. The chapter also highlights two key biochemical technologies used in the quantitative analysis of biochemical signaling: the radioimmunoassay developed by Rosalyn Yalow (Box 13-1) and receptor–ligand binding assays using Scatchard plots (Box 13-2).

Essential Concepts

Hormones

1. In general, every biochemical signaling pathway consists of a receptor protein that binds a hormone or some other signaling molecule. This binding results in conformational changes that either allow the receptor–hormone complex to directly affect gene expression or transduce the extracellular signal through a series or cascade of reactions, often involving protein kinases and phosphatases. The latter mechanism may result in dramatic 10^6-fold amplification of the original signal molecule's interaction with a single receptor.

2. Endocrine hormones are chemical signals that reach their target cells via the bloodstream. Each hormone binds to a specific receptor, and its binding results in the transduction of the hormone signal into a biochemical response inside the cell (often referred to as signal transduction). Endocrine hormones include soluble compounds such as amino acids and small polypeptides and insoluble lipids such as steroids. Many factors may modulate the specific response of a cell to a particular hormone, such as the presence or number of receptors, the synergistic or antagonistic activities of other hormone-bound receptors, and the specific isoforms of the kinases and phosphatases of the activated enzyme cascades.

3. The pancreatic islet and adrenal gland hormones represent some of the best studied hormone signaling systems. Energy metabolism is regulated largely by these hormones. The pancreas secretes the polypeptide hormones glucagon and insulin, which act antagonistically to each other to regulate blood glucose concentration. The adrenal medulla secretes the catecholamines epinephrine and norepinephrine, which bind to α- and β-

adrenoreceptors. These receptors often respond to the same ligand in opposite ways in different tissues.

4. Steroids are carried in the blood via the glycoprotein transcortin and albumin. As transcortins interact with cellular membranes, steroids pass across the membrane and bind to their cognate steroid receptors. Steroid–receptor complexes migrate to the nucleus, where they function as regulators of DNA transcription (see Chapter 28).

5. The receptor for growth hormone (a polypeptide hormone) shows features characteristic on many membrane-bound receptors. The binding of growth hormone to the extracellular domain of its receptor results in dimerization of the receptor. This allows it to bind to protein kinases that phosphorylate Tyr residues on target proteins (via transfer of the γ-phosphate from ATP), and are therefore called tyrosine kinases. Growth hormone and insulin were among the first polypeptides to be produced in industrial quantities by recombinant DNA technology.

Receptor Tyrosine Kinases

6. A large family of cell surface receptors, known as receptor tyrosine kinases (RTKs), includes protein kinase activity that specifically phosphorylates tyrosine residues on target proteins. The insulin receptor is a receptor tyrosine kinase. RTKs usually dimerize upon binding ligand, which leads to autophosphorylation of the receptor protein. This autophosphorylation in turn may lead to the phosphorylation of target proteins. These proteins may bind directly to the RTK via a protein domain called the Src homology 2 (SH2) domain, which recognizes the phospho-Tyr residues, or they may bind to integrator proteins such as the insulin receptor substrate proteins (IRS-1 and IRS-2), which are directly phosphorylated by the RTK.

7. The human genome encodes at least 90 protein tyrosine kinases (PTKs) and 388 Ser/Thr kinases (representing >2% of human genes). PTK domains undergo dramatic conformational changes close to the active site, presumably bringing substrates in close juxtaposition, as well as excluding water from the active site, which could result in wasteful ATP hydrolysis.

8. Many RTKs interact with proteins that activate the monomeric G protein Ras, which in turn activates a protein kinase cascade. In the best-known pathway, the activation of Ras leads to the activation of Raf (MAPKKK), which is a Ser/Thr protein kinase that phosphorylates and activates MEK (MAPKK). MEK is both a Tyr kinase and a Ser/Thr kinase that activates MAP kinase (MAPK), another Ser/Thr kinase with many targets including other kinases and transcription factors. Many MAPK cascade proteins are organized into complexes by scaffold proteins that are essential for the rapid amplification of signal transduction in these signaling pathways.

9. G proteins are an important family of regulatory proteins that bind and hydrolyze GTP, resulting in a cycle of conformational changes reminiscent of the ATP-binding-and-hydrolysis cycle of myosin.

    (a) Monomeric Ras interacts with a guanine nucleotide exchange factor (GEF; Sos in some cases), which stimulates Ras to exchange of its bound GDP for GTP. Ras also interacts with a GTPase activating protein (GAP), which stimulates the inherent GTPase of Ras by a factor of 10^5.

    (b) An activated RTK binds to an adapter protein–GEF complex, which then interacts with Ras·GDP, resulting in a rapid exchange of GDP for GTP to produce active Ras·GTP. Activated Ras·GTP binds to and stimulates other effectors such as Raf, becoming inactivated by the hydrolysis of GTP.

10. Receptor tyrosine kinase family members also bind proteins called growth factors that regulate cell growth and differentiation. Many genes that encode elements of growth-factor signaling pathways are known as proto-oncogenes, because their mutation may yield an oncogene. The product of an oncogene may disrupt normal cell growth and lead to cancer. Many cancer-causing viruses bear oncogenes. For example, the v-*erbB* oncogene, first identified in an animal virus, codes for the epidermal growth factor receptor without its ligand-binding domain, which results in constitutively active phosphorylation of protein substrates by this RTK. Other oncogenes include mutated forms of Ras and the transcription factors Fos and Jun (see Box 13-3).

11. Many receptors do not have intrinsic kinase activity (e.g., the growth hormone receptor and many growth factor receptors). Dimerization or trimerization usually leads to activation of nonreceptor tyrosine kinases (NRTKs) that are complexed to one of the receptors. One of the best studied NRTKs is Src (named after the sarcoma virus wherein it was first discovered), which is one member of a large family of NRTKs. Many of these have common structural domains that allow the protein to dock or interact with other signaling proteins, such as SH2 (Src homology-2) domain, which binds to phosphorylated receptors, and the SH3 domain, which binds to proline-rich domains in downstream effector proteins.

12. Protein kinases that phosphorylate serine and threonine residues make up another large class of signaling enzymes, the best known of which are the mitogen-activated protein kinases (MAPKs). The MAPK cascade is often tethered by scaffold proteins that organize combinations of kinase isoforms. These signaling pathways initiate a variety of cellular responses including cell growth and differentiation, proliferation, apoptosis (programmed cell death), and cell survival.

13. The activity of protein kinases is antagonized by the activity of protein phosphatases that hydrolyze the phosphoryl groups added to proteins by protein kinases and thereby inhibit or reverse the effects of the signaling pathway that originally activated the protein kinase. The phosphatases are specific for phospho-Tyr or phospho-Ser/Thr. Protein phosphatases in general have broader specificity than the protein kinases and so they form a smaller family of proteins.

14. Receptor-ligand binding interactions can be quantified using a variety of technologies, including radioimmunoassays (Box 13-1). This quantification can provide insights into the physiological hypersensitivity of some cancers to endogenous growth factors, or help distinguish the binding affinities of different isoforms of receptors (e.g., adrenoreceptor isoforms). This quantification is based on the dissociation constant for the receptor and its ligand at equilibrium, which can be graphically represented (Scatchard plot) to reveal both the dissociation constant and number of receptors (see Box 13-2).

Heterotrimeric G Proteins

15. Some signal transduction systems that involve heterotrimeric G proteins include a G protein–coupled receptor (GPCR) and adenylate cyclase, which produces the second messenger cAMP. Production of the second messenger leads to activation of many target proteins. Thus, the initial hormone signal is amplified, although its duration is usually limited.

16. GPCRs transmit the binding of an extracellular signal (or ligand) to the cell interior by alternating between two discrete conformations, one with the ligand bound and another without. The glucagon receptor and the α- and β-adrenoreceptor s are GPCRs. All GPCRs contain seven transmembrane α helices.

17. GPCR and other signaling systems respond to *changes* in stimulation rather than to specific values of extracellular signals, such that they reduce their response to long-term stimuli in a process called desensitization.

18. G proteins form a large class of GTP-hydrolyzing proteins, of which there are two families involved in signal transduction: heterotrimeric G proteins and small monomeric G proteins (e.g., the Ras family). Heterotrimeric G proteins consist of a G_α, G_β, and G_γ subunit, in which GTP binds to and is hydrolyzed in the G_α subunit. The inactive state consists of the trimer with GDP bound to the G_α subunit. The binding of this complex to a ligand-bound receptor results in a conformational change that stimulates the exchange of GDP for GTP. The γ phosphate of GTP in turn induces a conformation change in the three switch regions of the G_α subunit, which causes the dissociation of the G_α subunit from the $G_{\beta\gamma}$ dimer. Both the G_α and the $G_{\beta\gamma}$ subunits can act as allosteric effectors of downstream protein targets, although targets of the G_α subunit are the best characterized. A G_α subunit self-inactivates through the slow hydrolysis of its bound GTP to GDP and P_i.

19. The exchange of GTP for GDP and GTP hydrolysis in G proteins can be augmented by accessory proteins.
    (a) GTPase-activating protein (GAP) stimulates GTP hydrolysis.
    (b) Guanine nucleotide exchange factor (GEF) stimulates the dissociation of GDP from G proteins.
        These accessory proteins are most commonly associated with the small monomeric G proteins (e.g., Ras).

20. Signal transduction via the adenylate cyclase system involves activation of a heterotrimeric G protein. The G_α subunit of a stimulatory G protein ($G_{s\alpha}$) activates adenylate cyclase, an integral membrane protein, whereas the G_α subunit of an inhibitory G protein ($G_{i\alpha}$) inhibits the enzyme.

21. The second messenger 3',5'-cyclic AMP (cAMP), the product of the adenylate cyclase reaction, activates protein kinase A (PKA). cAMP binds to the regulatory subunits of PKA so as to release the catalytic subunits in active form. The effects of cAMP are attenuated by the action of a phosphodiesterase that hydrolyzes cAMP to AMP, and by protein phosphatases that reverse the phosphorylation catalyzed by PKA. Many drugs and toxins interfere with components of the adenylate cyclase signaling system (see Boxes 13-4 and 13-5).

The Phosphoinositide Pathway

22. The phosphoinositide signaling pathway produces two second messengers: inositol trisphosphate (IP_3) and diacyglycerol (DAG), which are hydrolysis products of the plasma membrane phospholipid phosphatidylinositol-4,5-bisphosphate (PIP_2). Activation of a heterotrimeric G protein by a ligand–receptor complex activates phospholipase C (PLC), which hydrolyzes PIP_2. The resulting IP_3 binds to IP_3-gated Ca^{2+} channels in the endoplasmic reticulum, thereby releasing Ca^{2+} into the cytosol. This Ca^{2+} can then bind to calmodulin and activate calmodulin-sensitive kinases. DAG activates protein kinase C (PKC), which activates other kinases and proteins.

23. Ca^{2+}-calmodulin dependent kinases (CaM-dependent kinases) are autoinhibited kinases wherein the N- or C- terminal sequence serves as a "pseudosubtrate" that occludes the kinase active site. Calmodulin, by binding to the inhibitory "pseudosubstrate" sequence of the CaM-dependent kinase, causes a conformational change that bends the pseudosubstrate domain away from the active site, thereby relieving the intrinsic autoinhibition of the enzyme.

24. Different signal transduction pathways may antagonize each other or act synergistically via cross talk. For instance, some RTKs activate a PKC isoform that contains SH2 domains. The insulin receptor phosphorylates key target proteins called insulin receptor substrates (IRS-1 and IRS-2), which serve as scaffold proteins to recruit members from multiple signaling pathways. Moreover, other sites on the insulin receptor can bind proteins activating other signaling pathways; hence, the insulin receptor is capable of stimulating numerous signaling pathways depending on which are present in the stimulated cell.

25. Nearly every step of a signal transduction pathway is subject to regulation such that the magnitude of the ultimate biochemical response reflects the presence and degree of activation or inhibition of multiple signaling components.

26. Insulin signaling is an example of a complex signaling system in which cellular responses to insulin depend on the biochemical phenotype, which requires a larger systems-level approach to understand the physiological response to insulin in different cell types. Similarly, some drugs or toxins (e.g., anthrax toxin; discussed in Box 13-5) may affect several signaling pathways.

Questions

Hormones

1. The α- and β-adreonoreceptors stimulate different cellular effects in response to their ligands or agonists; however, the ultimate physiological response is the same. What is the net result of the activation of both kinds of receptors?

2. In what way are steroid receptors unique in their action once they have bound their cognate ligand?

Receptor Tyrosine Kinases

3. How are SH2 and SH3 domains important in the activation of Ras by receptor tyrosine kinases?

4. How does the binding of ligand to a receptor tyrosine kinase lead to changes in gene expression?

Heterotrimeric G Proteins

5. Compare the overall structure and mode of signal transduction by G protein–coupled receptors and receptor tyrosine kinases.

6. During G protein activation, which subunit exchanges GDP for GTP? What is the immediate consequence of this exchange?

7. Suggest three mechanisms by which the β-adrenergic signaling pathway utilizing the second messenger cAMP can be "turned off."

8. You are studying protein kinase A (PKA)–mediated cell signaling in a mammalian cell line, in which PKA levels are sensitive to two different hormones, XGF and YGF. Examine the results shown below and suggest a mechanism for how XGF and YGF affect PKA.

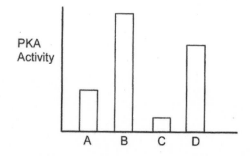

Key: A: Controls with no XGF or YGF treatment
B: Treatment of cells with XGF only
C: Treatment of cells with YGF only
D: Treatment with both XGF and YGF

9. The actions of caffeine, theobromine, and theophylline result in relative decreases in cAMP. Suggest a biochemical rationale based on the information given in Box 13-4.

The Phosphoinositide Pathway

10. Describe the intracellular signaling cascade mediated by phospholipase C activity. How is phospholipase C activated?

11. The insulin receptor can interact with several proteins. What are these proteins and what signaling pathways are they linked to?

12. Li^+ ions inhibit the recycling of PIP_2. Is Li^+ likely to counteract or enhance the effects of phorbol-13-acetate?

13. How is the regulation of the steady-state level of IP_3 analogous to that of cAMP?

Answers to Questions

1. Both receptor systems lead to the mobilization of energy reserves, shunting them to key tissues. The physiological responses include relaxation of smooth muscle for rapid blood delivery as well as stimulation of skeletal muscle contraction to increase mobility, most commonly for the flight or fight response.

2. Steroid–receptor complexes act as transcriptional regulators, while soluble hormones act more indirectly by activating intermediary proteins that trigger cascades of enzymatic activity affecting multiple pathways.

3. The SH2 and SH3 domains function in the assembly of multiprotein signaling complexes that linkphosphorylated receptor tyrosine kinases to the monomeric G protein Ras. For example, the SH2 domain in the Grb2/Sem-5 protein binds the phosphorylated cytoplasmic domain of the receptor tyrosine kinase, while its SH3 domains recognize and bind proline-rich sequences in Sos protein. This complex induces Ras to release GDP and bind GTP in order to become active (see Figure 13-7).

4. Ligand binding to a receptor tyrosine kinase leads to the activation of Ras via the formation of a complex of SH2- and SH3-containing proteins. Activated Ras triggers a kinase cascade consisting of Raf, MEK, and MAPK, which are activated in series. Activated MAPK migrates to the nucleus, where it phosphorylates a variety of proteins, including transcription factors. The activated transcription factors stimulate gene expression.

5. G protein–coupled receptors are transmembrane proteins that contain seven membrane-spanning segments. Ligand binding leads to interactions between the receptor and an associated G protein, which is anchored to the membrane by its lipid groups. Activation of the G protein leads to production of second messengers such as cAMP (if the G protein stimulates adenylate cyclase) and IP_3 and Ca^{2+} (if the G protein stimulates phospholipase C).

Receptor tyrosine kinases are usually monomeric proteins with a single membrane-spanning segment. Ligand binding induces dimerization of the receptor, which leads to autophosphorylation at one or more tyrosine residues in its cytoplasmic domain. These phosphorylated residues provide binding sites for target proteins with SH2 domains, which are generally activated upon binding to the receptor tyrosine kinase.

6. When the G protein binds to a hormone–receptor complex, the G_α subunit exchanges GDP for GTP. Binding of GTP induces a conformational change in G_α that causes it to dissociate from $G_{\beta\gamma}$. GTP binding also increases the affinity of G_α for adenylate cyclase, which is thereby activated.

7. Among possible mechanism by which a β-adrenergic signaling pathway can be "turned off" include:
    (a) Internalization of the receptor by endocytosis
    (b) Dissociation of the ligand from the receptor
    (c) Increased cAMP phosphodiesterase activity
    (d) Increased rate of GTP hydrolysis or inhibition of GTP exchange for GDP
    (e) Increased protein phosphatase activity to counteract the actions of PKA
    (f) Proteolytic degradation of key enzymes of the signaling pathway

    Note that (a) and (b) "turn off" the pathway because the activated heterotrimeric G protein is short-lived. In general, however, regulatory mechanisms result in the cell becoming refractory to stimulatory or inhibitory effects on a signaling pathway.

8. Both hormones probably act via the adenylate cyclase signaling pathway: XGF activates a stimulatory G protein to release a $G_{s\alpha}$ subunit, while YGF activates an inhibitory G protein that that releases a $G_{i\alpha}$ subunit. These activated G proteins both affect the activity of adenylate cyclase to modulate the amount of cAMP that is synthesized and available to stimulate PKA activity.

9. All of these drugs antagonize adenosine receptors that act through inhibitory G proteins. This creates an imbalance in the regulation of adenylate cyclase activity that is modulated in drug-responsive cells by signaling systems generating stimulatory and inhibitory G proteins.

10. Activation of phospholipase C leads to the hydrolysis of phosphatidylinositol-4,5-bisphosphate (PIP$_2$) to inositol-1,4,5-trisphosphate (IP$_3$) and 1,2-diacyglycerol (DAG). Both of these products are second messengers. IP$_3$ moves through the cytosol and binds to the IP$_3$ receptor, a Ca^{2+} channel on the endoplasmic reticulum. Binding of IP$_3$ to its receptor triggers the release of Ca^{2+} from the ER, increasing the intracellular $[Ca^{2+}]$. Ca^{2+} can stimulate many cellular activities either by interacting directly with target proteins or through the formation of the Ca^{2+}-calmodulin complex, which regulates many other target proteins. The DAG is retained within the membrane, where it activates protein kinase C. This enzyme then phosphorylates and activates additional target proteins.

Members of the phospholipase C family can be activated in two ways. Some forms are activated by interacting with a G_α–GTP complex derived from a G protein following ligand binding to a G protein–coupled receptor. Other isoforms are activated by binding, via their SH2 domains, to a receptor tyrosine kinase that has undergone ligand binding, dimerization, and autophosphorylation.

11. When it binds insulin, the insulin receptor undergoes a conformational change that activates its tyrosine kinase domains, resulting in autophosphorylation. The RTK can then phosphorylate the target proteins IRS-1 and IRS-2. These proteins can interact with SH2-containing proteins, which in turn may activate G proteins, act as kinases, or act as phosphatases. Phosphorylation of IRS proteins may also lead to activation of protein kinase C. The insulin receptor is thereby linked to kinase cascades, which is typical of RTKs, and to elements of the phosphoinositide pathway.

12. All other things being equal, Li^+ is likely to counteract the action of phorbol-13-acetate, a DAG mimic that activates protein kinase C, as the pool of available IP_3 is hydrolyzed to IP_2.

13. Both IP_3 and cAMP synthesis are stimulated by stimulatory G proteins and both second messengers are inactivated by hydrolysis of a phosphoester bond.

14
Introduction to Metabolism

This chapter provides a brief overview of the biological strategies and thermodynamics of metabolism. Metabolism is the overall biochemical processes that living systems use to acquire and use free energy. Organisms break down macromolecules to a common set of smaller molecules, or metabolites, which then serve as precursors for new biosynthesis. This chapter introduces some basic thermodynamic features of metabolic pathways and the mechanisms that have evolved to allow an organism to control the flow of a few common metabolites through different metabolic pathways.

Organisms harness the free energy from the degradation of macromolecules by trapping it in certain nucleotides (ATP, NAD^+, and FAD) and certain thioesters, which then make free energy available to energy-requiring pathways. The chapter describes these energy transmitters in terms of their thermodynamic features and their chemical properties. Some phosphorylated compounds have significant negative free energies of hydrolysis, which are described as their phosphoryl group-transfer potentials, their tendency to transfer their phosphoryl group to another compound. ATP, with its intermediate phosphoryl group-transfer potential, is the principal energy currency of life. Thioesters are also "high-energy" compounds. One of these, coenzyme A, shuttles acyl groups in metabolic processes.

Oxidation–reduction reactions are the most important process through which living organisms acquire and use free energy. The chapter reviews the principles of redox reactions, including the Nernst equation, and describes the mathematical relationship between free energy and reduction potentials. The chapter concludes with a brief look at the methods used to map the labyrinth of biochemical pathways in living cells and to understand the regulation of biochemical pathways.

Essential Concepts

1. Metabolism, the network of all biochemical reactions in cells, can be divided into two parts:
   (a) Catabolism (degradation), in which free energy is released as organic molecules are broken down into smaller constituents.
   (b) Anabolism (biosynthesis), in which biological molecules are synthesized from smaller, simpler molecules.

2. In general, catabolic reactions release free energy, which can then be used to drive endergonic synthetic reactions. This coupling requires free energy transmitters including nucleotides (ATP, NADH, NADPH, and $FADH_2$) and thioesters (e.g., coenzyme A).

Overview of Metabolism

3.   Organisms use different strategies for capturing free energy from their environment:
    (a) Autotrophs synthesize all their macromolecules from simple molecules obtained from their environment. The chemolithotrophs oxidize inorganic compounds, and the photoautotrophs use light to drive synthetic reactions.
    (b) Heterotrophs obtain free energy from the oxidation of organic compounds (usually produced by autotrophs).

4.   Organisms can be further classified by their requirements for oxygen. Obligate aerobes require oxygen for the oxidation of nutrients. Obligate anaerobes are poisoned by oxygen and must use another electron acceptor for the oxidation of nutrients. Facultative anaerobes can oxidize nutrients both in the absence and presence of oxygen.

5.   Some oxidation–reduction reactions do not involve oxygen and are most easily recognized by examining the oxidation state of the carbon atom(s) involved in the chemical reaction. The oxidation state of a carbon atom does not reflect its atomic charge (which is close to neutral) but is a convenient tool for following oxidation–reduction reactions.

6.   Vitamins and many mineral nutrients participate in metabolic reactions as coenzymes (though most must first be chemically modified from the ingested form) and enzyme cofactors, respectively. The water-soluble vitamins are key chemical participants in the active sites of enzymes involved in energy metabolism and biosynthesis. Fat-soluble vitamins are involved in light reception (vitamin A), Ca^{2+} absorption (vitamin D), blood clotting reactions (vitamin K), and protection from oxidation (vitamin E). Minerals serve not only as enzyme cofactors (e.g., magnesium and zinc) but as key components of small molecules (e.g., sulfur in amino acids), and structural elements (e.g., calcium phosphate in bones and teeth).

7.   Metabolic pathways are compartmentalized in the cytosol of prokaryotic and eukaryotic cells. Eukaryotic cells further compartmentalize metabolic pathways in membrane-bounded organelles. In multicellular organisms, many metabolic pathways are compartmentalized in different tissues.

8.   In any given metabolic pathway, most reactions are near equilibrium, such that the law of mass action largely dictates the flow rate (flux) of metabolites.

9.   In each metabolic pathway, there is at least one reaction that is far from equilibrium, in which the reactants accumulate above their equilibrium values and $\Delta G \lll 0$. Such reactions are referred to as rate-determining steps since they control flux in the pathway. The flux changes only with a change in the enzyme's ability to increase the reaction rate.

10. Metabolic pathways have three key characteristics:
    (a) They are irreversible.
    (b) They have an exergonic step that serves as the first committed step and ensures irreversibility.
    (c) Catabolic and anabolic pathways involving the interconversion of two metabolites differ in key exergonic reactions.

11. Control of flux in the rate-determining step requires control of the enzyme catalyzing it by one or more of the following mechanisms:
     (a) Allosteric control by feedback regulation from an end product of the pathway.
     (b) Covalent modification of the enzyme, which may increase or decrease its ability to accelerate a reaction.
     (c) Substrate cycles in which interconversion of two substrates utilizes different rate-determining enzymes.
     (d) Genetic control, which regulates the steady-state levels of the enzyme.
         Mechanisms (a)–(c) respond quickly to changes in physiological states (seconds to minutes), whereas mechanism (d) involves slower, long-term adaptive changes to new physiological states (minutes to days).

"High-Energy" Compounds

12. The free energy derived from the degradation of organic compounds is transiently captured in "high-energy" compounds whose subsequent breakdown provides the free energy to drive otherwise endergonic reactions.

13. ATP is the primary energy currency of cells. Its energy resides in the thermodynamic instability of its two phosphoanhydride bonds. The free energy of hydrolysis of ATP, in which the phosphate group is transferred to water, is called its phosphoryl group-transfer potential. ATP has an intermediate phosphoryl group-transfer potential, making it a conduit for the transfer of free energy from higher-energy compounds to lower-energy compounds.

14. The high-energy character of the phosphoanhydride bonds results from:
     (a) increased resonance stabilization of the hydrolysis products.
     (b) the destabilizing effect of electrostatic repulsions between the charged phosphates at neutral pH.
     (c) increased solvation energy of the hydrolysis products.

15. Many endergonic reactions in cells are coupled to the hydrolysis of ATP or pyrophosphate (PP_i) so that the net reaction is exergonic. In some biosynthetic reactions, the transfer of a nucleotidyl group "activates" the substrate for further reaction (e.g., the polymerization reactions of polysaccharides and the formation of aminoacyl–tRNA for protein synthesis).

16. ATP can be replenished by transfer of a phosphoryl group to ADP from a compound with a higher phosphoryl group-transfer potential. Such a transfer is called substrate-level phosphorylation. The concentrations of ATP and other nucleotides are maintained in part by the activity of kinases.

17. The transfer of acyl groups requires their "activation" by the formation of a thioester bond to a sulfur-containing compound such as coenzyme A. The hydrolysis of thioesters is about as exergonic as the hydrolysis of ATP. Hence, thioester cleavage drives the otherwise endergonic transfer of the acyl group.

Oxidation–Reduction Reactions

18. Oxidation–reduction reactions are the principal source of free energy for life. The oxidation of organic compounds is coupled to the reduction of the nucleotide cofactors NAD^+ (and $NADP^+$) and FAD.

19. A measure of the potential electrical energy (electromotive force or reduction potential) in an electrochemical cell is described by the Nernst equation:

$$\Delta\mathcal{E} = \Delta\mathcal{E}^{\circ\prime} - \frac{RT}{n\mathcal{F}}\ln\left(\frac{\left[A_{red}\right]\left[B^{n+}_{ox}\right]}{\left[A^{n+}_{ox}\right]\left[B_{red}\right]}\right)$$

Here, $\Delta\mathcal{E}$ is the reduction potential, $\Delta\mathcal{E}^{\circ\prime}$ is the reduction potential when all the components are in their biochemical st2andard states, $\mathcal{F}$ is the faraday ($96{,}485$ J·V^{-1}·mol^{-1}), n is the number of moles of electrons transferred per mole of reactants reduced, and R is the gas constant.

20. $\Delta\mathcal{E}$ is related to the free energy change in a redox reaction by the following relationship:

$$\Delta G = -n\mathcal{F}\Delta\mathcal{E}$$

Electrons flow spontaneously from a compound with the lower reduction potential to a compound with the higher reduction potential.

Experimental Approaches to the Study of Metabolism

21. Metabolites can be traced by labeling them with isotopes of certain atoms that can be detected by their radioactivity or through nuclear magnetic resonance spectroscopy. Such labels are useful for establishing precursor–product relationships and for examining the rates of biochemical transformation in living cells and tissues.

22. Other methods for analyzing a metabolic pathway include the use of metabolic inhibitors, which inhibit specific enzymes, and genetic mutations in enzymes involved in the pathway. In recent years, genetic engineering has become a powerful new tool for studying metabolism, as the gene for a specific enzyme can be added, deleted, or specifically altered.

23. An organism's metabolic capabilities can be assessed through the approaches of systems biology, which may include genomic sequencing, DNA microarray analysis, and techniques for protein identification and quantitation.

24. A picture of gene expression, as represented by all of the different species of a cell's mRNA (transcriptome), can be obtained using DNA microarrays. These "DNA chips" can be used to determine allelic differences between individuals or to assess how transcriptional activity of various genes differs between tissues or at different developmental stages.

25. An organism's overall metabolism can also be assessed through proteomics, the identification and quantitation of all of a cell's proteins, including their posttranslational modifications, in a given tissue at a given time.

Key Equations

$$\Delta G = -n\mathcal{F}\Delta\mathcal{E}$$

$$\Delta\mathcal{E} = \Delta\mathcal{E}^{\circ\prime} - \frac{RT}{n\mathcal{F}} \ln \left(\frac{[A_{red}][B_{ox}^{n+}]}{[A_{ox}^{n+}][B_{red}]} \right)$$

$$\Delta\mathcal{E}^{\circ} = \mathcal{E}^{\circ}_{(e^- \text{ acceptor})} - \mathcal{E}^{\circ}_{(e^- \text{ donor})}$$

Questions

Overview of Metabolism

1. In searching for life on Mars in the 1970s, the Viking spacecraft tested the Martian soil for rapid oxidation–reduction reactions. Explain why such reactions might indicate the presence of life.

2. Use the following terms to fill in the diagram below: ADP, P_i, $NADP^+$, ATP, NADPH, carbohydrates, proteins, lipids, acetyl-CoA, catabolism, anabolism.

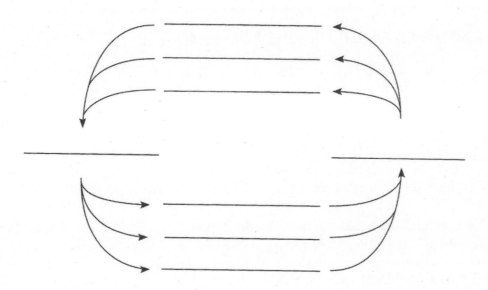

3.   Phosphofructokinase (PFK) catalyzes the reaction

ATP + fructose-6-phosphate (F6P) $\rightleftarrows$ fructose-1,6-bisphosphate (FBP) + ADP

Explain why the reaction rate is relatively insensitive to changes in the concentrations of F6P or FBP. What does this tell you about PFK?

"High-Energy" Compounds

4.   In the cell, divalent cations such as Mg^{2+} bind to the anionic phosphate groups of ATP. How would the removal of the divalent cations affect the ΔG for the hydrolysis of ATP?

5.   What processes maintain the cellular concentration of ATP?

6.   Why don't "high-energy" compounds such as phosphoenolpyruvate and phosphocreatine (Figure 14-6) break down quickly under physiological conditions?

7.   Explain how phosphocreatine acts as an ATP "buffer."

8.   Many metabolic reactions are actually coupled reactions. A common coupled reaction is substrate phosphorylation by ATP. For example, the oxidation of glucose begins with its phosphorylation to glucose-6-phosphate (G6P).

(a)   For the reaction

$$\text{Glucose} + P_i \rightleftharpoons \text{glucose-6-phosphate} + H_2O \qquad \Delta G^{\circ\prime} = 14 \text{ kJ·mol}^{-1}$$

Calculate the ratio of $[G6P]/[glucose][P_i]$ at equilibrium at 25°C.

(b)   In muscle cells at 37°C, the steady-state ratio of $[ATP]/[ADP]$ is 12. Assuming that glucose and G6P achieve equilibrium values in muscle, what is the ratio of $[G6P]$ to $[glucose]$?

9.   Aldolase catalyzes the reaction

Fructose-1,6-bisphosphate (FBP) $\rightleftharpoons$
Glyceraldehyde-3-phosphate (GAP) + Dihydroxyacetone phosphate (DHAP)

$\Delta G^{\circ\prime}$ for this reaction is 22.8 kJ·mol^{-1}. In the cell at 37°C, the ΔG for this reaction is −5.9 kJ·mol^{-1}. What is the ratio $[GAP][DHAP]/[FBP]$?

Oxidation–Reduction Reactions

10.   Using the curved-arrow convention, show the transfer of electrons in the reduction of pyruvate to lactate in the presence of $NADH + H^+$.

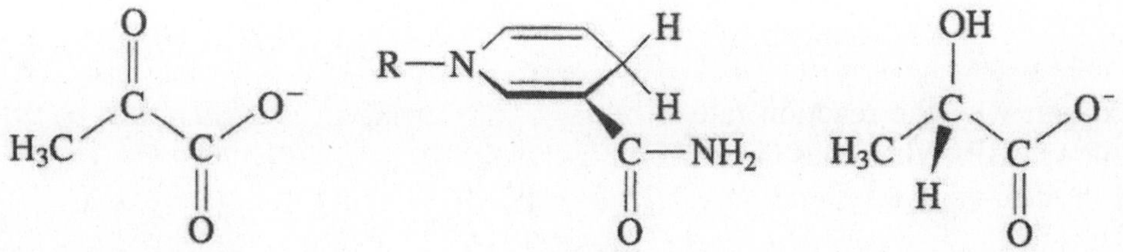

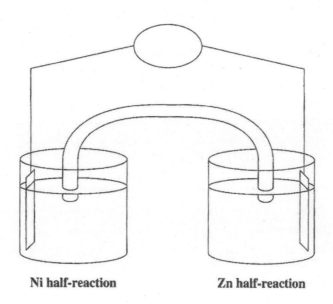

pyruvate   NADH   lactate

11. In the diagram below, indicate the direction of flow of electrons and the voltage (on the meter) for the following half-reactions under standard conditions:

$$Zn^{2+} + 2e^- \rightarrow Zn \qquad E° = -0.763 \text{ V}$$
$$Ni^{2+} + 2e^- \rightarrow Ni \qquad E° = -0.250 \text{ V}$$

Ni half-reaction   Zn half-reaction

12. At pH 0,

$$\tfrac{1}{2} O_2 + 2 H^+ + 2 e^- \rightleftharpoons H_2O \qquad \mathscr{E}° = 1.23 \text{ V}$$

Is oxygen reduction more favored at pH 0 or at pH 7? Explain in electrochemical terms as well as in terms of chemical equilibria.

13. Consider the reaction in which acetoacetate is reduced by NADH to β-hydroxybutyrate:

$$\text{Acetoacetate} + \text{NADH} + \text{H}^+ \rightleftharpoons \text{β-hydroxybutyrate} + \text{NAD}^+$$

Calculate ΔG for this reaction at 25°C when [acetoacetate] and [NADH] are 0.01 M, and [β-hydroxybutyrate] and [NAD$^+$] are 0.001M.

Experimental Approaches to the Study of Metabolism

14. Radioactive isotope tracers and metabolic inhibitors have been essential to the elucidation of metabolic pathways. Can both kinds of agents be used to determine the order of metabolites in a metabolic pathway?

15. Many biosynthetic pathways have been elucidated by the analysis of genetic mutations in organisms such as *Neurospora crassa* (a mold) and *Escherichia coli*. How would you elucidate the steps of a hypothetical biosynthetic pathway in *N. crassa* in which compound A leads to compound Z?

16. You have isolated four mutants in amino acid metabolism of the mold *N. crassa*. Mutant 1 requires two compounds for growth, X and Z. Mutant 2 requires only X. Mutants 3 and 4 require only Z. Mutant 3 accumulates a compound, W, that supports the growth of Mutant 4 but not that of Mutants 1 or 2. Mutant 4 accumulates a compound, Y, that alone supports the growth of Mutant 1.
    (a) Diagram the biosynthetic pathway connecting compounds W, X, Y, and Z, indicating the step at which each mutant is blocked.
    (b) According to the diagram in (a), what is the first committed step in the synthesis of Z?

17. Evaluate the accuracy of the following two statements on the interpretation of DNA chip results such as those presented in Figures 14-18 and 14-19.
    (a) DNA chip data allow one to determine the number of mRNA copies present in the cell.
    (b) DNA chip data provide an accurate assessment of the relative protein diversity and content of cells under different physiological states.

Answers to Questions

1. The fundamental processes by which living organisms acquire and use free energy depend on oxidation–reduction reactions. For example, the oxidation of a metabolic fuel generates a reduced cofactor, such as NADH, whose subsequent reoxidation generates ATP. Rapid (i.e., enzyme-catalyzed) oxidation–reduction reactions therefore would be consistent with (but would not prove) the presence of life on Mars.

2.

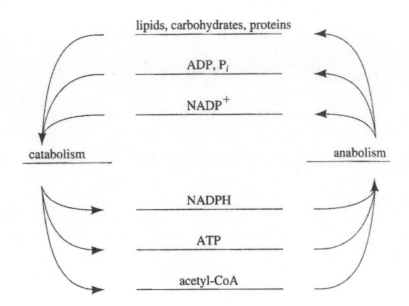

3. PFK operates far from equilibrium, so it is not sensitive to changes in substrate concentrations. It most likely catalyzes a rate-determining step of a metabolic pathway. (In fact, it is part of the glycolytic pathway, and its activity is under allosteric control.)

4. ΔG would become more negative because, without the shielding effect of the divalent cations, the phosphate groups in ATP would experience more repulsion.

5. Processes that maintain the cellular concentration of ATP are oxidative phosphorylation (the synthesis of ATP from ADP + P_i as driven by the free energy of dissipation of a transmembrane proton concentration gradient); substrate-level phosphorylation (direct transfer of a phosphoryl group from a "high-energy" compound such as phosphoenolpyruvate to ADP); and reactions catalyzed by kinases such as nucleoside diphosphate kinase and adenylate kinase.

6. These "high-energy" compounds have large negative values for ΔG of hydrolysis. Therefore, their breakdown is thermodynamically spontaneous (exergonic). However, the kinetics of their breakdown depends on the availability and activity of enzymes to catalyze such reactions. In the absence of the appropriate enzymatic activity, the compounds are quite kinetically stable.

7. The reaction catalyzed by creatine kinase

$$\text{ATP + creatine} \rightleftharpoons \text{phosphocreatine + ADP}$$

is at equilibrium and is freely reversible in the cell. Hence, the concentration of phosphocreatine is directly sensitive to changes in [ATP]. When ATP is plentiful, phosphocreatine is formed by the forward reaction. When [ATP] drops, the reaction proceeds in reverse, so that phosphocreatine can transfer its phosphoryl group to ADP to produce more ATP. The extent of this reaction depends on the decrease in [ATP].

8. (a) At equilibrium, $\Delta G^{o\prime} = -RT \ln K_{eq}$,

Hence, $K_{eq} = \dfrac{[G6P]}{[Glucose][P_i]} = e^{-\Delta G^{o\prime}/RT}$

$= e^{-(14,000 \, J\cdot mol-1)/(8.3145 \, J\cdot K-1\cdot mol-1)(298K)}$

$= e^{-5.65}$

$= 0.0035$

(b) The phosphorylation of glucose coupled to ATP hydrolysis is the sum of the following reactions:

ATP + H_2O $\rightleftharpoons$ ADP + P_i	$\Delta G^{o\prime}$ = −30.5 kJ·mol^{-1}	(Table 13-2)
glucose + P_i $\rightleftharpoons$ glucose-6-phosphate + H_2O	$\Delta G^{o\prime}$ = 14 kJ·mol$^-$	

ATP + glucose $\rightleftharpoons$ glucose-6-phosphate + ADP $\qquad \Delta G^{o\prime}$ = −16.5 kJ·mol^{-1}

At equilibrium, $\Delta G = 0$ and $\Delta G^{o\prime} = -RT \ln K_{eq}$. Thus,

$K_{eq} = e^{-\Delta G^{o\prime}/RT} = \dfrac{[ADP][G6P]}{[ATP][Glucose]}$

Since [ATP]/[ADP] = 12,

$\dfrac{[G6P]}{[Glucose]} = e^{-\Delta G^{o\prime}/RT}$

$= 12e^{-(-16,500 J\cdot mol^{-1})/(8.3145 K\cdot K^{-1}\cdot mol^{-1})(310K)}$

$= 12e^{6.40} = 72.33$

9. Use Equation 14-1:

$$\Delta G = \Delta G^{o\prime} + RT \ln\left(\dfrac{[GAP][DHAP]}{[FBP]}\right)$$

$$\dfrac{\Delta G - \Delta G^{o\prime}}{RT} = \ln\left(\dfrac{[GAP][DHAP]}{[FBP]}\right)$$

$$\dfrac{(-5900 kJ\cdot mol^{-1})(22,800 kJ\cdot mol^{-1})}{(8.3145 J\cdot K^{-1}\cdot mol^{-1})(310K)} = \ln\left(\dfrac{[GAP][DHAP]}{[FBP]}\right)$$

$$-11.13 = \ln\left(\dfrac{[GAP][DHAP]}{[FBP]}\right)$$

$$e^{-11.13} = \dfrac{[GAP][DHAP]}{[FBP]}$$

$$1.46 \times 10^{-5} = \dfrac{[GAP][DHAP]}{[FBP]}$$

10.

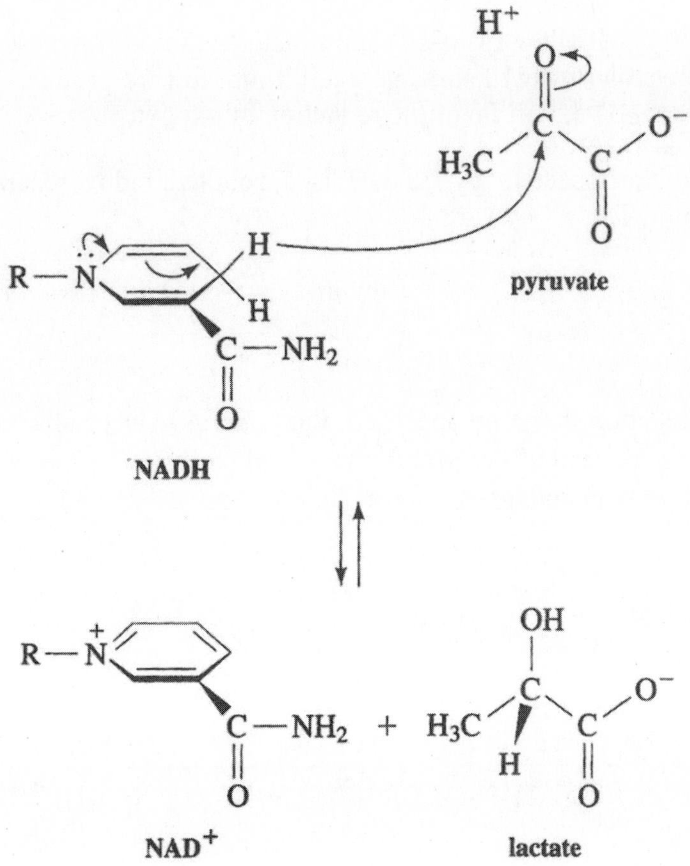

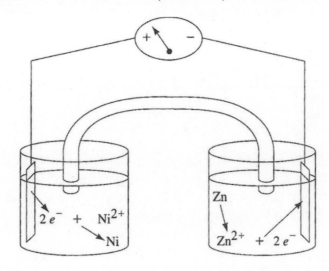

NADH

pyruvate

NAD⁺   lactate

11.   Electrons flow from substances of lower reduction potential to substances of higher reduction potential. $\Delta \mathscr{E}^\circ = -0.250\ \text{V} - (-0.763\ \text{V}) = 0.513\ \text{V}$.

12. At pH 7.0, $\mathscr{E}^{\circ\prime} = 0.815$ V (Table 14-5). Therefore, reduction is more favorable at pH 0 (where $\mathscr{E}^{\circ} = 1.23$ V) since the more positive the reduction potential, the more negative the ΔG (Equation 14-7). The law of mass action dictates that an increase in the concentration of one of the reactants (H^+) will shift the equilibrium toward product. Thus, decreasing the pH, which increases $[H^+]$, will favor the reduction of oxygen.

13. The reduction of acetoacetate by NADH is a coupled redox reaction, where the half-reactions (shown in Table 14-5) are

$$\text{Acetoacetate} + 2\,H^+ + 2\,e^- \rightleftharpoons \beta\text{-hydroxybutyrate} \qquad \mathscr{E}^{\circ\prime} = -0.346\ \text{V}$$
$$NAD^+ + H^+ + 2\,e^- \rightleftharpoons NADH \qquad \mathscr{E}^{\circ\prime} = -0.315\ \text{V}$$

For the overall reaction direction specified, the NAD^+/NADH half-reaction is the electron donor, and the acetoacetate/β-hydroxybutyrate half-reaction is the electron acceptor. Use the Nernst equation to calculate ΔE:

$$\Delta\mathscr{E} = \Delta\mathscr{E}^{\circ\prime} - \frac{RT}{n\mathscr{F}}\ln\left(\frac{[\beta\text{-Hydroxybutyrate}][NAD^+]}{[\text{Acetoacetate}][NADH]}\right)$$

$$= \mathscr{E}^{\circ\prime}_{e^-\ acceptor} - \mathscr{E}^{\circ\prime}_{e^-\ donor} - \frac{(8.3145\ \text{J}\cdot\text{K}^{-1}\cdot\text{mol}^{-1})(298\text{K})}{(2)(96,485\ \text{J}\cdot\text{V}^{-1}\cdot\text{mol}^{-1})}\ln\left[\frac{(0.001)(0.001)}{(0.01)(0.01)}\right]$$

$$= (-0.346\ \text{V} + 0.315\ \text{V}) - (0.01284\ \text{V})\ln(0.01)$$
$$= -0.031\ \text{V} - (0.01284\ \text{V})(-4.605)$$
$$= 0.028\ \text{V}$$

Next, use Equation 14-7 to calculate ΔG:

$$\begin{aligned}
\Delta G &= n\mathscr{F}\Delta\mathscr{E} \\
&= -(2)(96,485\ \text{J}\cdot\text{V}^{-1}\cdot\text{mol}^{-1})(0.028\ \text{V}) \\
&= -5403\ \text{J}\cdot\text{mol}^{-1} \\
&= -5.4\ \text{kJ}\cdot\text{mol}^{-1}
\end{aligned}$$

14. Radioactive tracers can be used to determine the order of metabolic transformations in a pathway. An inhibitor that blocks a step of the pathway causes earlier intermediates to accumulate but does not necessarily reveal their order, and cannot reveal information about the order of metabolites following the blocked step.

15. To study the metabolic steps between compounds A and Z, first generate mutants by irradiating the cells or treating them with a chemical mutagen. Next, screen cells for their inability to grow in the absence of compound Z, the end product of the pathway. This would yield cells with defects in the A→Z pathway. To identify mutations in enzymes that catalyze individual steps of the pathway, screen the mutant cells for their inability to grow in the absence of each of the compounds suspected to be intermediates in the transformation of A to Z.

16.

(a) Since Mutant 1 requires two compounds for growth, the pathway for their synthesis is probably branched, such that both are derived from a common precursor lacking in Mutant 1. Mutant 2 is blocked in the pathway leading to X, while Mutants 3 and 4 are blocked in the pathway leading to Z. The step blocked in Mutant 4 precedes that blocked in Mutant 3, since Mutant 3 accumulates compound W, which supports the growth of Mutant 4. Since compound Y, which accumulates in Mutant 4, supports the growth of Mutant 1, it must be the common precursor of X and Z. Therefore, the most likely pathway is

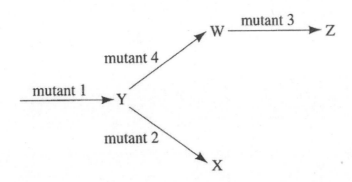

(b) The first committed step in the synthesis of Z is the step that converts Y to W, the step missing in Mutant 4.

17.

(a) DNA chip data only allow investigators to assess the *relative* concentrations of specific mRNAs from cells under two different physiological states, since mRNA populations from each source are quantitatively compared to each other.

(b) DNA chip data only show gene expression at the point of mRNA synthesis. The relative amounts of proteins made from these mRNAs cannot be accurately assessed, since the rates of translation of different mRNAs may differ under different physiological states. The protein diversity and content (i.e., the proteome) at a systems level requires the tools of proteomics, such as the ICAT method for quantitative proteome analysis.

15
Glucose Catabolism

Glycolysis, or the biochemical conversion of glucose to pyruvate, is one of the best understood metabolic pathways. The mechanisms and structures of the ten glycolytic enzymes are known in some detail, and they serve as models for the study of other enzymes with similar reaction mechanisms. The regulation of metabolic pathways, a topic introduced in Chapter 14, is illustrated here. This chapter also describes the fate of pyruvate (the end product of glycolysis), the entry of other sugars into the glycolytic pathway, and the pentose phosphate pathway that also catabolizes glucose.

Essential Concepts

Overview of Glycolysis

1   Glucose is a major source of metabolic energy in many cells. The energy released during its conversion (oxidation) to pyruvate is conserved in the form of ATP and the reduced coenzyme NADH. Ten enzymes catalyze the glycolytic pathway, which occurs in both prokaryotes and eukaryotes and is almost universal.

2.   Stage I of glycolysis is a preparatory stage in which glucose is "activated" by phosphorylation by ATP and broken down into two C_3 sugars. Stage II produces "high-energy" intermediates that phosphorylate ADP to form ATP, for a net gain of 2 ATP. One glucose molecule yields two pyruvate molecules and requires the oxidizing power of two NAD^+. The overall equation for glycolysis is

$$\text{Glucose} + 2\,NAD^+ + 2\,ADP + 2\,P_i \rightarrow$$
$$2\text{ pyruvate} + 2\,NADH + 2\,ATP + 2\,H_2O + 4\,H^+$$

The Reactions of Glycolysis

3.   Hexokinase, the first enzyme in the pathway, transfers a phosphoryl group from ATP to the C6-OH group of glucose to produce glucose-6-phosphate (G6P). Hexokinase undergoes a large conformational change on binding glucose, which excludes water from the active site and promotes the specific transfer of the phosphoryl group from ATP to glucose.

4.   Phosphoglucose isomerase catalyzes the conversion of glucose-6-phosphate to fructose-6-phosphate (F6P). This reaction proceeds via an enediolate intermediate.

5.   Phosphofructokinase (PFK) converts fructose-6-phosphate to fructose-1-6-bisphosphate (FBP) by another phosphoryl-group transfer from ATP. This reaction, which is irreversible, is the first committed step of the pathway. The phosphofructokinase reaction is the rate-determining step of glycolysis and the principal regulatory point.

6.   Aldolase cleaves fructose-1,6-bisphosphate to two C_3 compounds: glyceraldehyde-3-phosphate (GAP) and dihydroxyacetone phosphate (DHAP). In animals and plants, aldolase operates via Schiff base and enamine intermediates.

7.   Triose phosphate isomerase catalyzes the interconversion of glyceraldehyde-3-phosphate and dihydroxyacetone phosphate through an enediolate intermediate. This α/β barrel enzyme has achieved catalytic perfection. Its activity allows the products of the first stage of glycolysis to proceed through the second stage as glyceraldehyde-3-phosphate.

8.   Glyceraldehyde-3-phosphate dehydrogenase (GAPDH) catalyzes the formation of the first glycolytic intermediate that has sufficient free energy to synthesize ATP from ADP. Glyceraldehyde-3-phosphate is converted to 1,3-bisphosphoglycerate (1,3-BPG) by the reduction of NAD^+ and the addition of inorganic phosphate. The sulfhydryl group of an enzyme Cys residue attacks GAP, which is then reduced by NAD^+ to form an acyl thioester intermediate that is attacked by P_i.

9.   1,3-Bisphosphoglycerate, a "high-energy" mixed anhydride of a phosphate and a carboxylic acid, is the substrate for phosphoglycerate kinase, which transfers the phosphoryl group at C1 to ADP to generate 3-phosphoglycerate (3PG) and the first ATP product of glycolysis.

10.   3-Phosphoglycerate is converted to 2-phosphoglycerate (2PG) by phosphoglycerate mutase. The active form of this enzyme contains a phosphohistidine residue whose phosphoryl group is transferred to 3PG to produce 2,3-bisphosphoglycerate (2,3-BPG), which then transfers the phosphoryl group at the 3 position back to the histidine, yielding 2PG.

11.   Enolase dehydrates (removes water from) 2-phosphoglycerate to form phosphoenolpyruvate (PEP). This reaction produces the second "high-energy" intermediate of glycolysis.

12.   The free energy of phosphoenolpyruvate is released in the reaction catalyzed by pyruvate kinase. The transfer of the phosphoryl group to ADP produces the second ATP product of glycolysis and an enol product whose tautomerization to the keto form yields pyruvate. Most of the free energy of the pyruvate kinase reaction is supplied by this tautomerization step. Keep in mind that because the initial substrate of glycolysis is a C_6 compound that is converted to two C_3 compounds, the first stage of glycolysis consumes 2 ATP but the second stage generates 4 ATP (2 for each GAP), for a net yield of 2 ATP.

Fermentation: The Anaerobic Fate of Pyruvate

13.   The NADH produced by glycolysis must be converted back to NAD^+ in order for the glyceraldehyde-3-phosphate dehydrogenase reaction to proceed. Consequently, pyruvate is not the end product of glucose metabolism but can undergo one of three processes to regenerate NAD^+: homolactic fermentation, alcoholic fermentation, or oxidative metabolism. In oxidative metabolism, pyruvate is oxidized to CO_2 via the citric acid cycle.

14. In muscle cells, when O_2 is in short supply, lactate dehydrogenase reduces pyruvate to lactate, with the concomitant oxidation of NADH to NAD^+. The lactate that builds up in muscles upon strenuous exertion can either be reconverted to pyruvate later or carried by the blood to the liver, where it can be converted back to glucose by a process called gluconeogenesis.

15. Under anaerobic conditions, yeast carry out alcoholic fermentation. In the first reaction of this process, pyruvate is decarboxylated to acetaldehyde and carbon dioxide. Pyruvate decarboxylase catalyzes this reaction with the aid of the coenzyme thiamine pyrophosphate, which stabilizes the reaction's carbanion intermediate. In the second reaction, acetaldehyde is reduced with NADH to form ethanol and NAD^+, as catalyzed by alcohol dehydrogenase. These two reactions have been known for thousands of years: The released CO_2 leavens (raises) bread and the ethanol is used to make alcoholic beverages.

16. The anaerobic catabolism of glucose can be 100 times faster than the catabolism of glucose in the presence of oxygen. However, fermentation produces 2 ATP per glucose, whereas oxidative metabolism (via the citric acid cycle and oxidative phosphorylation) generates 38 ATP for each glucose molecules that is converted to CO_2 and H_2O.

Regulation of Glycolysis

17. The reaction catalyzed by phosphofructokinase, which has a large negative ΔG, is the first committed step in glycolysis and the primary regulatory point for the pathway. PFK is allosterically inhibited by ATP, which binds to an inhibitory site and stabilizes PFK's T (less active) state. This is an example of feedback inhibition, since ATP is a product of the pathway. AMP, ADP, and fructose-2,6-bisphosphate (F2,6P) relieve the inhibition of PFK by ATP by preferentially binding to the R (more active) state. PFK thereby senses the energy state of the cell and adjusts the flux through glycolysis accordingly.

18. Additional control of glycolytic flux is provided by a substrate cycle. PFK catalyzes the reaction F6P + ATP $\rightarrow$ FBP + ADP, whereas fructose-1,6-bisphosphatase (FBPase) catalyzes the opposing reaction FBP + H_2O $\rightarrow$ F6P + P_i. The sum of these reactions is the hydrolysis of ATP. Both enzymes exist in the same cell and their relative activity is under hormonal and neuronal control. Substrate cycling, which produces heat, can provide a form of nonshivering thermogenesis.

Metabolism of Hexoses Other than Glucose

19. Three other sugars—fructose, galactose, and mannose—are major sources of cellular energy. In muscle, fructose can be directly phosphorylated by hexokinase to F6P. However, liver glucokinase cannot directly phosphorylate fructose. In the liver, fructokinase phosphorylates C1 to generate fructose-1-phosphate. Fructose-1-phosphate aldolase generates dihydroxyacetone phosphate and glyceraldehyde by an aldol cleavage. Glyceraldehyde kinase phosphorylates C3 of glyceraldehyde to produce the glycolytic intermediate GAP. Three other enzymes (alcohol dehydrogenase, glycerol kinase, and glycerol phosphate dehydrogenase) convert glyceraldehyde to dihydroxyacetone phosphate, which is then converted to GAP by triose phosphate isomerase. Individuals who

have defective fructose-1-phosphate aldolase have fructose intolerance and quickly develop a strong distaste for anything sweet.

20. Galactose, which differs from glucose in the configuration of the OH group at C4, requires four reactions to enter glycolysis. First, galactose is converted to galactose-1-phosphate by galactokinase. Next, galactose-1-phosphate uridylyl transferase transfers the UMP group of UDP–glucose to produce UDP–galactose and glucose-1-phosphate. An epimerase converts UDP–galactose to UDP–glucose. Finally, phosphoglucomutase converts glucose-1-phosphate to glucose-6-phosphate.

21. Mannose, the C2 epimer of glucose, is recognized by hexokinase. The resulting mannose-6-phosphate is then converted to the glycolytic intermediate fructose-6-phosphate by phosphomannose isomerase.

The Pentose Phosphate Pathway

22. Besides ATP, cells require the reducing power of NADPH for the biosynthesis of macromolecules (anabolism). NADPH is used in biosynthesis, whereas NADH is used in oxidative metabolism (catabolism). Cells keep the $[NAD^+]/[NADH]$ ration near 1000 (which favors metabolite oxidation) and the $[NADP^+]/[NADPH]$ ratio near 0.01 (which favors reductive biosynthesis). The oxidation of glucose by the pentose phosphate pathway generates NADPH.

23. The first stage of the pentose phosphate pathway consists of three steps, its oxidative reactions:
    (a) Glucose-6-phosphate is oxidized to 6-phosphoglucono-δ-lactone by glucose-6-phosphate dehydrogenase, producing the first NADPH.
    (b) 6-Phosphogluconolactonase hydrolyzes the lactone (cyclic ester) to yield 6-phosphogluconate.
    (c) 6-Phosphogluconate dehydrogenase then catalyzes the oxidative decarboxylation of 6-phosphogluconate by $NADP^+$ to yield ribulose-5-phosphate, CO_2, and the second NADPH.

24. Stage two of the pentose phosphate pathway is catalyzed by two enzymes that act on ribulose-5-phosphate (Ru5P): Ribulose-5-phosphate epimerase converts Ru5P to xylulose-5-phosphate (Xu5P), and ribulose-5-phosphate isomerase converts Ru5P to ribose-5-phosphate (R5P). The R5P can be used to produce nucleosides for RNA and DNA synthesis.

25. In the third stage of the pentose phosphate pathway, 3 five-carbon sugars are converted to 2 fructose-6-phosphate and 1 glyceraldehyde-3-phosphate. First, transketolase transfers a two-carbon unit from Xu5P to R5P yielding the seven-carbon sugar sedoheptulose-7-phosphate (S7P) and glyceraldehyde-3-phosphate (GAP). Next, transaldolase transfers a three-carbon unit from S7P to GAP to form fructose-6-phosphate (F6P) and the four-carbon sugar erythrose-4-phosphate (E4P). Finally, another transketolase reaction converts Xu5P and E4P to F6P and GAP.

26. The reversible nature of the second and third stages of the pentose phosphate pathway permits the cell to meet its need for R5P (a nucleic acid precursor) and NADPH. For

example, if the need for NADPH is greater than that for R5P, the excess R5P is converted to F6P and GAP for consumption via glycolysis. Conversely, if the demand for R5P outstrips the need for NADPH, the glycolytic intermediates F6P and GAP can be diverted to the pentose phosphate pathway to synthesize R5P.

27. The flux of glucose-6-phosphate through the pathway is regulated by the rate of the glucose-6-phosphate dehydrogenase (G6PD) reaction. This enzyme is controlled by the availability of its substrate $NADP^+$ so that the pathway flux increases in response to increasing levels of $NADP^+$ (which indicates increased cellular demand for NADPH).

28. A deficiency of G6PD is the most common human enzyme deficiency. The resulting shortage of NADPH increases the sensitivity of red blood cells to oxidative stress, since NADPH is required to maintain the supply of reduced glutathione, which removes organic hydroperoxides that occasionally form. Certain compounds such as the antimalarial drug primaquine stimulate peroxide formation, which induces hemolytic anemia in G6PD-deficient individuals. However, mutations in G6PD confer resistance to malaria.

Questions

Overview of Glycolysis

1. Describe the two-stage "chemical strategy" of glycolysis and write a balanced equation for each phase.

2. The two "high-energy" compounds produced in glycolysis are _____ and _____.

The Reactions of Glycolysis

3. Examine the following five glycolytic intermediates:

    (a) Name each intermediate.
    (b) Write the order in which they appear in glycolysis.
    (c) Which intermediate is a reactant in substrate-level phosphorylation?
    (d) List the glycolytic enzyme for which each intermediate is a substrate.
    (e) Which phosphorylated intermediates of glycolysis are not shown above?

4.   List the four kinases of glycolysis.

5.   There are _____ isomerization reactions in glycolysis. The enzymes that catalyze them are _____.

6.   For the reaction catalyzed by triose phosphate isomerase, what is the equilibrium ratio of reactants and products under standard biochemical conditions? How does this ratio differ from the ratio observed in the cell at 37°C?

7.   How does the GAPDH-catalyzed exchange of ^{32}P between P_i and 1,3-bisphosphoglycerate corroborate the existence of an acyl–enzyme intermediate in the GAPDH reaction?

8.   2,3-BPG is an intermediate in the reaction catalyzed by phosphoglycerate mutase. Why does the cell require trace amounts of 2,3-BPG?

9.   If the cytosolic $[NAD^+]/[NADH]$ ratio is 100 and the $[ATP]/[ADP][P_i]$ ratio is 10, what is the actual (not equilibrium) ratio of $[GAP]/[3PG]$ at 37°C in the cell? Assume that $[H^+] = 1$, its value in the biochemical standard state.

10.   Fluoride ions specifically inhibit enolase in the presence of P_i in cell extracts.
    (a) Explain why both 2PG and 3PG accumulate in the presence of F^- and P_i.
    (b) Explain why 1,3-BPG does not accumulate.

11.   Match each ^{14}C-labeled glucose with the ^{14}C-labeled pyruvate produced by glycolysis.

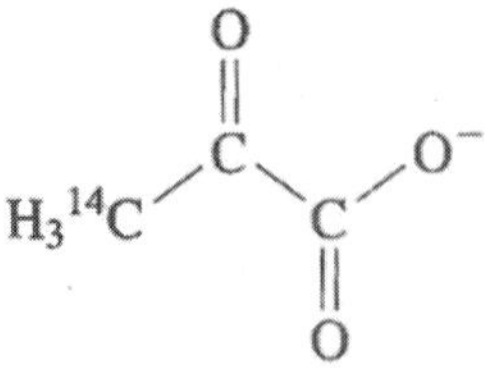

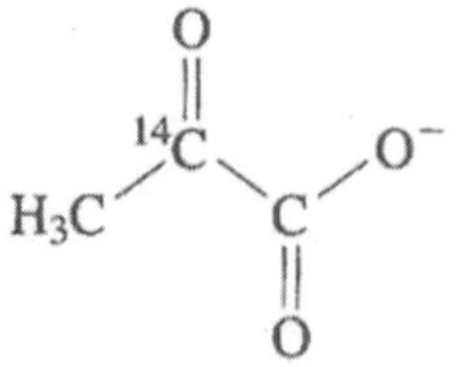

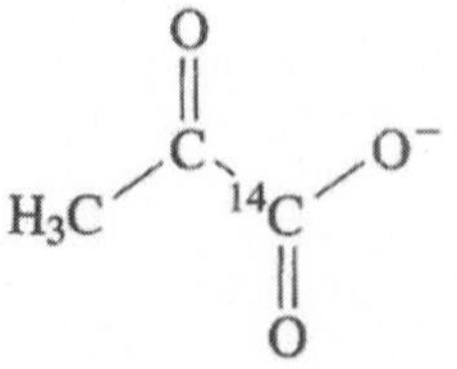

(a) $HOCH_2$

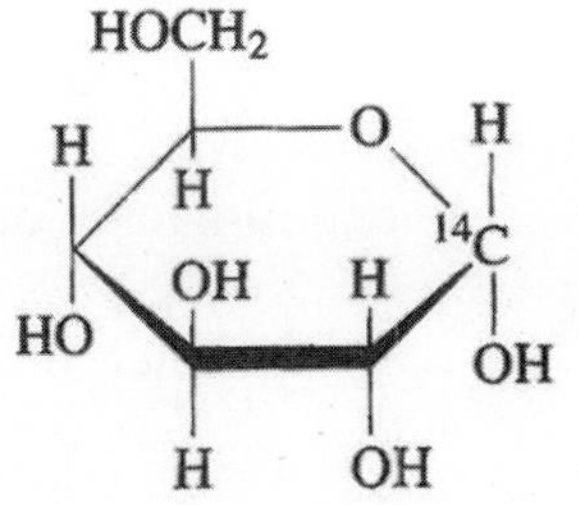

H
H
OH   H
HO
^{14}C
OH
H   OH
H

(c) $HO^{14}CH_2$

H
H
OH   H
HO
H   OH
H   OH

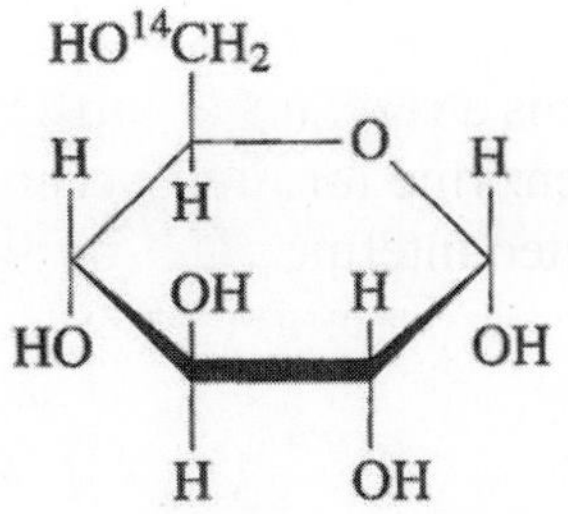

(b) $HOCH_2$

H
H
^{14}C
OH   H
HO
OH
H   OH
H

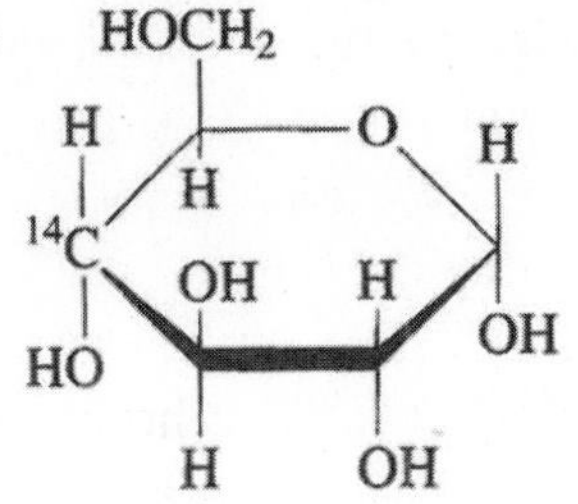

(d) $HOCH_2$

H
H
OH   H
HO ^{14}C
OH
H   OH
H

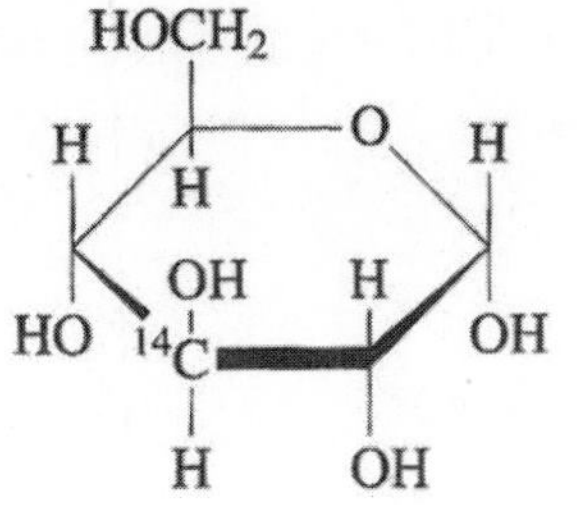

Fermentation: The Anaerobic Fate of Pyruvate

12. How does a muscle cell maintain the $[NAD^+]/[NADH]$ ratio during the catabolic breakdown of glucose?

13. Which of the carbon(s) of glucose must be labeled with ^{14}C for the end products of alcoholic fermentation to be unlabeled ("cold") ethanol and $^{14}CO_2$?

14. Compare the rates of ATP production in fermentation versus oxidative phosphorylation. Which process is utilized in rapid bursts of muscular activity?

Regulation of Glycolysis

15. In many metabolic pathways the first reaction is the rate-determining step of the pathway.
    (a) What is the rate-determining step of glycolysis?
    (b) What rationale can you offer for this "unusual" regulation?

16. What is meant by a futile cycle?

Metabolism of Hexoses Other than Glucose

17. The catabolism in the liver of which sugar—mannose, fructose, or galactose—bears the closest similarity to the first two reactions of glycolysis?

18. Defend or refute the following statement: The association of NAD^+ with UDP–galactose-4-epimerase suggests the generation of an additional NADH in the oxidation of galactose to pyruvate.

19. Why is galactosemia especially dangerous to nursing infants?

20. Some organisms use glycerol as a carbon energy source, and it is also an intermediate in fructose metabolism.
    (a) Write equations for the reactions required to oxidize glycerol to pyruvate.
    (b) Compare the ATP yield of two molecules of glycerol versus one molecule of glucose in glycolysis.
    (c) Does the anaerobic fermentation of glycerol maintain the redox balance of the cell? Explain.

The Pentose Phosphate Pathway

21. The diagram below shows the interconversions of the nonoxidative reactions of the pentose phosphate pathway.

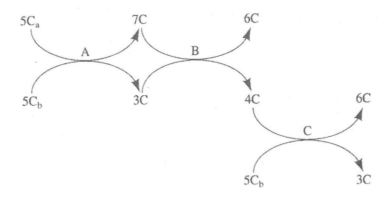

    (a) Which sugar phosphates correspond to the 5C compounds?
    (b) Which sugar phosphate corresponds to the 6C compound?
    (c) Which sugar phosphate corresponds to the 3C compound?
    (d) Which reaction(s) involves transketolase? Transaldolase?

22. The nonoxidative reactions of the pentose phosphate pathway convert pentose phosphates into hexose phosphates.
    (a) For every 3 glucose phosphates that enter the pentose phosphate pathway, how many fructose-6-phosphates are recovered?
    (b) For every 3 glucose-6-phosphates that enter the pentose phosphate pathway, how many glyceraldehyde-3-phosphates are recovered?

23. Ribulose-5-phosphate is converted to xylulose-5-phosphate and ribose-5-phosphate by an epimerase and an isomerase, respectively. What distinguishes these isomerizations?

24. You obtain a mutant transketolase from yeast that binds R5P and E4P poorly. What unique side product of the reaction catalyzed by this enzyme might you find in these cells?

25. Indicate the appropriate choices in parentheses: Transaldolase transfers a (1, 2, 3)-carbon unit from a(n) (ketose / aldose) to a(n) (ketose / aldose) to form a(n) (ketose / aldose).

26. Indicate the appropriate choices in parentheses: Transketolase transfers a (1, 2, 3)-carbon unit from a(n) (ketose / aldose) to a(n) (ketose / aldose) to form a(n) (ketose / aldose).

27. Which step commits glucose to oxidation via the pentose phosphate pathway? How is this enzyme regulated?

28. The conversion of G6P to R5P via the reactions of glycolysis and the pentose phosphate pathway, *without* the production of NADPH, can be summarized as

$$5 \text{ G6P} + \text{ATP} \rightleftharpoons 6 \text{ R5P} + \text{ADP} + \text{H}^+$$

What reaction requires ATP? How do you account for the stoichiometry?

Answers to Questions

1. In the energy investment phase, glucose is phosphorylated and cleaved to yield two molecules of glyceraldehyde-3-phosphate. Two ATPs are consumed in this stage so that the second stage can produce "high-energy" compounds whose breakdown drives ATP synthesis. The initial hexose is split, in the first stage, to two trioses. These undergo the reactions of the second stage, thereby generating four ATPs for a net "profit" of two ATPs per glucose. The equation for Stage I is

$$\text{Glucose } + \text{ 2 ATP } \rightarrow \text{ 2 GAP } + \text{ 2 ADP } + \text{ 2 H}^+$$

The equation for Stage II is

$$2 \text{ GAP } + \text{ 2 NAD}^+ + \text{ 4 ADP } + \text{ 2 HPO}_4^{2-} \rightarrow$$
$$2 \text{ pyruvate } + \text{ 2 NADH } + \text{ 4 ATP } + \text{ 2 H}_2\text{O } + \text{ 2 H}^+$$

2. ATP and NADH

3.

   (a) **A** Fructose-1,6-bisphosphate
        B Glucose-6-phosphate
        C Phosphoenolpyruvate
        D 3-Phosphoglycerate
        E Glyceraldehyde-3-phosphate
   (b) **B, A, E, D, C**
   (c) **C**
   (d) **A** Aldolase
        B Phosphoglucose isomerase
        C Pyruvate kinase
        D Phosphoglycerate mutase
        E Glyceraldehyde-3-phosphate dehydrogenase
   (e) Fructose-6-phosphate, dihydroxyacetone phosphate, 1,3-bisphosphoglycerate, and 2-phosphoglycerate.

4. Hexokinase, phosphofructokinase, phosphoglycerate kinase, and pyruvate kinase.

5. There are <u>three</u> isomerization reactions in glycolysis. The enzymes that catalyze them are <u>phosphoglucose isomerase, triose phosphate isomerase, and phosphoglycerate mutase</u>.

6. Triose phosphate isomerase catalyzes the interconversion of GAP and DHAP. Its $\Delta G^{\circ\prime} =$ 7.9 kJ·mol^{-1} and $\Delta G = 4.4$ kJ·mol^{-1} (Table 15-1). Since $\Delta G = -RT \ln K$, $\ln K = -\Delta G/RT$ and $K = e^{-\Delta G/RT}$. Under standard conditions,

$$K = \frac{[\text{GAP}]}{[\text{DHAP}]} = e^{-(7900\ \text{J·mol}^{-1})/(8.3145\ \text{J·K}^{-1}\text{mol}^{-1})(310\text{K})}$$

$$= 0.047$$

In the cell, where the reaction is not at equilibrium,

$$\Delta G = \Delta G^{\circ\prime} + RT \ln \frac{[\text{GAP}]}{[\text{DHAP}]}$$

$$\frac{[\text{GAP}]}{[\text{DHAP}]} = e^{(\Delta G - \Delta G^{\circ\prime}/RT)}$$

$$\frac{[\text{GAP}]}{[\text{DHAP}]} = e^{(4400 - 7900\ \text{J·mol}^{-1})/(8.3145\ \text{J·K}^{-1}\cdot\text{mol}^{-1})(310\ \text{K})}$$

$$= 0.26$$

Therefore, the cellular ratio [GAP]/[DHAP] is nearly 6 times larger than under standard conditions, so that in the cell, formation of GAP is favored.

7. This isotope exchange reaction is consistent with the existence of an acyl–enzyme intermediate through the following series of events: For exchange to occur, 1,3-bisphosphoglycerate (1,3-BPG) must react with GAPDH to form an acyl–enzyme intermediate and eliminate P_i. (Figure 15-9, Reaction 5 in reverse). In this reaction, the ^{32}P-labeled P_i reacts with the acyl-enzyme intermediate to yield ^{32}P-labeled 1,3-BPG.

8. 2,3-BPG is occasionally released by bisphosphoglycerate mutase. The resulting enzyme is inactive because it requires a phospho-His residue to phosphorylate 2PG. The presence of trace amounts of 2,3-BPG permits this substance to rebind to the enzyme and thereby activate it by forming the phospho-His residue.

9. The combined glyceraldehyde-3-phosphate dehydrogenase/phosphoglycerate kinase reaction is

$$\text{GAP} + \text{NAD}^+ + \text{ADP} + P_i \rightarrow 3\text{PG} + \text{NADH} + \text{ATP} + \text{H}^+$$

$\Delta G^{\circ\prime} = -16.7$ kJ·mol^{-1} and $\Delta G = -1.1$ kJ·mol^{-1} (Table 15-1).

Since $\Delta G = \Delta G^{\circ\prime} + RT \ln Q$, where

$$Q = \frac{[3PG][NADH][ATP][H^+]}{[GAP][NAD^+][ADP][P_i]}$$

$$Q = e^{(\Delta G - \Delta G^{\circ\prime}/RT)}$$
$$= e^{(-1100 + 16{,}700 \text{ J·mol}^{-1})/(8.3145 \text{ J·K}^{-1}\text{·mol}^{-1})(310 \text{ K})}$$
$$= 425$$

Since $Q = \dfrac{[3PG][NADH][ATP][H^+]}{[GAP][NAD^+][ADP][P_i]} = \dfrac{[3PG]}{[GAP]} \times \dfrac{1}{100} \times 10 \times 1 = 425$

$$\frac{[GAP]}{[3PG]} = \frac{1}{4250} = 2.35 \times 10^{-4}$$

10.

    (a) Inhibition of enolase causes its substrate, 2PG, to accumulate. Because the preceding reaction (catalyzed by phosphoglycerate mutase) is near equilibrium, 2PG equilibrates with 3PG so that [3PG] increases as [2PG] increases.

    (b) The formation of 3PG from 1,3-BPG is endergonic, so this reaction (catalyzed by phosphoglycerate kinase) does not proceed in reverse. Therefore, an increase in [3PG] does not cause [1,3-BPG] to increase.

11.

    (a) A
    (b) C
    (c) A
    (d) C

12. During the catabolic breakdown of glucose, NAD^+ is reduced to NADH, thereby lowering the $[NAD^+]/[NADH]$ ratio. Under anaerobic conditions, pyruvate can be converted to lactate with the concomitant oxidation of NADH to NAD^+. Under aerobic conditions, oxidative phosphorylation regenerates the NAD^+. Both of these processes will increase the $[NAD^+]/[NADH]$ ratio.

13. C3 or C4 of glucose must be labeled with ^{14}C.

14. The rate of ATP production during anaerobic fermentation can be as much as 100 times greater than during oxidative phosphorylation. Therefore, anaerobic fermentation provides the bulk of ATP during rapid bursts of muscle activity, even though the yield of ATP per glucose is much lower than in oxidative phosphorylation.

15.
(a) The phosphofructokinase reaction is the rate-determining step.
(b) The first reaction of glycolysis, catalyzed by hexokinase, is not a suitable control point for the overall pathway because significant amounts of sugar enter glycolysis as glucose-6-phosphate, the product of this reaction. For example, galactose is converted to glucose-6-phosphate to enter glycolysis. Not all this glucose-6-phosphate proceeds through glycolysis: Some is shunted to the pentose phosphate pathway. Furthermore, mannose and, in muscle, fructose enter glycolysis as fructose-6-phosphate, which freely interconverts with glucose-6-phosphate. Therefore, the production of fructose-1,6-bisphosphate by PFK is the first committed step of glycolysis and the most effective point for regulation.

16. Substrate cycles were originally referred to as futile cycles because of their apparent futile waste of free energy.

17. Mannose.

18. No additional NADH is produced. The NAD^+ is probably reduced and then reoxidized in the sequential oxidation and reduction of C4 during epimerization of the hexose.

19. The primary sugar in human milk is lactose, a disaccharide of glucose and galactose.

20.
(a) Glycerol can be converted to pyruvate through reactions catalyzed by glycerol kinase, glycerol phosphate dehydrogenase, triose phosphate isomerase, and the enzymes of Stage II of glycolysis (see Figure 15-27):

$$\text{Glycerol} + \text{ATP} \rightarrow \text{glycerol-3-phosphate} + \text{ADP}$$
$$\text{Glycerol-3-phosphate} + NAD^+ \rightarrow \text{DHAP} + \text{NADH} + H^+$$
$$\text{DHAP} \rightarrow \text{GAP}$$
$$\text{GAP} + NAD^+ + 2\,\text{ADP} + P_i \rightarrow \text{pyruvate} + \text{NADH} + 2\,\text{ATP} + H_2O + 2\,H^+$$

$$\text{Glycerol} + 2\,NAD^+ + \text{ADP} + P_i \rightarrow \text{pyruvate} + 2\,\text{NADH} + \text{ATP} + H_2O + 3\,H^+$$

(b) The net yield of ATP is the same for the oxidation of 1 glucose or 2 glycerol to pyruvate.
(c) For every pyruvate formed from glycerol, 2 NAD^+ are reduced to NADH. Alcoholic fermentation regenerates only 1 NAD^+ per pyruvate, so this pathway does not maintain the redox balance of the cell.

21.
(a) $5C_a$ is ribose-5-phosphate, and $5C_b$ is xylulose-5-phosphate.
(b) 6C is fructose-6-phosphate.
(c) 3C is glyceraldehyde-3-phosphate.
(d) A and C are transketolase reactions, and B is a transaldolase reaction.

22.

    (a) Two

    (b) One directly, although four more can be produced via glycolysis from the two F6P molecules also produced.

23. The conversion of Ru5P to Xu5P is a racemization that inverts the configuration at the C3 chiral center. The isomerization of Ru5P to R5P converts a ketose to an aldose by shifting the position of a double bond.

24.

2-hydroxy-ethanol (hydroxy-acetaldehyde)

As in the pyruvate decarboxylase reaction, an aldehyde might be expected to be released from the TPP cofactor of transketolase in the absence of the second substrate.

25. (3); (ketose); (aldose); (ketose and an aldose)

26. (2); (ketose); (aldose); (ketose and an aldose)

27. The committed step is the exergonic reaction mediated by G6PD, which is regulated by the availability of $NADP^+$.

28.

    (1) The 5 G6P are converted to 5 F6P by phosphoglucose isomerase.

    (2) One F6P is converted to FBP, which requires ATP for the reaction catalyzed by phosphofructokinase.

    (3) The FBP is converted to 2 GAP by aldolase and triose phosphate isomerase.

    (4) The transaldolase and transketolase reactions of the pentose phosphate pathway operate in reverse to convert 2 F6P + 1 GAP to 2 Xu5P and 1 R5P. Since the starting materials are 4 F6P and 2 GAP, the result is 4 Xu5P and 2 R5P.

    (5) The 4 Xu5P can be isomerized to 4 R5P, for a total yield of 6 R5P from the original 5 G6P.

16

Glycogen Metabolism and Gluconeogenesis

This chapter discusses glycogen breakdown and synthesis, gluconeogenesis, and oligosaccharide synthesis, with an emphasis on the mechanisms that regulate these metabolic pathways. Six major enzymes are involved in the breakdown and synthesis of glycogen: glycogen phosphorylase, glycogen debranching enzyme, and phosphoglucomutase for glycogen breakdown, and UDP–glucose pyrophosphorylase, glycogen synthase, and glycogen branching enzyme for glycogen synthesis. Box 16-2 shows how inherited metabolic diseases contribute to our understanding of glycogen metabolism. The chapter explains how the synthesis of glycogen from glucose-1-phosphate requires the free energy of nucleotide hydrolysis in a reaction that is the opposite of the exergonic breakdown of glycogen. The mechanisms of glycogen phosphorylase and glycogen synthase are examined, including the role of the oxonium ion transition state in each case. Box 16-3 explores the strategies for optimizing the branched structure of glycogen. The chapter then discusses the regulation of glycogen phosphorylase and glycogen synthase by allosteric effectors and covalent modification. Regulation involves an enzyme cascade where extracellular signals (hormones), acting via the second messengers cyclic AMP and Ca^{2+}, initiate the activation of successive kinases, resulting in the reciprocal activation of glycogen phosphorylase and inactivation of glycogen synthase. . The role of phosphoprotein phosphatase-1 in modulating the ratio of phosphorylated to dephosphorylated enzymes is also explored as an additional layer of complexity in the regulation of the enzymes involved in glycogen breakdown and synthesis.

The chapter then moves on to gluconeogenesis, the process by which pyruvate and related metabolites can be converted to glucose. Gluconeogenesis uses many of the same enzymes as glycolysis but requires other enzymes to bypass the exergonic steps of glycolysis. The chapter discusses the interesting role of the allosteric effector fructose-2,6-bisphosphate in the regulation of gluconeogenesis and glycolysis in liver and heart muscle. The last section of this chapter discusses the roles of nucleotide sugars and dolichol pyrophosphate in oligosaccharide synthesis.

Essential Concepts

1. The mobilization of glucose from glycogen stores in the liver provides a constant supply of glucose to the central nervous system and red blood cells, which use glucose as their sole energy source. Under fasting conditions, amino acids (mainly from muscle protein degradation) serve as precursors for new glucose synthesis (gluconeogenesis).

2. Glucose-6-phosphate (G6P) is a key branch point in glucose metabolism in the liver, as it can be polymerized to glycogen, degraded to pyruvate, converted to ribose-5-phosphate, or hydrolyzed to glucose.

Glycogen Breakdown

3. Glycogen breakdown (glycogenolysis) utilizes three enzymes:
    (a) Glycogen phosphorylase, which catalyzes the phosphorolysis of the glucose residues at the nonreducing ends of glycogen to yield glucose-1-phosphate (G1P).
    (b) Glycogen debranching enzyme, which transfers a tri- or tetrasaccharide and hydrolyzes the $\alpha(1{\rightarrow}6)$ linkage at branch points.
    (c) Phosphoglucomutase, which converts G1P to G6P.

4. Phosphorylase utilizes the cofactor pyridoxal-5′-phosphate (PLP) in the general acid–base catalytic mechanism in glycogen phosphorolysis. Inorganic phosphate attacks the terminal glucose residue, which passes through an oxonium ion intermediate before being released as G1P. The activity of the dimeric enzyme is regulated by allosteric effectors and by phosphorylation/dephosphorylation of the protein at Ser 14.

5. Glycogen debranching enzyme acts as an $\alpha(1{\rightarrow}4)$ transglycosylase by transferring a tri- or tetrasaccharide from a limit branch (a four- or five-glucose-residue segment that phosphorylase cannot further degrade) to the nonreducing end of another branch. The remaining glucosyl residue, which is attached to glycogen by an $\alpha(1{\rightarrow}6)$ linkage, is hydrolyzed at a separate active site on the enzyme to release free glucose.

6. Phosphoglucomutase converts G1P to G6P by way of a G1,6P intermediate. A phosphate group from a Ser residue is transferred to C6 of G1P, followed by transfer of the C1 phosphate back to the Ser residue, in a manner similar to the phosphoglycerate mutase reaction.

7. The liver expresses glucose-6-phosphatase, an enzyme that hydrolyzes G6P. The resulting free glucose equilibrates with glucose in the blood so that the breakdown of liver glycogen leads to an elevation of blood glucose levels.

Glycogen Synthesis

8. Glycogen synthesis requires three enzymes to covert G1P to glycogen. First, UDP–glucose pyrophosphorylase catalyzes the transfer of UMP from UTP to the phosphate group of G1P to form UDP–glucose and PP_i. PP_i is eventually hydrolyzed to P_i by inorganic pyrophosphatase, which provides the exergonic push for this reaction.

9. Glycogen synthase catalyzes a transfer reaction in which the glucosyl residue of UDP–glucose is added to the nonreducing end of glycogen through an $\alpha(1{\rightarrow}4)$ bond. Glycogen synthase can only extend a pre-existing $\alpha(1{\rightarrow}4)$-linked chain. The glycogen molecule originates through the action of the protein glycogenin, which assembles a seven-residue glycogen "primer" for glycogen synthase to act on.

10. Glycogen branching enzyme transfers a seven-residue segment from the end of an $\alpha(1{\rightarrow}4)$-linked glucan chain to the C6-hydroxyl group of a glucosyl residue on the same chain or another chain, thereby forming an $\alpha(1{\rightarrow}6)$-linked branch.

Control of Glycogen Metabolism

11. The opposing processes of glycogen breakdown and synthesis are coordinately controlled by allosteric regulation and covalent modification of key enzymes. The major allosteric regulators are ATP, AMP, and G6P. Covalent modification occurs with the transfer of P_i from ATP to certain enzymes by the action of specific kinases.

12. Glycogen phosphorylase exists in two forms: phosphorylase *a* (the phosphorylated, more active enzyme) and phosphorylase *b* (the dephosphorylated, less active enzyme). Each form also has two conformations: the T (relatively inactive) state and the R (relatively active) state. AMP promotes the T $\rightarrow$ R conformational change and thereby activates phosphorylase *b*, whereas ATP and G6P inhibit this conformational change. Phosphorylase *a* is less sensitive to allosteric effectors and is mainly in the R state. However, high concentrations of glucose promote the R $\rightarrow$ T transition.

13. Covalent modification regulates glycogen phosphorylase and glycogen synthase in a reciprocal fashion. Glycogen phosphorylase tends to be more active when it is phosphorylated, so the ratio of phosphorylase *a* to phosphorylase *b* largely determines the rate of glycogen phosphorolysis. This ratio is set by the activity of phosphorylase kinase, which is regulated by protein kinase A (PKA), and by phosphoprotein phosphatase-1 (PP1). In contrast, glycogen synthase is activated by dephosphorylation so that PKA promotes glycogen breakdown while inhibiting glycogen synthesis, and PP1 inhibits glycogen breakdown while promoting glycogen synthesis.

14. Phosphoprotein phosphatase-1 (PP1) can dephosphorylate glycogen phosphorylase, phosphorylase kinase, and glycogen synthase. In muscle, insulin-stimulated protein kinase activates PP1 by phosphorylating its G_M subunit, which binds to glycogen in muscle . The PKA-mediated phosphorylation of another site on the G_M subunit causes the catalytic subunit of PP1 to be released in the cytoplasm, where it cannot dephosphorylate glycogen-bound enzymes. In addition, phosphoprotein phosphatase inhibitor 1 (inhibitor-1) inhibits PP1. Inhibitor-1 activity is stimulated by phosphorylation by PKA, which helps preserve the phosphorylated (active) forms of phosphorylase kinase and phosphorylase *a*.

15. Glycogen synthase is inactivated by phosphorylation by the same enzyme system that phosphorylates glycogen phosphorylase. Hence, activation of phosphorylase kinase, which activates phosphorylase *a*, inactivates glycogen synthase. This regulatory mechanism provides a rapid and large-scale control of flux in the substrate cycle that links glycogen and G1P. Glycogen synthase activity is also controlled by other kinases.

16. Hormones, including glucagon, insulin, epinephrine, and norepinephrine, ultimately control glycogen metabolism. These hormones bind to transmembrane protein receptors and initiate a series of reactions that lead to the production of molecules called second messengers (e.g., cAMP and Ca^{2+}), which modulate the activities of numerous intracellular proteins. Glucagon binding to its receptor in the liver results in an elevation of cAMP, which favors glycogen breakdown. Epinephrine and norepinephrine bind to α- and β-adrenoreceptors in the liver and to β-adreno receptors in muscle. The binding of these

hormones to β-adrenoreceptors increases [cAMP], whereas their binding to α-adrenoreceptors increases cytosolic [Ca^{2+}]. Insulin binding to its receptor in tissues other than the liver decreases [cAMP] and promotes glycogen synthesis.

Gluconeogenesis

17.   The principal noncarbohydrate precursors of glucose are lactate, pyruvate, and amino acids. In animals, these compounds (except for leucine and lysine) are converted, at least in part, into oxaloacetate, which is required for gluconeogenesis.

18.   The conversion of pyruvate (or lactate) to glucose follows a pathway that is the reverse of glycolysis except where it bypasses the exergonic steps catalyzed by pyruvate kinase, phosphofructokinase, and hexokinase.

19.   To bypass the pyruvate kinase reaction, pyruvate is first carboxylated by pyruvate carboxylase in a reaction that is driven by ATP hydrolysis. The enzyme's biotin prosthetic group is converted to a carboxybiotinyl group in order to transfer CO_2 to pyruvate. The product of this reaction is oxaloacetate, which is subsequently decarboxylated to phosphoenolpyruvate (PEP). The HCO_3^- added to pyruvate therefore leaves as CO_2. This second reaction, catalyzed by PEP carboxykinase (PEPCK) is driven by the hydrolysis of GTP.

20.   Oxaloacetate is produced in the mitochondria, while the reactions that convert PEP to glucose occur in the cytosol. In species with mitochondrial PEPCK, the PEP formed in the mitochondria is exported to the cytosol via a specific transporter. In species with cytosolic PEPCK, oxaloacetate is first converted to malate or aspartate, each of which has a transporter that allows mitochondrial–cytosolic exchange, and, in the cytosol, is reconverted to oxaloacetate.

21.   PEP is converted to fructose-1,6-bisphosphate (FBP) by the enzymes of glycolysis operating in reverse. FBP is then hydrolyzed to fructose-6-phosphate and P_i by the action of fructose bisphosphatase (FBPase-1). Similarly, G6P is hydrolyzed by glucose-6-phosphatase, yielding glucose and P_i. These reactions therefore bypass the exergonic hexokinase and phosphofructokinase reactions.

22.   Gluconeogenesis and glycolysis are reciprocally regulated in the liver. The principal allosteric regulator is the metabolite fructose-2,6-bisphosphate (F2,6P), which is a potent activator of PFK-1 and inhibitor of FBPase-1. F2,6P is formed and degraded by a bifunctional enzyme referred to as PFK-2/FBPase-2. PKA phosphorylates this enzyme, thereby activating FBPase-2 and inactivating PFK-2, which results in a net decrease in F2,6P (and favors gluconeogenesis). In heart muscle, the situation is reversed, so that phosphorylation activates PFK-2, which facilitates the muscle's ability to extract energy from glucose via glycolysis.

23.   Glucose metabolism is also regulated by other mechanisms:
(a) Acetyl-CoA activates pyruvate carboxylase.

(b) Alanine inhibits pyruvate kinase. The amino group of alanine is transferred to an α-keto acid by transamination to yield pyruvate and a new amino acid. The resulting pyruvate then serves as a substrate for gluconeogenesis.

(c) Long-term regulation of gluconeogenesis occurs via changes in gene expression. Prolonged low concentrations of insulin or high concentrations of cAMP stimulate the transcription of the genes for PEPCK, FBPase, and glucose-6-phosphatase, and repress the transcription of the genes for glucokinase, PFK, and the PFK-2/FBPase-2 bifunctional enzyme.

Other Carbohydrate Biosynthetic Pathways

24. The formation of glycosidic bonds in oligo- and polysaccharides is facilitated by nucleotide sugars, as in the polymerization of glucose by glycogen synthase. The principal nucleotides used are ADP and CDP. Nucleotide sugars are glycosyl donors in the synthesis of *O*-linked oligosaccharides and in the processing of *N*-linked oligosaccharides.

25. *N*-linked oligosaccharides are initially built on dolichol, an isoprenoid lipid carrier in the endoplasmic reticulum (ER). This process begins in the cytosol but finishes in the lumen of the ER, as the dolichol-oligosaccharide flips back and forth across the ER membrane. After the oligosaccharide reaches 14 residues, it is transferred to a protein, leaving dolichol pyrophosphate.

26. Lactose synthesis in mammals involves the mammary gland protein α-lactalbumin, which changes the substrate specificity of a galactosyltransferase so that it synthesizes lactate rather than *N*-acetyllactosamine.

Questions

Glycogen Breakdown

1. List the three enzymes required for the breakdown of glycogen.

2. What feature of the structure of glycogen phosphorylase is consistent with the observation that the enzyme cannot cleave a glycosidic bond within five residues of a branch point?

3. What role does pyridoxal phosphate (PLP) play in the mechanism of glycogen phosphorylase?

4. What would be the product of the nonenzymatic phosphorolysis of glycogen?

5. In the diagram of glycogen shown below, circle the substrates for glycogen debranching enzyme.

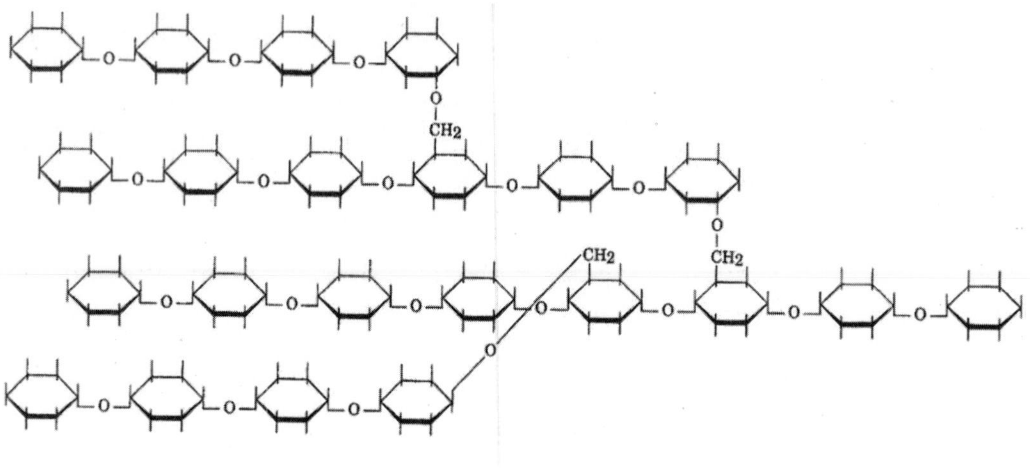

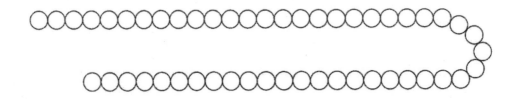

6. The rate of debranching is much slower than that of phosphorolysis. Explain how highly branched glycogen molecules release glucose-1-phosphate at a greater rate than relatively unbranched ones.

7. Write an equation for the equilibrium constant for the phosphorylase reaction. $\Delta G^{\circ\prime}$ for this reaction is $+3.1$ kJ·mol^{-1}, yet glycogen breakdown in the liver and muscle is thermodynamically favored at 37°C. What is the minimum ratio of $[P_i]/[G1P]$ required to make the phosphorylase reaction exergonic? Assume that the concentration of glycogen does not change significantly.

Glycogen Synthesis

8. How is the thermodynamic barrier to glycogen synthesis overcome by cells?

9. In the diagram below, each circle represents a glycosyl unit. By the action of branching enzyme, show the most branched structure this molecule can assume. Indicate the reducing and nonreducing ends of the molecule and the position where a glycogenin molecule would be found. There are 50 glycosyl residues.

10. Below is a diagram showing the synthesis and breakdown of glycogen [(glucose)]. Fill in the blanks and identify the enzyme that catalyzes each numbered reaction.

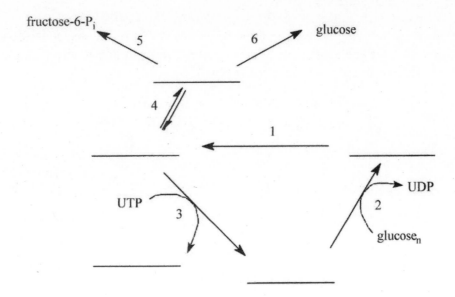

Control of Glycogen Metabolism

11. What structural features distinguish the T conformation of glycogen phosphorylase from the R conformation of the enzyme? (*Hint:* See Figure 12-15.)

12. Glycogen phosphorylase is activated in vigorously active muscle without significant changes in intracellular [cAMP]. Explain.

13. The presence of epinephrine results in the stimulation of PFK-2 in heart muscle. How does epinephrine affect glycolysis in this organ?

14. Which target enzyme in glycogen metabolism requires both α- and β-adrenoreceptors to be activated for full enzyme activity?

15. Which of the hereditary glycogen storage diseases results in a pronounced decrease of stored glycogen?

Gluconeogenesis

16. What is the fate of $H^{14}CO_3^-$ added to a liver homogenate that is active in gluconeogenesis?

17. Write a balanced equation for the formation of PEP from pyruvate and compare it with the reverse reaction of glycolysis.

18. While acetyl-CoA, the end product of fatty acid oxidation, cannot be converted into glucose, another product of fat degradation, glycerol, can be converted to glucose. Where does glycerol enter the gluconeogenic pathway?

Other Carbohydrate Biosynthetic Pathways

19. UDP–galactose can donate is sugar residue to glucose to form the disaccharide lactose. For the structures of UDP–galactose and glucose below, indicate the reactive electrophilic and nucleophilic centers and the products of the reaction.

UDP–galactose   Glucose

Answers to Questions

1. Glycogen phosphorylase, glycogen debranching enzyme, and phosphoglucomutase.

2. A narrow crevice on the surface of glycogen phosphorylase, which connects the glycogen storage site and the active site, can accommodate up to five residues of an unbranched glycogen chain but cannot accommodate a branched chain.

3. The phosphoryl group of PLP participates in general acid–base catalysis by donating a proton to the anionic P_i that reacts with glycogen to release G1P.

4. The reaction product would be a mixture of α and β anomers of G1P because the reaction intermediate is an oxonium ion whose C1 can react with phosphate approaching either face of the sugar residue.

5.

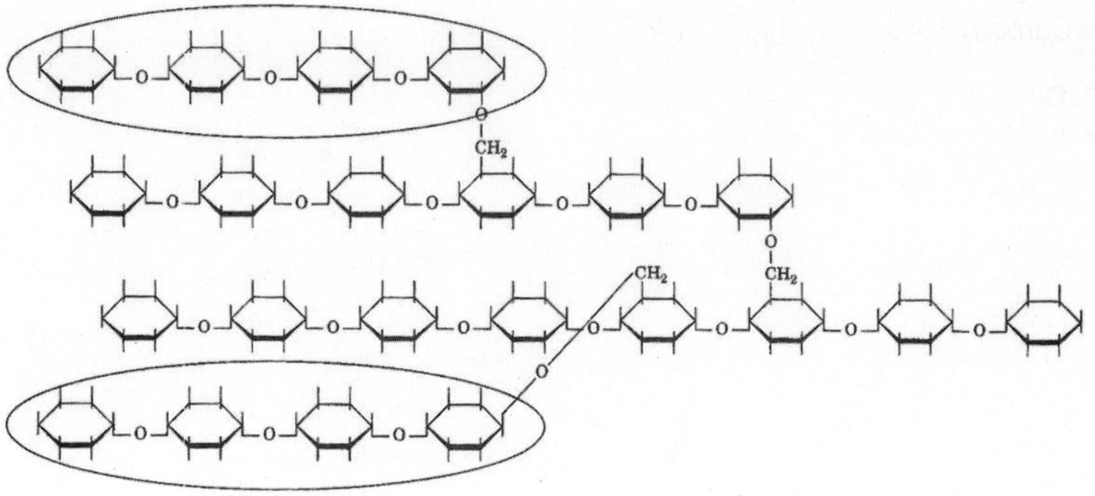

Glycogen debranching enzyme has two catalytic functions: (1) It transfers three or four $\alpha(1\rightarrow4)$-linked glucose residues from a "limit branch" of glycogen to the nonreducing end of another branch, and (2) it hydrolyzes the $\alpha(1\rightarrow6)$ bond of the remaining residue to form free glucose. In the diagram above, the top and bottom branches serve as substrates in which the trisaccharide is transferred and the remaining glucose residue is released. Once these branches have been eliminated, continued phosphorolysis would produce an additional "limit branch" substrate for debranching enzyme.

6.   Highly branched glycogen molecules have more nonreducing ends available for phosphorylase to produce G1P, so the rate of G1P release remains high until a limit branch is encountered.

7.

$$K_{eq} = \frac{[\text{glycogen}_{n-1}][\text{G1P}]}{[\text{glycogen}_n][\text{P}_i]} \approx \frac{[\text{G1P}]_{eq}}{[\text{P}_i]_{eq}}$$

$$\Delta G = \Delta G^{o\prime} + RT \ln \frac{[\text{G1P}]}{[\text{P}_i]}$$

In order for ΔG to be less than zero, $-RT \ln ([\text{G1P}]/[\text{P}_i])$ must be at least as great as $\Delta G^{o\prime}$.

$-RT \ln ([\text{G1P}]/[\text{P}_i]) \geq \Delta G^{o\prime}$

$-\ln ([\text{G1P}]/[\text{P}_i]) \geq \Delta G^{o\prime}/RT$

$\ln ([\text{G1P}]/[\text{P}_i]) \leq -\Delta G^{o\prime}/RT$

$[\text{G1P}]/[\text{P}_i] \leq e^{-\Delta G^{o\prime}/RT}$

$[\text{G1P}]/[\text{P}_i] \leq e^{-(3100 \text{ J·mol}^{-1})/(8.3145 \text{ J·K}^{-1}\text{·mol}^{-1})(310 \text{ K})}$

$[\text{G1P}]/[\text{P}_i] \leq 0.30$

Therefore, $[\text{G1P}]/[\text{P}_i]$ can be no greater than 0.30, so $[\text{P}_i]/[\text{G1P}]$ must be at least $1/0.30 = 3.33$.

8.   A "high-energy" compound is required to bypass the exergonic glycogen phosphorylase reaction. The formation of UDP–glucose from G1P and UTP yields PP_i, whose hydrolysis provides the thermodynamic "pull" to add another glucose residue to glycogen.

9.

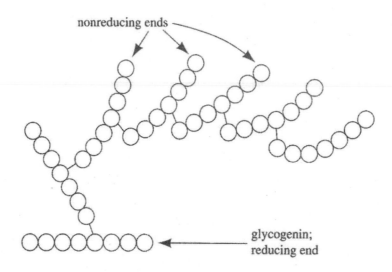

Branching enzyme has three constraints: (1) Branches must be separated by four glucosyl residues; (2) seven residues are transferred at a time to make a branch; and (3) the donating chain must be at least 11 residues long. The first branch is made by moving a seven-residue segment from the reducing end. The most highly branched structure results when branches are added to branches for a final product with six $\alpha(1{\to}6)$ branch points, as shown above. There are seven nonreducing ends (three are labeled).

10.

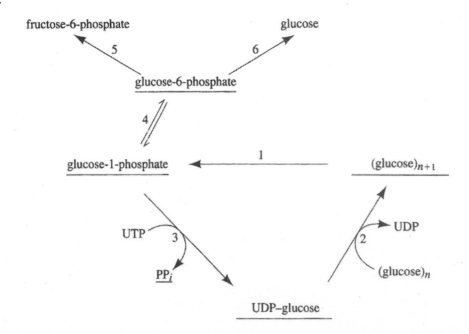

1   Glycogen phosphorylase
2   Glycogen synthase
3   UDP–glucose pyrophosphorylase
4   Phosphoglucomutase
5   Phosphoglucose isomerase
6   Glucose-6-phosphatase

11. In the T form, the active site is less accessible to its substrates compared to the R form due to the presence of a loop that covers the T state active site so as to prevent access of substrate to it. In the R state, the tower helices have tilted and pulled apart relative to their positions in the T state, thereby inducing a ~10° counter-rotation of the two subunits. This also displaces and disorders the loop covering the active site, thus making the active site accessible to substrate. In addition, the side chain of Arg 569, which is located in the active site, rotates in such a way that it increases the R-state enzyme's affinity for its P_i substrate.

12. Ca^{2+} released during muscle contraction activates phosphorylase kinase via binding of Ca^{2+} to calmodulin (the δ subunit of phosphorylase kinase). Phosphorylase kinase is therefore active (although not maximally so) even without PKA-catalyzed phosphorylation of its α and β subunits.

13. Epinephrine-dependent stimulation of PFK-2 leads to an increase in [F2,6P], which activates PFK-1. The result is increased flux through glycolysis.

14. The full activity of phosphorylase kinase requires both the presence of Ca^{2+} (whose concentration increases in response to hormone binding to α-adrenoreceptors) and phosphorylation by PKA (which is activated by the binding of the cAMP second messenger whose synthesis is stimulated by hormone binding to β-adrenoreceptors).

15. Only Type 0 (glycogen synthase deficiency) causes a decrease in stored glycogen.

16. The $H^{14}CO_3^-$ is added to pyruvate to yield oxaloacetate via the pyruvate carboxylase reaction. The ^{14}C is then released as CO_2 by the PEP carboxykinase reaction, which converts oxaloacetate to PEP + CO_2.

17.

gluconeogenesis:

$$pyruvate + HCO_3^- + ATP \rightleftharpoons oxaloacetate + ADP + P_i$$
$$oxaloacetate + GTP \rightleftharpoons PEP + GDP + CO_2$$
$$\overline{pyruvate + HCO_3^- + ATP + GTP \rightleftharpoons PEP + ADP + GDP + P_i + CO_2}$$

glycolysis:

$$H^+ + PEP + ADP \rightleftharpoons ATP + pyruvate$$

The formation of PEP from pyruvate requires the investment of two ATP equivalents ("high-energy" phosphoanhydride bonds), whereas PEP's reaction to form pyruvate yields only one ATP.

18. Glycerol enters the gluconeogenic pathway as dihydroxyacetone phosphate (DHAP; see Figure 15-27 on p. 517). Glycerol is a substrate for glycerol kinase, which catalyzes the reaction

Glycerol + ATP $\rightleftharpoons$ glycerol-3-phosphate + ADP

Glycerol phosphate dehydrogenase then catalyzes the reaction

Glycerol-3-phosphate + NAD$^+$ $\rightleftharpoons$ DHAP + NADH + H$^+$

19.

lactose

17
Citric Acid Cycle

This chapter discusses the citric acid cycle as both a catabolic and an anabolic process. The citric acid cycle was elucidated by several investigators, but the key insights were provided by Hans Krebs, so the cycle is often referred to as the Krebs cycle. It is also called the tricarboxylic acid (TCA) cycle, which refers to its first intermediate, citrate. The chapter first provides an overview of the key features of the citric acid cycle. It then takes a closer look at each step of the cycle, beginning with a discussion of the pyruvate dehydrogenase complex that converts pyruvate to acetyl-CoA, the fuel molecule that enters the cycle. In this chapter, you will encounter several coenzymes that are critical to the functioning of the citric acid cycle, all derived from water-soluble vitamins. A discussion of the regulation of the citric acid cycle then ensues, in which you will encounter familiar regulatory mechanisms as well as new ones. The chapter then turns to the anabolic features of the citric acid cycle. The citric acid cycle is one of the key metabolic hubs in the cell: Its intermediates are generated by a variety of degradative reactions, and they provide precursors for several biosynthetic pathways. As citric acid cycle intermediates are diverted to anabolic pathways, they are replaced through anaplerotic reactions. The final section of the chapter introduces the glyoxylate cycle found in plants. This pathway is the only known pathway for the net conversion of acetyl-CoA to glucose. The chapter also discusses arsenic poisoning in the citric acid cycle (Box 17-2). Box 17-3 discusses the evolution of the citric acid cycle from a series of reductive reactions.

Essential Concepts

Overview of the Citric Acid Cycle

1. The citric acid cycle is a central pathway for recovering energy from the three major metabolic fuels: carbohydrates, fatty acids, and amino acids. These fuels are broken down to yield acetyl-CoA, which enters the citric acid cycle by condensing with the C_4 compound oxaloacetate. The citric acid cycle is a series of reactions in which 2 CO_2 are released for every acetyl-CoA that enters the cycle, so that oxaloacetate is always reformed. Hence, the cyclical series of reactions acts catalytically to process acetyl-CoA continuously.

2. The oxidation of the acetyl carbon skeleton in the citric acid cycle is coupled to the reduction of NAD^+ and FAD. Oxidation of the resulting NADH and $FADH_2$ by the electron-transport chain supplies free energy for ATP synthesis and regenerates NAD^+ and FAD for the oxidation of additional acetyl-CoA. In aerobic respiration, O_2 serves as the terminal acceptor of the acetyl group's electrons. In anaerobic respiration, this function is carried out by molecules such as NO_3^-, SO_4^{2-}, and Fe^{3+}.

3. One complete round of the cycle yields 2 CO_2, 3 NADH, 1 $FADH_2$, and 1 GTP (which is the equivalent of 1 ATP). Hence, the net reaction of the citric acid cycle is

$$3 \text{ NAD}^+ + \text{FAD} + \text{GDP} + P_i + \text{acetyl-CoA} + 2\text{ H}_2\text{O} \rightarrow$$
$$3 \text{ NADH} + \text{FADH}_2 + \text{GTP} + \text{CoA} + 2\text{ CO}_2 + 3\text{ H}^+$$

The carbon atoms of the acetyl group entering the cycle do not exit the cycle as CO_2 in the first round but do so in subsequent rounds.

Synthesis of Acetyl-Coenzyme A

4. In eukaryotes, all the enzymes of the citric acid cycle (and the pyruvate dehydrogenase complex) occur in the matrix (inner compartment) of the mitochondrion. The pyruvate produced by glycolysis in the cytosol is transported into the mitochondrion via a pyruvate–H^+ symport protein.

5. The pyruvate dehydrogenase complex is a large multienzyme complex containing multiple copies of three enzymes (E_1, E_2, and E_3) organized in a polyhedral array. For example, the *E. coli* pyruvate dehydrogenase complex consists of 24 copies of E_1 and 24 copies of E_2 arranged about the corners of concentric cubes, and 12 copies of E_3. In mammals, the pyruvate dehydrogenase complex has dodecahedral symmetry.

6. The pyruvate dehydrogenase complex catalyzes five reactions in the oxidative decarboxylation of pyruvate. Five different coenzymes are involved.
    (a) Pyruvate dehydrogenase (E_1) decarboxylates pyruvate in a reaction identical to that catalyzed by pyruvate decarboxylase in alcoholic fermentation (Figure 15-20). However, in the pyruvate dehydrogenase reaction, the hydroxyethyl group remains linked to the thiamine pyrophosphate (TPP) prosthetic group rather than being released as acetaldehyde.
    (b) Dihydrolipoyl transacetylase (E_2) transfers the hydroxyethyl group from TPP to a lipoic acid residue that is tethered to an enzyme Lys side chain via an amide linkage (lipoamide) to form acetyl-dihydrolipoamide.
    (c) E_2 then transfers the acetyl group to CoA to form acetyl-CoA and dihydrolipoamide.
    (d) Dihydrolipoyl dehydrogenase (E_3) oxidizes the dihydrolipoamide group of E_2 via the reduction of E_3's reactive Cys—Cys disulfide bond.
    (e) E_3 is then reoxidized by NAD^+, in a reaction involving the transfer of electrons via the enzyme's FAD group, to regenerate the reactive disulfide bond, thereby preparing the pyruvate dehydrogenase complex for another round of hydroxyethyl transfer and oxidation.

Enzymes of the Citric Acid Cycle

7. Citrate synthase catalyzes the condensation of acetyl-CoA and oxaloacetate to form citrate in a highly exergonic reaction. The enzyme undergoes a large conformational change as part of an Ordered Sequential reaction mechanism.

8. Aconitase catalyzes the isomerization of citrate to isocitrate via stereospecific dehydration and rehydration.

9. Isocitrate dehydrogenase catalyzes the oxidative decarboxylation of isocitrate to produce α-ketoglutarate and the first CO_2 and NADH of the citric acid cycle.

10. α-Ketoglutarate dehydrogenase (a multienzyme complex similar to the pyruvate dehydrogenase complex) catalyzes the oxidative decarboxylation of α-ketoglutarate by a mechanism identical to that of the pyruvate dehydrogenase complex. The reaction products are succinyl-CoA and the second CO_2 and second NADH of the cycle. Note that the two carbons released as CO_2 in this round of the citric acid cycle are not the carbons that entered the cycle as acetyl-CoA.

11. Succinyl-CoA synthetase catalyzes a coupled reaction in which thioester hydrolysis is coupled to the phosphorylation of GDP via succinyl-phosphate and phospho-His intermediates. This enzyme therefore catalyzes substrate-level phosphorylation. The GTP produced in this step is easily interconverted with ATP by nucleoside diphosphate kinase.

12. The remaining reactions of the citric acid cycle regenerate oxaloacetate.
    (a) Succinate dehydrogenase catalyzes the stereospecific dehydrogenation of succinate to fumarate. The enzyme's FAD prosthetic group is reduced in this redox reaction and is reoxidized when it gives up its electrons to the respiratory electron-transport chain.
    (b) Fumarase catalyzes the hydration of fumarate to malate.
    (c) Malate dehydrogenase catalyzes the oxidation of malate, which regenerates oxaloacetate and produces the third NADH of the cycle. This highly endergonic reaction is driven by the exergonic reaction that follows, namely the condensation of oxaloacetate with acetyl-CoA, which begins the next round of the citric acid cycle.

Regulation of the Citric Acid Cycle

13. The citric acid cycle generates ATP through substrate-level phosphorylation (GTP production) and by the subsequent reoxidation of its 3 NADH and $FADH_2$ products by the electron-transport chain. Each NADH generates ~2.5 ATP and $FADH_2$ generates ~1.5 ATP, so each round of the citric acid cycle yields ~10 ATP.

14. The pyruvate dehydrogenase complex is regulated by product inhibition by NADH and acetyl-CoA, and, in eukaryotes, by covalent modification via the phosphorylation/dephosphorylation of E_1. Pyruvate dehydrogenase kinase represses E_1 activity by phosphorylating it at a specific Ser residue, and pyruvate dehydrogenase phosphatase stimulates E_1 activity by dephosphorylating it.

15. The citric acid cycle is regulated principally at the steps catalyzed by citrate synthase, isocitrate dehydrogenase, and α-ketoglutarate dehydrogenase. Regulatory mechanisms include:
    (a) Substrate availability. The most critical regulators are acetyl-CoA, oxaloacetate, and the ratio of [NADH] to [NAD^+].
    (b) Product inhibition. Citrate competes with oxaloacetate in the citrate synthase reaction, and succinyl-CoA and NADH inhibit α-ketoglutarate dehydrogenase.
    (c) Competitive feedback inhibition. Succinyl-CoA competes with acetyl-CoA in the citrate synthase reaction.

(d) Allosteric regulation. ADP activates isocitrate dehydrogenase, while ATP allosterically inhibits it. Ca^{2+} activates pyruvate dehydrogenase, isocitrate dehydrogenase, and α-ketoglutarate dehydrogenase.

Reactions Related to the Citric Acid Cycle

16. The citric acid cycle is amphibolic (both anabolic and catabolic). As an anabolic cycle, the citric acid cycle provides intermediates for gluconeogenesis (oxaloacetate, which must be first converted to malate for export from the mitochondrion), amino acid biosynthesis (oxaloacetate, α-ketoglutarate), porphyrin synthesis (succinyl-CoA), and lipid biosynthesis (citrate). The citric acid cycle also functions catabolically to complete the degradation of carbohydrates and fatty acids (which yield acetyl-CoA) and amino acids that are converted to fumarate, succinyl-CoA, α-ketoglutarate, and oxaloacetate.

17. Anaplerotic reactions replenish citric acid cycle intermediates that have been siphoned off into anabolic reactions. The most important of these reactions is catalyzed by pyruvate carboxylase:

$$Pyruvate + CO_2 + ATP + H_2O \rightarrow oxaloacetate + ADP + P_i$$

18. Acetyl-CoA cannot serve as a precursor for gluconeogenesis in animals, but it can do so in plants via the glyoxylate pathway. This pathway is a variation of the citric acid cycle that takes place in two organelles, the mitochondrion and the glyoxysome (a specialized peroxisome found in plants). The glyoxysome contains isocitrate lyase (which cleaves isocitrate to succinate and glyoxylate) and malate synthase (which condenses glyoxylate with acetyl-CoA to form malate). The net reaction for the glyoxylate pathway is

$$2\,Acetyl\text{-}CoA + 2\,NAD^+ + FAD + 3\,H_2O \rightarrow$$
$$oxaloacetate + 2\,CoA + 2\,NADH + FADH_2 + 4\,H^+$$

Note that in the glyoxylate pathway, no carbons are lost as CO_2.

Questions

Overview of the Citric Acid Cycle

1. The citric acid cycle can be divided into two phases with respect to the oxidation of acetyl-CoA. Describe each phase and write its balanced equation.

2. Early experiments showed that malonate, which inhibits succinate dehydrogenase, blocks cellular respiration. This led to the idea that succinate participates in oxidative metabolism as an intermediate and not as just another metabolic fuel. Using isotopically-labeled reagents, what observation(s) would demonstrate the validity of this interpretation?

Synthesis of Acetyl-Coenzyme A

3. Name the three enzymes that form the pyruvate dehydrogenase complex.

4. For each reaction listed below, indicate the appropriate enzyme(s) in the pyruvate dehydrogenase complex and the relevant cofactor(s), if applicable.

Reaction	Enzyme	Cofactor
(a) Oxidative formation of an enzymatic disulfide bond	_____	_____
(b) Transfer of hydroxyethyl group bound to TPP	_____	_____
(c) Liberation of CO_2	_____	_____
(d) Oxidation of dihydrolipoamide	_____	_____
(e) Formation of acetyl-CoA	_____	_____

5. What are the roles of FAD and NAD^+ in the pyruvate dehydrogenase catalytic mechanism?

6. You are given two preparations of purified pyruvate dehydrogenase complex enzymes with all the required cofactors. You add pyruvate to each preparation and then measure the rate of production of acetyl-CoA and acetaldehyde under aerobic conditions.

		Acetyl-CoA *(molecules/s)*	*Free acetaldehyde* *(molecules/s)*
Preparation	A	10^5	10^{-6}
Preparation	B	10^{-2}	10^3

How might the pyruvate dehydrogenase complex enzymes differ in each preparation?

7. Which of the following labeled glucose molecules would yield $^{14}CO_2$ following glycolysis and the pyruvate dehydrogenase reaction?
    (a) 1-$[^{14}C]$-glucose
    (b) 3-$[^{14}C]$-glucose
    (c) 4-$[^{14}C]$-glucose
    (d) 6-$[^{14}C]$-glucose

Enzymes of the Citric Acid Cycle

8. What is the energetic function of the thioester bond of CoA in the citrate synthase reaction?

9. For the half-reaction

$$FAD + 2\,e^- + 2\,H^+ \rightleftharpoons FADH_2$$

$\mathscr{E}^{\circ\prime} \approx 0.00$ V for FAD bound to succinate dehydrogenase.
    (a) Calculate $\Delta G^{\circ\prime}$ for the oxidation of succinate to fumarate by enzyme-bound FAD.
    (b) How does this compare to the $\Delta G^{\circ\prime}$ for the oxidation of succinate to fumarate by free NAD^+?
    (c) Based on these results, explain why nature chose FAD rather than NAD^+ as the oxidizing agent in the succinate dehydrogenase reaction.

10. What is the fate of C4 (the carboxyl that is β to the carbonyl) of oxaloacetate?

Regulation of the Citric Acid Cycle

11. How would the rapid accumulation of succinyl-CoA affect the rate of glucose oxidation?

Reactions Related to the Citric Acid Cycle

12. List four anabolic pathways that utilize citric acid cycle intermediates as starting material.

13. How does acetyl-CoA affect the activity of pyruvate carboxylase? Why is this advantageous for the cell?

14. Write a balanced equation for the synthesis of glucose from acetyl-CoA via the glyoxylate cycle.

15. Which reactions of the glyoxylate cycle deviate from the citric acid cycle?

16. Which pathway intermediates pass between the plant mitochondrion and the glyoxysome during the functioning of the glyoxylate cycle?

Answers to Questions

1. The first phase involves the net oxidation of acetyl-CoA to two molecules of CO_2 and the regeneration of CoA (Reactions 1–5 of the cycle).

$$Acetyl\text{-}CoA + oxaloacetate + GDP + 2\,NAD^+ + H_2O + P_i \rightarrow$$
$$succinate + GTP + CoASH + 2\,NADH + 2\,CO_2 + 2\,H^+$$

    The second phase involves the regeneration of oxaloacetate from succinate (Reactions 6–8).

$$Succinate + FAD + NAD^+ + H_2O \rightarrow$$
$$oxaloacetate + NADH + FADH_2 + H^+$$

2. To demonstrate that succinate is an intermediate in glucose oxidation, show that, in the presence of malonate, cell extracts containing newly added ^{14}C-labeled acetyl-CoA (labeled at either of the acetyl C atoms) and any of the other citric acid cycle intermediates yield ^{14}C-labeled succinate.

3. Pyruvate dehydrogenase (E_1), dihydrolipoyl transacetylase (E_2), and dihydrolipoyl dehydrogenase (E_3).

4. Reaction | Enzyme | Cofactor

Reaction	Enzyme	Cofactor
(a) Oxidative formation of an enzymatic disulfide bond	E_3	FAD, NAD^+
(b) Transfer of hydroxyethyl group bound to TPP	E_2	lipoic acid
(c) Liberation of CO_2	E_1	TPP
(d) Oxidation of dihydrolipoamide	E_2, E_3	
(e) Formation of acetyl-CoA	E_2	CoA, lipoic acid

5. FAD functions as an electron conduit to NAD^+ as the enzyme reoxidizes the E_3 sulfhydryl groups that were reduced during the regeneration of lipoamide on E_2. The FAD and NAD^+ therefore participate in regenerating the functional groups of E_3.

6. The data suggest that in preparation B, E_1 (pyruvate dehydrogenase) is somehow unable to transfer its bound hydroxyethyl carbanion to the lipoamide cofactor of E_2 (dihydrolipoyl transacetylase). Instead, free acetaldehyde is released from the TPP group of E_1 (as occurs in the pyruvate decarboxylase reaction; Section 15-3B).

7. (b) and (c) would yield $^{14}CO_2$ after the pyruvate dehydrogenase reaction since the C3 and C4 atoms of glucose both become C1 of glyceraldehyde-3-phosphate and then pyruvate. Pyruvate dehydrogenase liberates C1 of pyruvate as CO_2.

8. The large, negative free energy of hydrolysis of the thioester bond provides the thermodynamic force for the otherwise endergonic condensation of the acetyl group and oxaloacetate under conditions of low oxaloacetate concentration. The oxaloacetate concentration is low because of the endergonic nature of the preceding malate dehydrogenase reaction.

9.

(a) For the reaction succinate + FAD → fumarate + $FADH_2$, succinate is the electron donor and FAD is the electron acceptor. $\mathscr{E}^{\circ\prime}$ for the succinate → fumarate half-reaction is 0.031 V (Table 14-5). Therefore,

$$\Delta \mathscr{E}^{\circ\prime} = \mathscr{E}^{\circ\prime}_{FAD} - \mathscr{E}^{\circ\prime}_{succinate} = 0.00 \text{ V} - 0.031 \text{ V} = -0.031 \text{ V}$$

$$\Delta G^{\circ\prime} = -n\mathscr{F}\Delta\mathscr{E}^{\circ\prime}$$
$$= -(2)(96,485 \text{ J·V}^{-1}\text{·mol}^{-1})(-0.031 \text{ V})$$
$$= 5982 \text{ J·mol}^{-1} = 6.0 \text{ kJ·mol}^{-1}$$

(b) When NAD^+ is the electron acceptor ($\mathscr{E}^{\circ\prime} = -0.315$ V),
$$\mathscr{E}^{\circ\prime} = -0.315 \text{ V} - 0.031 \text{ V} = -0.346 \text{ V}$$
$$\Delta G^{\circ\prime} = -(2)(96,485 \text{ J·V}^{-1}\text{·mol}^{-1})(-0.346 \text{ V})$$
$$= 66768 \text{ J·mol}^{-1} = 67 \text{ kJ·mol}^{-1}$$

Therefore, under standard biochemical conditions, the oxidation of succinate by FAD is slightly disfavored ($\Delta G^{\circ\prime} > 0$). However, the oxidation of succinate by NAD^+ is strongly disfavored ($\Delta G^{\circ\prime} \gg 0$).

(c) Succinate lacks the reducing power (has a reduction potential that is too high) to reduce NAD^+ to NADH but has sufficient reducing power to reduce FAD to $FADH_2$.

10. It is lost as CO_2 in the isocitrate dehydrogenase reaction in the first turn of the cycle (see Figure 17-2).

11. The rate of glucose oxidation would decrease because succinyl-CoA inhibits its own synthesis by the α-ketoglutarate dehydrogenase reaction, and it inhibits the citrate synthase reaction via feedback inhibition. The rapid accumulation of succinyl-CoA would also deplete the mitochondrial pool of CoA, thereby slowing the production of acetyl-CoA from glucose-derived pyruvate.

12. Glucose synthesis (gluconeogenesis) utilizes oxaloacetate; lipid biosynthesis utilizes acetyl-CoA derived from citrate; amino acid biosynthesis utilizes α-ketoglutarate and oxaloacetate, and porphyrin biosynthesis utilizes succinyl-CoA (Figure 17-17).

13. Acetyl-CoA activates pyruvate carboxylase. Increases in [acetyl-CoA] are indicative of an inability of the citric acid cycle to oxidize acetyl-CoA as fast as it is being produced (from glycolysis and fatty acid oxidation). Hence, activating pyruvate carboxylase, which adds more oxaloacetate to the pool of citric acid cycle intermediates will catalytically accelerate acetyl-CoA oxidation.

14. Two rounds of the glyoxylate cycle are necessary to yield two oxaloacetate, which give rise to one glucose by gluconeogenesis.

   glyoxylate cycle:
   $$4 \text{ Acetyl-CoA} + 4\,NAD^+ + 2\,FAD + 6\,H_2O \rightarrow$$
   $$2 \text{ oxaloacetate} + 4\,CoA + 4\,NADH + 2\,FADH_2 + 8\,H^+$$
   gluconeogenesis:
   $$2 \text{ Oxaloacetate} + 2\,GTP + 2\,ATP + 2\,NADH + 4\,H_2O + 6\,H^+ \rightarrow$$
   $$\text{glucose} + 2\,GDP + 2\,ADP + 2\,NAD^+ + 4\,P_i + 2\,CO_2$$
   net:
   $$4 \text{ Acetyl-CoA} + 2\,GTP + 2\,ATP + 2\,NAD^+ + 2\,FAD + 10\,H_2O \rightarrow$$
   $$\text{glucose} + 4\,CoA + 2\,CO_2 + 2\,GDP + 2\,ADP + 2\,NADH + 2\,FADH_2 + 2\,H^+ + 4\,P_i$$

15. *Isocitrate lyase*: isocitrate $\rightarrow$ succinate + glyoxylate
    Malate synthase: acetyl-CoA + glyoxylate $\rightarrow$ malate + CoA

16. Aspartate and α-ketoglutarate move from the mitochondrion to the glyoxysome, and succinate and glutamate move from the glyoxysome to the mitochondrion.

18

Electron Transport and Oxidative Phosphorylation

This chapter introduces you to the remarkable process by which cells harness the free energy of oxidation and use it to synthesize ATP. (You have already seen how ATP can be synthesized by the phosphorylation of ADP by a metabolite with a higher phosphoryl group-transfer potential.) The reduced cofactors generated in metabolic reactions, NADH and $FADH_2$, are reoxidized in the mitochondrion by a set of reactions in which electrons flow through a series of redox carriers, finally reducing oxygen to water. During electron transport, an electrochemical potential is developed across the inner mitochondrial membrane by the vectorial transfer of protons. This proton gradient is stable because the inner mitochondrial membrane is impermeable to ions. The free energy of the electrochemical proton gradient is utilized by an ATP synthase to catalyze the endergonic reaction $ADP + P_i \rightarrow ATP$. The coupling of the electrochemical gradient to ATP synthesis is described by the chemiosmotic hypothesis, which is supported by considerable evidence.

Essential Concepts

1. The complete oxidation of glucose carbons by glycolysis and the citric acid cycle can be written as

$$C_6H_{12}O_6 + 6\,H_2O \rightarrow 6\,CO_2 + 24\,H^+ + 24\,e^-$$

The reducing equivalents (electrons) are captured in the form of reduced coenzymes (NADH and $FADH_2$), which eventually transfer the electrons to molecular oxygen:

$$6\,O_2 + 24\,H^+ + 24\,e^- \rightarrow 12\,H_2O$$

This process regenerates NAD^+ and FAD and generates a proton concentration gradient across the inner mitochondrial membrane, whose dissipation provides the free energy for ATP synthesis. This process is known as oxidative phosphorylation.

The Mitochondrion

2. The mitochondrion is surrounded by a relatively porous outer membrane. An inner membrane is folded to form cristae and encloses the gel-like matrix, which contains the enzymes of the citric acid cycle and fatty acid oxidation. The matrix also contains genetic machinery (DNA, RNA, and ribosomes), reflecting the bacterial origin of this organelle. The proteins involved in electron transport and oxidative phosphorylation are located in the

inner mitochondrial membrane. The inner and outer membranes also contain proteins that mediate the transport of ions and metabolites.

3. NADH produced in the cytosol as a result of glycolysis must enter the mitochondrion to be aerobically oxidized. There are two shuttles for NADH.
    (a) The malate–aspartate shuttle allows NADH to be indirectly transported into the mitochondrion by reducing oxaloacetate to malate in the cytosol and transporting it into the mitochondrion, where it is reoxidized to produce oxaloacetate and NADH. The oxaloacetate is converted by transamination to aspartate and transported out again.
    (b) The glycerophosphate shuttle in insects first reduces cytosolic dihydroxyacetone phosphate to 3-phosphoglycerate and NAD^+. The 3-phosphoglycerate is oxidized by an inner mitochondrial membrane enzyme, flavoprotein dehydrogenase, which introduces electrons directly into the electron-transport pathway.

4. Most of the ATP generated in the mitochondria is used in the cytosol. The ADP–ATP translocator exports ATP out of the matrix while importing ADP. ATP has one more negative charge than ADP, so transport is electrogenic. Transport is driven by the electrochemical potential of the proton concentration gradient (positive outside). The proton gradient also favors the transport of P_i into the matrix by a P_i–H^+ symport system.

Electron Transport

5. In an electron transfer reaction, electrons flow from a substance with a lower reduction potential to a substance with a higher reduction potential. The standard reduction potential, $E^{\circ\prime}$, is a measure of a substance's affinity for electrons. For a redox reaction, $\Delta \mathscr{E}^{\circ\prime} = \mathscr{E}^{\circ\prime}_{(e-\text{acceptor})} - \mathscr{E}^{\circ\prime}_{(e-\text{donor})}$. When $\Delta \mathscr{E}^{\circ\prime}$ is positive, the reaction is spontaneous, since $\Delta G^{\circ\prime} = -n \mathscr{F} \Delta \mathscr{E}^{\circ\prime}$, where n is the number of electrons transported and F is the faraday (96,485 $J \cdot V^{-1} \cdot mol^{-1}$). The transfer of electrons from NADH to O_2 ($\Delta \mathscr{E}^{\circ\prime} = 1.13$ V and $\Delta G^{\circ\prime} = -218$ $kJ \cdot mol^{-1}$) provides enough free energy to synthesize three ATP molecules.

6. Four large protein complexes in the inner mitochondrial membrane are involved in transferring electrons from reduced coenzymes to O_2. Complexes I and II transfer electrons to the lipid-soluble electron carrier ubiquinone (coenzyme Q or CoQ), which transfers electrons to Complex III. From there, electrons pass to cytochrome *c,* a peripheral membrane protein with a heme prosthetic group, which transfers electrons to Complex IV. The reactions of Complexes I–IV are as follows:

(I)   NADH + CoQ (*ox*) → NAD^+ + CoQ (*red*)
       $\Delta \mathscr{E}^{\circ\prime} = 0.360$ V and $\Delta G^{\circ\prime} = -69.5$ $kJ \cdot mol^{-1}$

(II) $FADH_2$ + CoQ (*ox*) → FAD + CoQ (*red*)
       $\Delta \mathscr{E}^{\circ\prime} = 0.085$ V and $\Delta G^{\circ\prime} = -16.4$ $kJ \cdot mol^{-1}$

(III) CoQ (*red*) + cytochrome *c* (*ox*) → CoQ (*ox*) + cytochrome *c* (*red*)
       $\Delta \mathscr{E}^{\circ\prime} = 0.190$ V and $\Delta G^{\circ\prime} = -36.7$ $kJ \cdot mol^{-1}$

(IV) Cytochrome *c* (*red*) + ½ O_2 → cytochrome *c* (*ox*) + H_2O
       $\Delta \mathscr{E}^{\circ\prime} = 0.580$ V and $\Delta G^{\circ\prime} = -112$ $kJ \cdot mol^{-1}$

7. Complex I is an enormous protein complex containing flavin mononucleotide (FMN, which is FAD minus its AMP group) and multiple iron–sulfur clusters (which are one-electron carriers). The two electrons donated by NADH are transferred through these redox-active prosthetic groups and then to CoQ. As electrons are transferred, four protons are translocated from the matrix to the intermembrane space via a proton wire. Proton movement most likely occurs by protein conformational changes similar to those in bacteriorhodopsin.

8. Complex II, which contains the citric acid cycle enzyme succinate dehydrogenase, transfers electrons from succinate to FAD and then to CoQ. No protons are translocated by Complex II, which serves mainly to feed electrons into the electron transport chain.

9. Complex III (cytochrome bc_1 or cytochrome c reductase) contains two b-type cytochromes, cytochrome c_1, and an iron–sulfur protein, which contains a [2Fe–2S] cluster. Electron flow from CoQ through Complex III follows a bifurcated cyclic pathway known as the Q cycle. In the first round of the Q cycle, fully reduced ubiquinone (ubiquinol; QH_2) donates one electron to the iron–sulfur protein, which then transfers it to cytochrome c_1 and then to cytochrome c. This one-electron donation yields the ubisemiquinone anion (Q·⁻), which donates its remaining electron to the low potential cytochrome b (b_L), and then to the high potential cytochrome b (b_H). The resulting ubiquinone diffuses to the other side of the membrane, where it accepts the electron from b_L to reform Q·⁻. A second round of electron transfers completes the cycle: Another QH_2 donates its electrons, one to the iron–sulfur protein and one to cytochrome b_L. The net result is that two electrons are transferred, one at a time, to two cytochrome c molecules, and four protons are transferred from the matrix to the intermembrane space, two from each QH_2 that participates in the Q cycle.

10. Cytochrome c shuttles electrons between Complexes III and IV. Cytochrome c is a small water-soluble protein whose heme group is largely buried in a crevice surrounded by a ring of Lys residues. Both cytochrome c_1 and cytochrome c oxidase (Complex IV) have a corresponding patch of negatively charged amino acid residues to facilitate cytochrome c binding and electron transfer.

11. Complex IV (cytochrome c oxidase) has four redox centers [cytochrome a, cytochrome a_3, Cu_A (which contains two Cu ions), and Cu_B], and it carries out the following reaction:

$$4 \text{ Cytochrome } c \text{ (Fe}^{2+}) + 4 \text{ H}^+ + O_2 \rightarrow 4 \text{ cytochrome } c \text{ (Fe}^{3+}) + 2 \text{ H}_2\text{O}$$

O_2 reduction takes place at the cytochrome a_3–Cu_B binuclear complex, which mediates four one-electron transfer reactions. Four protons are consumed in the production of H_2O, and four additional proteins are pumped, most likely via a proton wire, from the matrix to the intermembrane space (two for each pair of electrons that enter the electron-transport chain).

Oxidative Phosphorylation

12. ATP synthase (Complex V) phosphorylates ADP by a mechanism driven by the free energy of electron transport, which is conserved in the formation of an electrochemical proton gradient across the inner mitochondrial membrane. The two processes are coupled as described by the chemiosmotic hypothesis. Four observations support this hypothesis:
    (a) Mitochondrial ATP formation requires an intact inner membrane.
    (b) The inner membrane is impermeable to ions, so an electrochemical gradient across the membrane can be sustained.
    (c) Electron transport pumps protons out of the mitochondrion to create a measurable electrochemical gradient.
    (d) Agents that increase the permeability of the inner mitochondrial membrane to protons inhibit ATP synthesis but not electron transport.

13. The protonmotive force results from the difference in concentration of protons (pH) in the matrix and the intermembrane space and from the difference in charge (membrane potential, $\Delta\Psi$) across the membrane. Thus, $\Delta G = 2.3\,RT\,[\text{pH}(side\ 1) - \text{pH}(side\ 2)] + Z\mathcal{F}\Delta\Psi$, where Z is the charge of the proton. $\Delta\Psi$ is positive when a proton is transported from negative to positive, or against its potential. Thus, pumping protons out of the matrix (against the gradient) is an endergonic process, whereas transporting them back in (with the gradient) is an exergonic process. About three protons are needed to supply sufficient energy to synthesize one ATP from ADP + P_i.

14. ATP synthase, also called F_1F_0-ATPase, has two functional units. The F_0 component includes the transmembrane proton channel, which, depending on the species, is a ring of 9 to 14 c subunits that rotates within the plane of the membrane such that a proton is picked up from the intermembrane space, completes about one rotation, and is released into the matrix. The F_0 component also includes a stator assembly that holds the F_1 component in place.

15. The F_1 component is a water-soluble protein of subunit composition $\alpha_3\beta_3\gamma\delta\varepsilon$ that associates with the membrane via F_0 to form a lollipoplike structure. The 3 α and 3 β subunits of F_1 form a structure of alternating α and β subunits with pseudo-threefold rotational symmetry. The elongated γ subunit, which also contacts F_0, rotates within the center of the $\alpha_3\beta_3$ ring. As the proton-translocating c-ring rotates, the γ subunit also rotates.

16. The binding change mechanism describes ATP synthesis in terms of three processes:
    (a) Translocation of protons carried out by F_0.
    (b) Formation of the phosphoanhydride bond of ATP, catalyzed by F_1.
    (c) Interaction of F_0 and F_1 to couple the dissipation of the proton gradient to formation of the phosphoanhydride bond.

17. According to the binding change mechanism, ATP is synthesized as each $\alpha\beta$ protomer shifts though three conformations in sequence. The three possible conformations are called open (O), loose (L), and tight (T). ADP and P_i bind to a protomer with the L conformation and are converted to ATP when the conformation shifts to the T state. The free energy of

the proton concentration gradient converts the T state to the O state (this is the rate-limiting step), thereby releasing ATP. This three-step mechanism is consistent with the pseudo-threefold axis of rotation. The γ subunit and c-ring rotate with respect to the $\alpha_3\beta_3$ assembly, and the geometric relationship of individual $\alpha\beta$ protomers to the γ subunit dictates their conformational state. Protons equal in number to the number of c subunits in the c-ring are required for one complete rotation, promoting three conformational shifts, each of which synthesizes one ATP.

18.   The ratio of the amount of ATP produced to the amount of substrate oxidized (measured as oxygen consumed) is called the P/O ratio. [The P/O ratio refers to atomic oxygen, O, rather than molecular oxygen, O_2, because each substrate (NADH or $FADH_2$) transfers two electrons, not four.] Depending on where a substrate's electrons enter the electron-transport chain, the P/O ratio is ~2.5, ~1.5, or ~1. For example, the two electrons transferred from NADH through Complexes I, III, and IV pump 10 protons, which would yield ~ 3 ATP, whereas the two electrons transferred from $FADH_2$ through Complexes II, III, and IV pump 6 protons, which would yield ~2 ATP. However, these numbers are lower due to leakage of protons and P_i transport. The complete oxidation of glucose therefore yields ~32 ATP. The P/O ratio is not necessarily a whole number, because protons are contributed to the gradient by more than one process and some protons leak back into the matrix.

19.   Electron transport and oxidative phosphorylation are normally strongly coupled processes; that is, neither process occurs in the absence of the other. This is because, if the rate of electron transport were to outpace the rate of ATP synthesis, the proton gradient would build up to the level that it would resist additional proton pumping by Complexes I, III, and IV and hence the rate of electron transport would be slowed. However, when uncoupling agents, which dissipate the proton gradient, are added to respiring mitochondria, electron transport proceeds unchecked while ATP synthesis stops. The free energy of electron transport is then redirected from ATP synthesis to generate heat. 2,4-Dinitrophenol is an uncoupling agent because it carries protons through the membrane from the intermembrane space to the matrix, thereby providing a route for dissipation of the gradient that bypasses F_0.

Control of Oxidative Metabolism

20.   Electron transfer from NADH to cytochrome c is nearly at equilibrium. In contrast, the cytochrome c oxidase reaction is irreversible and hence its rate depends on the concentration of its substrate, reduced cytochrome c. Increased NADH concentrations and decreased ATP concentrations lead to the production of more reduced cytochrome c and hence to increased electron transfer rates. Thus, the overall rate of oxidative phosphorylation depends on the ratios $[NADH]/[NAD^+]$ and $[ATP]/[ADP][P_i]$, which in turn may depend on the activities of the respective mitochondrial transporters.

21.   The coordinated control of oxidative metabolism centers on several key enzymes: hexokinase (HK), phosphofructokinase (PFK), pyruvate kinase (PK), pyruvate dehydrogenase (PDH), citrate synthase (CS), isocitrate dehydrogenase (IDH) and α-ketoglutarate dehydrogenase (KDH). High levels of ATP inhibit PFK and PK while high

[NADH]/[NAD$^+$] ratios inhibit PDH, IDH, and KDH. Citrate inhibits both PFK and CS. The need for ATP, represented by high concentrations of either AMP or ADP, activates PFK, PK, PDH, and IDH, while Ca^{2+} stimulates PDH, IDH, and KDH.

22. Anaerobic glycolysis produces 2 ATP per glucose consumed, whereas oxidative metabolism generates 32 ATP per glucose, a 16-fold increase. However, there are several drawbacks of oxygen-based metabolism. Many organisms depend on oxidative metabolism and would perish without a steady supply of oxygen. Reactive oxygen species generated by incomplete oxygen reduction are potentially dangerous.

23. Reactive oxygen species include the superoxide radical O$_2\cdot^-$ (produced by the reaction O$_2$ + e^- → O$_2\cdot^-$), which is a precursor of even more powerful oxidizing species such as HO$_2\cdot$ and ·OH. These free radicals readily extract electrons from other substances, creating a chain reaction. Neurodegenerative conditions such as Parkinson's, Alzheimer's, and Huntington's diseases are associated with mitochondrial oxidative damage. Free radical reactions arising from normal oxidative metabolism appear to be partially responsible for the aging process.

24. Antioxidants limit oxidative damage by destroying free radicals. Superoxide dismutase (SOD) catalyzes the production of water and hydrogen peroxide from superoxide:

$$2\ O_2\cdot^- + 2\ H^+ \rightarrow H_2O_2 + O_2$$

This enzyme electrostatically guides its substrate to the active site to catalyze a reaction near the diffusion-controlled limit. Mutations in Cu,Zn SOD are associated with amyotrophic lateral sclerosis (Lou Gehrig's disease).

25. Catalase and glutathione peroxidase degrade hydroperoxides. Some types of glutathione peroxidase require selenium, so Se also appears to be an antioxidant.

Questions

The Mitochondrion

1. Draw a cross-section of a mitochondrion and label the following structural features:

    Outer membrane (OM)   Inner membrane (IM)
    Matrix (M)   Intermembrane space (IMSP)
    ATP synthase complex (ASC)   Direction of proton flux
    ATP and P$_i$ transporters (T)   Cristae (CR)

2. Match the following enzyme or other molecule with its location:

    ___ Pyruvate dehydrogenase
    ___ 3-Phosphoglycerate dehydrogenase   A. Cytosol
    ___ Flavoprotein dehydrogenase

_____ Malate dehydrogenase   B. Mitochondrial outer membrane
_____ Cytochrome c
_____ Cytochrome c_1 C. Mitochondrial inner membrane
_____ Fatty acid oxidation enzymes
_____ Mitochondrial DNA   D. Mitochondrial intermembrane space
_____ ADP–ATP translocator
_____ Mitochondrial porin   E. Mitochondrial matrix

3.   About half the volume of the mitochondrial matrix is water, and the rest is protein. If a single protein of molecular mass 40,000 were as concentrated, what would be its molar concentration? Assume the protein's density is 1.37 $g \cdot mL^{-1}$.

4.   Oxidative phosphorylation requires the transfer of electrons donated by NADH. (a) Is NADH imported directly into the mitochondria? Explain. (b) Describe two import mechanisms that transfer cytosolic electrons from NADH into the mitochondrion. (c) Why is it important to maintain a relatively constant level of cytosolic NAD^+?

Electron Transport

5.   Which reactions of the citric acid cycle donate electron pairs to the mitochondrial electron-transport chain?

6.   The half-cell reduction potential is provided by the Nernst equation (Equation 14-8):

$$\mathscr{E}_A = \mathscr{E}°_A - \frac{RT}{n\mathscr{F}} \ln \left(\frac{[A_{red}]}{[A_{ox}^{n+}]} \right)$$

(a) On the graph below, plot the reduction potentials for the $FADH_2$/FAD half-cell ($E^{°\prime} = -0.219$ V) when the [$FADH_2$]/[FAD] ratios are 100, 10, 5, 2, 1, 0.5, 0.2, 0.1, and 0.01 at 25°C versus the percent reduction.

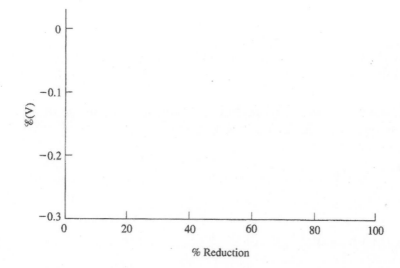

(b) Using the same [Reduced]/[Oxidized] ratios, plot the reduction potentials of cytochrome c ($E^{\circ\prime} = 0.235$ V).

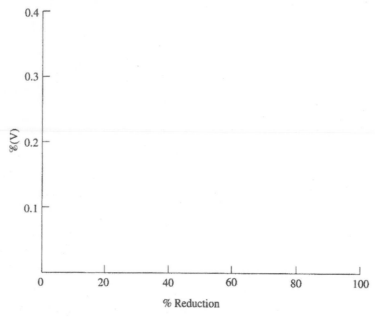

(c) What is ΔE for the oxidation of $FADH_2$ by cytochrome c when the $[FADH_2]/[FAD]$ ratio is 10 and the [cytochrome c (Fe^{2+})]/[cytochrome c (Fe^{3+})] ratio is 0.1?

7. Inhibitors of electron transport have been used to determine the order of electron carriers. What would be the expected redox states of cytochromes a, b_L, and c when (a) stigmatellin, (b) antimycin A, or (c) rotenone is added to succinate-driven respiring mitochondria?

8. What is the most abundant type of redox carrier in Complex I?

9. Which mitochondrial electron carriers are potentially proton carriers? Which are more abundant—proton carriers or electron carriers? What does this suggest about the mechanism of transmembrane proton transport?

10. What amino acid residues would you expect to find at the cytochrome c–binding sites of Complex III and Complex IV?

11. Which redox group(s) in Complex IV accepts electrons from cytochrome c? Which redox group(s) binds oxygen during its four-electron reduction?

Oxidative Phosphorylation

12. What key observations support the chemiosmotic hypothesis?

13. Can a pH gradient exist without $\Delta\Psi$? Can $\Delta\Psi$ exist without a ΔpH?

14. (a) What is a P/O ratio? (b) What happens to oxygen consumption when electron donors and inorganic phosphate are present in a suspension of mitochondria in the absence of ADP? (c) What happens when ADP is added?

15. (a) Valinomycin, an antibiotic ionophore, allows the free passage of only K^+ ions across a membrane. If K^+ and valinomycin are added to respiring, fully coupled ATP-synthesizing mitochondria, what happens to the pH gradient and the $\Delta\Psi$? (b) Nigericin, another ionophore, exchanges one K^+ for one H^+. How does this affect ATP synthesis and electron transport in mitochondria? (c) Gramicidin allows the free passage of many small molecules and ions across the membrane. What happens to ATP production and electron transport in the presence of gramicidin?

16. The free energy-requiring step in the synthesis of ATP is not the formation of ATP from ADP and P_i ($\Delta G \approx 0$), but the release of tightly bound ATP. Explain why this is not inconsistent with the $+30.5$ kJ·mol^{-1} free energy of formation of ATP in solution.

17. Dicyclohexylcarbodiimide (DCCD) reacts with Asp and Glu residues in the c subunits of F_0 and blocks ATP synthase activity. What happens to the rate of electron transport when DCCD is added to actively respiring mitochondria?

18. What does an H^+/P ratio measure? Why would it be impractical to determine an H^+/P ratio?

Control of Oxidative Metabolism

19. The conversion of glucose to 2 lactate has $\Delta G^{\circ\prime} = -196$ kJ·mol^{-1}. The complete oxidation of glucose to 6 CO_2 has $\Delta G^{\circ\prime} = -2823$ kJ·mol^{-1}. Compare the efficiencies of ATP synthesis by each of these processes under standard conditions.

20. On an average day, an adult dissipates about 7000 kJ of free energy. Assuming that this occurs under standard conditions, (a) how many moles of ATP must be hydrolyzed to provide this quantity of free energy? (b) What is the mass of this quantity of ATP? (c) If the amount of ATP in an adult is about 0.1 mole, how many times per day, on average, is a molecule of ADP recycled? The molecular mass of ATP is 507.

21. What is the irreversible step in electron transport and how is its rate controlled?

22. Cytochrome P450 (Section 12-4D) catalyzes a reaction in which two electrons supplied by NADPH reduce the heme Fe atom so that it can then reduce O_2 preparatory to the hydroxylation of a substrate molecule. Substrate binding to the enzyme displaces a water molecule that forms a ligand to the heme iron atom. This changes the reduction potential of the Fe from -0.300 V to -0.170 V. Why is this change necessary for efficient catalysis?

23. Rats that are fed a "cafeteria" diet (in which food is always available) tend to die sooner than rats whose dietary intake is limited. Propose an explanation for this observation.

Answers to Questions

1.

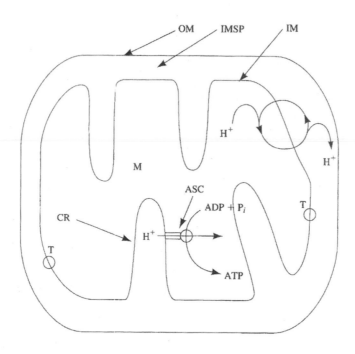

2.

    E Pyruvate dehydrogenase

    A 3-Phosphoglycerate dehydrogenase

    C Flavoprotein dehydrogenase

    E, A   Malate dehydrogenase

    D Cytochrome c

    C Cytochrome c_1

    E Fatty acid oxidation enzymes

    E Mitochondrial DNA

    C ADP–ATP translocator

    B Mitochondrial porin

3.   The protein's molar concentration would be

$$\frac{1}{40\,000 \text{ g·mol}^{-1}} \times \frac{1.37 \text{ g}}{\text{mL}} \times \frac{1000 \text{ mL}}{\text{L}} \times 50\%$$

$$= 0.0171 \text{ mol·L}^{-1}$$

$$= 17.1 \text{ mM}$$

4.

(a) The negatively charged phosphate groups of NADH (Figure 11-4) prevent its diffusion across the inner mitochondrial membrane, and there are no NADH transport proteins to facilitate its transport.

(b) The malate–aspartate shuttle allows the indirect import of NADH reducing equivalents (Figure 16-20). The reduction of oxaloacetate to malate by NADH followed by the facilitated transport of malate across the inner mitochondrial membrane yields NADH in the mitochondrial matrix when the malate is reoxidized to oxaloacetate. Oxaloacetate then returns to the cytosol by being converted to aspartate, for which there is a transporter. In insects, the glycerophosphate shuttle utilizes NADH to convert dihydroxyacetone phosphate to 3-phosphoglycerate by a flavoprotein dehydrogenase that donates electrons to the electron-transport chain in a manner similar to succinate dehydrogenase (Fig. 18-5).

(c) Cytosolic NAD^+ is required for the glyceraldehyde-3-phosphate dehydrogenase reaction of glycolysis. Limited $[NAD^+]$ would shut down glycolysis.

5. Isocitrate dehydrogenase, α-ketoglutarate dehydrogenase, and malate dehydrogenase produce NADH, which transfers its electrons to Complex I. Succinate dehydrogenase, whose FAD group is reduced in the oxidation of succinate to fumarate, is a component of Complex II.

6.
   (a) The percent reduction is

$$\frac{[\text{Reduced}]}{[\text{Reduced}] + [\text{Oxidized}]} \times 100 \quad \text{or}$$

$$\frac{[\text{FADH}_2]}{[\text{FADH}_2] + [\text{FAD}]} \times 100$$

For the $FADH_2/FAD$ half-cell, the Nernst equation is

$$\mathscr{E} = (-0.219V) - \left(\frac{(8.3145 J \cdot K^{-1} \cdot mol^{-1})(298K)}{(2)(96,486 J \cdot V^{-1} \cdot mol^{-1})}\right) \ln\left(\frac{[\text{FADH}_2]}{[\text{FAD}]}\right)$$

$$\mathscr{E} = (-0.219V) - (0.0218V) \ln\left(\frac{[\text{FADH}_2]}{[\text{FAD}]}\right)$$

$[\text{FADH}_2]/[\text{FAD}]$	% Reduction	E (V)
100	99.0	− 0.278
10	90.9	− 0.248
5	83.3	− 0.240
2	66.7	− 0.228
1	50.0	− 0.219
0.5	33.3	− 0.210
0.2	16.7	− 0.198
0.1	9.1	− 0.190
0.01	0.99	− 0.160

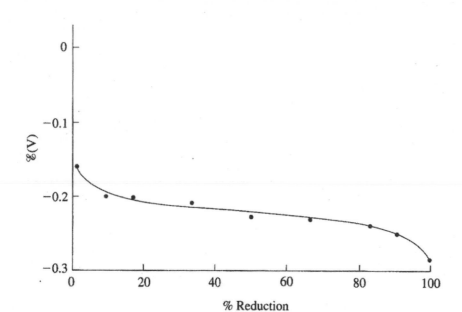

(b) For cytochrome c,

$$\mathscr{E} = (0.235 \text{ V}) - \left(\frac{(8.3145 \text{J} \cdot \text{K}^{-1} \cdot \text{mol}^{-1})(298\text{K})}{(1)(96,485 \text{ J} \cdot \text{V}^{-1} \cdot \text{mol}^{-1})} \right) \ln \left(\frac{\text{cyto } c \, [\text{Fe}^{2+}]}{\text{cyto } c \, [\text{Fe}^{3+}]} \right)$$

$$\mathscr{E} = (0.235 \text{ V}) - (0.0275 \text{ V}) \ln \left(\frac{\text{cyto } c \, [\text{Fe}^{2+}]}{\text{cyto } c \, [\text{Fe}^{3+}]} \right)$$

[cyto c (Fe^{2+})]/[cyto c (Fe^{3+})]	% Reduction	E (V)
100	99.0	0.117
10	90.9	0.176
5	83.3	0.194
2	66.7	0.217
1	50.0	0.235
0.5	33.3	0.253
0.2	16.7	0.276
0.1	9.1	0.294
0.01	0.99	0.353

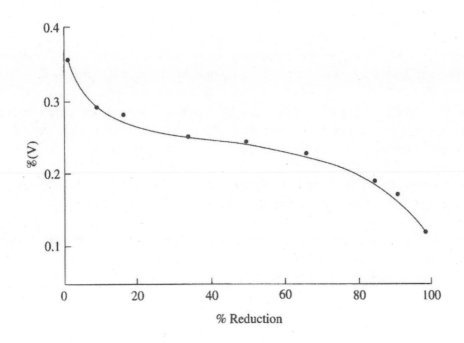

(c)   When $[FADH_2]/[FAD] = 10$, $\mathscr{E}_{FAD} = -0.248$ V
When $[cyto\ c\ (Fe^{2+})]/[cyto\ c\ (Fe^{3+})] = 0.1$, $\mathscr{E}_{cyto\ c} = 0.294$ V
According to Equation 13-10,

$$
\begin{aligned}
\Delta\mathscr{E} &= \mathscr{E}_{(e-\ acceptor)} - \mathscr{E}_{(e-\ donor)} \\
&= \mathscr{E}_{cyto\ c} - \mathscr{E}_{FAD} \\
&= 0.294\ V - (-0.248\ V) \\
&= 0.542\ V
\end{aligned}
$$

7.

(a)   Stigmatellin inhibits the electron flow from $CoQH_2$ to Complex III. Hence cytochrome a (a component of Complex IV) cannot obtain electrons and would be largely in its oxidized state. Similarly, cytochrome b_L (part of Complex II) and cytochrome c (which links Complexes III and IV) would be largely in their oxidized states.

(b)   Antimycin A blocks electron transport in Complex III from heme b_H to CoQ or $CoQ\cdot^-$. Thus, cytochrome b_L would be reduced since it could not pass its electrons on to cytochrome b_H. However, cytochromes a and c would be oxidized since they are both downstream of the block.

(c)   Rotenone blocks electron transport in Complex I. However, since electrons are being introduced into the electron-transport chain at Complex II, electron transport can proceed all the way to O_2. Therefore, all three cytochromes would be predominantly in their reduced forms.

8.   The most abundant type of electron carrier in Complex I is iron–sulfur clusters.

9. Only two electron carriers in the mitochondrial electron-transport chain can also carry protons, FMN and ubiquinone (CoQ). The paucity of proton carriers compared to electron carriers suggests that transmembrane proton transport involves protein-mediated proton pumping (a proton wire) rather than only direct transport by the proton carriers.

10. Since cytochrome c contains positively charged Lys residues around its heme crevice, its redox partners must have negatively charged residues such as Glu and Asp at their cytochrome c–binding sites.

11. The Cu_A center accepts the first electron from cytochrome c. Oxygen binds to the partially reduced Fe(II)–Cu(I) form of the cytochrome a_3–Cu_B binuclear complex.

12. Key observations that support the chemiosmotic hypothesis are: (a) Oxidative phosphorylation requires an intact inner membrane; (b) the inner mitochondrial membrane is impermeable to ions such as H^+, OH^-, K^+, and Cl^-; (c) electron transport results in the transport of H^+ out of the mitochondrial matrix; and (d) compounds that increase proton permeability across the membrane uncouple phosphorylation from electron transport and inhibit ATP synthesis.

13. A pH gradient can exist without $\Delta\Psi$. If a counterion such as Cl^- moved in the same direction as the H^+, or if K^+ moved in the opposite direction, a $[H^+]$ gradient could form without altering the net charge on either side of the membrane. Similarly, the transmembrane movement of ions other than H^+ could generate $\Delta\Psi$ without generating a proton concentration gradient.

14.
   (a) The P/O ratio is a measure of how many ATPs are synthesized for every O atom reduced.
   (b) Oxygen consumption stops when the production of ATP from ADP cannot be carried out. This is because electron transport and proton translocation cannot take place independently. Hence, when the proton gradient has built up to the point that electron transport has insufficient free energy to translocate additional protons across the inner mitochondrial membrane, electron transport must stop.
   (c) When ADP is added, ATP can be synthesized. O_2 consumption then resumes because the dissipation of the proton concentration gradient by ATP synthase reduces the free energy of proton translocation across the inner mitochondrial membrane to the point that electron transport can continue.

15.
   (a) The pH gradient would remain unchanged. However, the $\Delta\Psi$ would collapse because K^+ would equilibrate across the membrane in response to the $\Delta\Psi$.
   (b) Nigericin would dissipate the proton gradient by exchanging protons for K^+ and hence arrest ATP synthesis. The rate of electron transport would increase because there would be no buildup of a proton gradient to hold electron transport in check. However, $\Delta\Psi$ would remain intact because there would be no net change in charge across the membrane.

(c) Gramicidin would cause the collapse of both the $\Delta\Psi$ and the pH gradient, so ATP synthesis would stop but electron transport would accelerate.

16. In solution, the ATP-forming reaction is $ADP + P_i \rightarrow ATP + H_2O$. Recall that an enzymatic reaction may progress through several steps, but it cannot alter the ΔG for a reaction. For the ATP synthase–catalyzed reaction, the binding of substrates and release of products are just two of the steps in the overall process. Letting E represent the enzyme, the reaction can be written as

$$ADP + P_i + E \rightarrow ADP{\cdot}P_i{\cdot}E \rightarrow ATP{\cdot}H_2O{\cdot}E \rightarrow ATP + H_2O + E$$

where each step has a different ΔG value. However, the overall reaction remains

$$ADP + P_i + E \rightarrow ATP + H_2O + E$$

and hence the overall free energy change is still 30.5 kJ·mol^{-1}.

17. The reaction of DCCD with ATP synthase inhibits ATP formation by reacting with Asp 61, preventing its protonation and deprotonation, actions required for inducing the rotation of the c subunit. The proton gradient cannot be dissipated and therefore builds up to the point that it arrests further electron transport.

18. The H^+/P ratio is a measure of the number of protons transported across the inner membrane for each ATP molecule synthesized. This could be difficult to determine since the measurement of pH would not take into account $\Delta\Psi$, which also contributes to the protonmotive force. In addition, the measurement would be highly sensitive to cytosolic pH changes that were not involved in mitochondrial electron transport.

19. $\Delta G^{\circ\prime} = +30.5$ kJ·mol^{-1} for the synthesis of ATP from $ADP + P_i$. The conversion of glucose to lactate by glycolysis is accompanied by the synthesis of 2 ATP. Thus, the efficiency of ATP production under standard conditions is [(2 × 30.5 kJ·mol^{-1})/(196 kJ·mol^{-1})] × 100 = 31%. When glucose is completely oxidized, the yield is 32 ATP. The efficiency of this process is [(32 × 30.5 kJ·mol^{-1})/(2823 kJ·mol^{-1})] × 100 = 35%. Therefore, not only does oxidative phosphorylation yield more ATP than glycolysis, but at least under standard conditions, it does so with greater efficiency.

20.
   (a) The energy is supplied by 7000 kJ/30.5 kJ·mol^{-1} = 230 moles of ATP.
   (b) 230 moles × 507 g·mol^{-1} = 11700 g = 117 kg
   (c) Since 230 moles are needed each day, 0.1 mol of ADP must recycled 230/0.1 = 2300 times.

21. The irreversible step in electron transport is the formation of water from oxygen. The rate of cytochrome c oxidase is controlled by the ratio of reduced to oxidized cytochrome c, which is in turn controlled by the [NADH]/[NAD$^+$] and [ATP]/[ADP][P$_i$] ratios.

22. The change in redox potential of the Fe atom promotes its ability to accept the electrons from NADPH ($\mathscr{E}°' = -0.320$). Recall that electrons flow spontaneously from a substance with a more negative redox potential to a substance with a more positive redox potential.

23. Rats that consume large quantities of metabolic fuels have a higher rate of oxidative metabolism and hence generate more oxygen radicals than rats that consume less food and have lower rates of oxidative metabolism. The cumulative oxidative damage would be greater in the cafeteria-fed rats, which therefore die sooner.

19

Photosynthesis

In the preceding chapters, you learned that free energy is derived from reduced foodstuffs such as glucose and that the energy produced from catabolic pathways is used to generate both ATP and NAD(P)H. Photosynthesis, an ancient and important process, allows energy to be harvested directly from the most abundant and renewable source, the sun. Photosynthesis is a light-driven process in which carbon dioxide is "fixed" to produce carbohydrates. This occurs in two phases: (1) The light reactions (requiring light) produce ATP and NADPH, and (2) the dark reactions (not requiring light) use ATP and NADPH to synthesize carbohydrates. This chapter describes how different pigments (e.g., chlorophylls in plants and bacteria) efficiently capture light energy and redistribute it to specific reaction centers. Purple photosynthetic bacteria contain one photosystem that recycles its electrons, whereas higher plants have two photosystems that use water as a source of electrons to reduce NADPH. The oxidation of water in higher plants generates O_2 as a by-product of photosynthesis. As in mitochondria, the topology of chloroplasts is central to the biochemistry of photosynthesis, starting with the light-driven reactions in the thylakoid membrane and finishing with the dark reactions in the stroma. The dark reactions occur via the Calvin cycle, a set of reactions that synthesize glyceraldehyde-3-phosphate from 3 CO_2. The chapter also discusses the control of the Calvin cycle along with a variant called the C_4 pathway.

Essential Concepts

1. Photosynthesis is divided into two processes:
    (a) In the light reactions, organisms capture light energy to synthesize ATP and generate reducing equivalents in the form of NADPH.
    (b) In the dark reactions, carbon dioxide is converted to carbohydrates using the ATP and NADPH generated in the light reactions. Although the dark reactions are not light-driven, they occur only when it is light and hence are better described as light-independent.

Chloroplasts

2. Plants differ from bacteria by providing a separate organelle for the photosynthetic machinery, the chloroplast. A chloroplast is enveloped by a highly permeable outer membrane and a nearly impermeable inner membrane. The inner membrane encloses the stroma, which contains the soluble enzymes of carbohydrate synthesis, and the thylakoid membrane, which is organized in stacks of pancake-like disks (grana) that enclose the thylakoid compartment and that are linked by unstacked stromal lamellae. The proteins that capture light energy and mediate electron-transport processes are embedded in the thylakoid membrane.

3. Various pigment molecules absorb light of different wavelengths. The principal photosynthetic pigment is chlorophyll, a cyclic tetrapyrrole that ligands a central Mg^{2+} ion. Photosynthetic organisms also contain other pigments, such as carotenoids, phycoerythrin, and phycocyanin, which together with chlorophyll absorb most of the visible light in the solar spectrum.

4. Multiple pigment molecules are arranged in light-harvesting complexes (LHCs), which are proteins that act as antennae to gather light energy and redirect it to photosynthetic reaction centers, where the light energy is converted to chemical energy in the form of ATP and NADPH. The accessory pigments in the LHCs boost light absorption at wavelengths at which chlorophyll does not absorb strongly.

The Light Reactions

5. Photons propagate as discrete energy packets called quanta, whose energy, E, is given by Planck's law: $E = h\nu = hc/\lambda$, where h is Planck's constant (6.626×10^{-34} J·s) c is the speed of light (2.998×10^8 m·s^{-1}), λ is the wavelength of light, and ν is the frequency of the radiation.

6. When a molecule absorbs a photon, one of its electrons is promoted to a higher energy orbital. The excited electron can return to the ground state in several ways:
    (a) In internal conversion, electronic energy is converted to kinetic energy (heat).
    (b) Fluorescence results when the molecule emits a photon at a lower wavelength.
    (c) The excitation energy can be transferred to another molecule by exciton transfer (resonance energy transfer). This occurs in the transfer of light energy from LHCs to the photosynthetic reaction center.
    (d) The molecule may undergo photooxidation by transfer of an electron to another molecule. The excited chlorophyll at the reaction center transfers electrons in this manner.

7. The bacterial photosynthetic reaction center of *Rps. viridis* consists of a series of prosthetic groups arranged with nearly twofold symmetry: 2 closely associated bacteriochlorophyll *a* (BChl *a*) molecules known as the special pair, 2 bacteriopheophytin *a* (BPheo *a*; BChl *a* that lacks an Mg^{2+} ion), 2 additional BChl *a* molecules, a menaquinone, a ubiquinone, and an Fe(II) ion.

8. In purple photosynthetic bacteria, the special pair undergoes photooxidation virtually every time it absorbs a photon. The transferred electron is first passed to the BPheo *a* on the "right" side of the photosynthetic reaction center and then to the menaquinone and then the ubiquinone to yield a semiquinone radical anion (Q_B^-). A second photon absorption then transfers a second electron to yield Q_B^{2-}, which picks up two protons from the cytoplasm to form QH_2 and then exchanges with the membrane-bound pool of ubiquinone.

9. The electrons ejected from the special pair return to the photosynthetic reaction center via an electron-transport chain consisting of a cytochrome bc_1 complex and cytochrome c_2. Electrons flow follows a Q cycle in cytochrome bc_1, which translocates four protons to the

periplasmic space for every two electrons transferred. The free energy of the resulting transmembrane proton gradient drives ATP synthesis.

10. In plants and cyanobacteria, photooxidation occurs at two reaction centers, and electron transport is noncyclical. The path of electrons from water to NADPH is described by the Z-scheme, in which photosystem II (PSII) passes its electrons to the cytochrome b_6f complex via the mobile electron carrier ubiquinol (QH_2), and cytochrome b_6f then transfers these electrons to photosystem I (PSI) via the mobile Cu-containing protein plastocyanin. Since PSII and PSI are thereby "connected in series," the energy of each electron is boosted by two photon absorptions.

11. PSII includes the Mn-containing oxygen-evolving center (OEC), which cycles through five electronic states (S_0–S_4) in the conversion of H_2O to O_2 and is driven by four consecutive excitations of the PSII reaction center (called P680). The four electrons released from H_2O follow a path similar to that of the bacterial photosynthetic reaction center, eventually reaching the membrane plastoquinone pool.

12. Electron transport through the cytochrome b_6f complex (which resembles the mitochondrial Complex III) generates a transmembrane proton gradient via a Q cycle. Eight protons are translocated to the thylakoid lumen for the four electrons released from each H_2O. Plastocyanin, a peripheral membrane protein, has a Cu redox center than ferries one electron at a time from cytochrome b_6f to PSI.

13. PSI contains multiple pigments and redox groups, including chlorophylls, carotenoids, [4Fe–4S] clusters, and phylloquinone. Photooxidation of the PSI special pair (called P700) allows the electron received from plastocyanin to pass through a series a electron carriers in one of two routes:
    (a) In the noncyclic pathway, electrons flow through PSI to the [2Fe–2S]-containing one-electron carrier ferredoxin (Fd), which is a soluble stromal protein. Two Fd's deliver their electrons to ferredoxin–$NADP^+$ reductase, which thereupon carries out the two-electron reduction of $NADP^+$ to NADPH.
    (b) In the cyclic pathway, electrons return from PSI through cytochrome b_6f to the plastoquinone pool and thereby participate in the Q cycle. This pathway augments the proton gradient across the thylakoid membrane and hence contributes additional free energy for the synthesis of ATP but does not yield NADPH.

14. The free energy of the proton gradient is tapped by chloroplast CF_1CF_0-ATP synthase, which closely resembles the mitochondrial F_1F_0-ATPase. Approximately 12 protons enter the thylakoid lumen for each O_2 generated in noncyclic electron transport (4 H^+ from the OEC reaction and 8 H^+ from the Q cycle). The synthesis of ATP requires the transport of ~3 protons from the thylakoid lumen to the stroma.

The Dark Reactions

15. CO_2 is incorporated into carbohydrates by carboxylation of a 5-carbon sugar, ribulose-5-phosphate (R5P). The resulting 6-carbon compound is split into two molecules of 3-phosphoglycerate (3PG), which is then converted to glyceraldehyde-3-phosphate (GAP). Some of the GAP is diverted to carbohydrate synthesis, and the rest is converted back to Ru5P. This set of 13 reactions, called the Calvin cycle, has two stages:
    (a) In the production phase, 3 Ru5P react with 3 CO_2 to yield 6 GAP (for a net yield of 1 GAP from 3 CO_2), at a cost of 9 ATP and 6 NADPH.
    (b) In the recovery phase, the carbons of 5 GAP are shuffled via aldolase- and transketolase-catalyzed reactions to reform 3 Ru5P, without consuming ATP or NADPH.
    The GAP product of the cycle is then converted to glucose-1-phosphate, a precursor of sucrose and starch.

16. Ribulose bisphosphate carboxylase, which accounts for up to 50% of leaf proteins, catalyzes the carboxylation of ribulose-1,5-bisphosphate (RuBP). Enzymatic abstraction of a proton from RuBP generates an enediolate that attacks CO_2. H_2O then attacks the resulting β-keto acid to yield two 3PG.

17. The activity of RuBP carboxylase is controlled in several ways so that the dark reactions proceed only when the light reactions are able to provide the ATP and NADPH necessary to drive them:
    (a) RuBP carboxylase is most active at pH 8.0, which occurs when protons are pumped out of the stroma during the light reactions.
    (b) The Mg^{2+} that enters the stroma to compensate for the efflux of H^+ stimulates RuBP carboxylase.
    (c) Plants synthesize 2-carboxyarabinitol-1-phosphate, an inhibitor of RuBP carboxylase, only in the dark.

18. The Calvin cycle enzymes fructose bisphosphatase and sedoheptulose bisphosphatase are also activated by increases in pH, $[Mg^{2+}]$, and [NADPH]. The redox state of ferredoxin is sensed by a thiol-exchange cascade involving ferredoxin–thioredoxin reductase, thioredoxin, and disulfide groups on the bisphosphatases, so that the Calvin cycle enzymes are stimulated when Fd is reduced (i.e., when the light reactions are occurring).

19. RuBP carboxylase can also react with oxygen, which competes with CO_2 at the carboxylase active site. This process, called photorespiration, converts RuBP into 3PG and 2-phosphoglycolate, a two-carbon compound. A series of reactions in the chloroplast and peroxisome convert two 2-phosphoglycolate to two glycine, which are converted in the mitochondria to serine + CO_2. The serine is converted back to the Calvin cycle intermediate 3PG by reactions that require NADH and ATP. Thus, photorespiration consumes O_2 and produces CO_2, at the expense of ATP and NADH, thereby reversing the results of photosynthesis. All known RuBP carboxylases have this activity, which may protect chloroplasts from photoinactivation when high light intensity has greatly reduced the local CO_2 concentration.

20. The rate of photorespiration becomes significant on hot bright days when photosynthesis has depleted the level of CO_2 at the chloroplast and raised the concentration of O_2. However, C_4 plants such as corn and sugar cane prevent photorespiration by concentrating CO_2 at the chloroplast. They do so by using phosphoenolpyruvate (PEP) carboxylase to make oxaloacetate. Oxaloacetate is converted to malate in the mesophyll cells (which lack RuBP carboxylase) and transported to the bundle-sheath cells, where the Calvin cycle operates. There, malic enzyme cleaves malate into pyruvate and CO_2. The CO_2 is thereby delivered to RuBP carboxylase at a high enough concentration to essentially eliminate photorespiration. The pyruvate is transported back to the mesophyll cells and converted to PEP at the expense of two "high-energy" bonds. In the tropics, C_4 plants grow faster than so-called C_3 plants. In more temperate climates, where the rate of photorespiration is reduced, C_3 plants have an advantage because they require less energy to fix CO_2.

21. Many desert succulent plants conserve water by opening their stomata only at night to acquire CO_2. The CO_2 is converted to PEP at night and is released as CO_2 in the day, to be fixed by RuBP carboxylase. This process is called Crassulacean acid metabolism (CAM) because it was discovered in the family Crassulaceae.

Key equations

$$E = h\nu = \frac{hc}{\lambda}$$

Questions

1. Define the terms light reactions and dark reactions. Do the dark reactions occur in the dark? Explain.

Chloroplasts

2. Draw a cross-section of a chloroplast and indicate the locations of the following proteins and other structural features:

Outer membrane	Thylakoid lumen
Stromal lamella	Photosystem I
Inner membrane	Photosystem II
Intermembrane space	Cytochrome b_6f
Grana	CF_1CF_0-ATP synthase
Stroma	Direction of proton pumping

The Light Reactions

3. Calculate the energy of a mole of photons with wavelengths of (a) 400 nm, (b) 500 nm, (c) 600 nm, and (d) 700 nm.

4. Purple photosynthetic bacteria have different pigments than higher plants. Why is this an advantage for these bacteria?

5. What distinguishes the chlorophyll in a reaction center from the antennae chlorophyll?

6. The initial electron transfers in the bacterial photosynthetic reaction center are extremely rapid, but the lifetime of the terminal semiquinone is relatively long. (a) Why is it essential for the electrons to quickly leave the vicinity of the special pair? (b) Why does the terminal semiquinone persist?

7. Compare the electron flow in purple photosynthetic bacteria to that in higher plant chloroplasts. What is the origin of the electrons and their eventual fates? How many protons are translocated?

8. What is the standard reduction potential for the oxidation of H_2O (see Table 14-5)? Can this value be obtained from purple bacterial photosynthesis? Compare this to two-center photosynthesis.

9. The number of O_2 molecules released per photon absorbed by a suspension of algae can be measured at different wavelengths. When algae are illuminated by 700 nm light, very little O_2 is produced. However, when they are also illuminated by 500 nm light, the O_2 production is well in excess of that produced with only the 500 nm light. Explain.

10. What is the fate of water-derived electrons in chloroplasts treated with DCMU? What simple screening method could be used to identify plants that had been exposed to DCMU?

11. What do the S states represent in the oxygen-evolving center (OEC)? What S state predominates in dark-adapted chloroplasts?

12. What chloroplast component generates the majority of the proton gradient used for ATP production? Does it have a mitochondrial counterpart?

13. Why does the Cu atom of plastocyanin have an unusually high standard reduction potential?

14. What are the similarities and differences between photosystem I and photosystem II?

15. Estimate the minimum reduction potential for P680. Estimate the maximum reduction potential for ferredoxin. Explain your answers.

16. Describe the distribution of LHCs between grana (stacked membranes) and stromal lamellae (unstacked membranes) in chloroplasts under a bright sun and in shady light.

The Dark Reactions

17. What is the first stable radioactive sugar intermediate seen when $^{14}CO_2$ is added to algae such as *Chlorella*? When the supply of $^{14}CO_2$ is cut off, what compound accumulates? What do these results suggest about the pathway of CO_2 incorporation into carbohydrates?

18. Using Figure 19-26, write out the 13 reactions of the Calvin cycle. What is the main difference between Stage I and Stage II reactions? Write a net equation for each stage.

19. Glycolysis and the Calvin cycle are opposing pathways. Which reactions form a potential futile cycle? How are these reactions controlled in plant chloroplasts?

20. What is photorespiration? What is the ultimate result of this process?

21. The concentration of atmospheric CO_2 has been increasing for many decades. If this trend continues, how might it affect the relative abundance of C_3 and C_4 plants?

Answers to Questions

1. The light reactions depend on light for activity and include the reactions of the photosystems and the electron transport chain. The dark reactions are those of the Calvin cycle, which convert CO_2 to GAP.

    The dark reactions do not require light for their mechanisms. However, the light reactions regulate the dark reactions to ensure that the cell maintains adequate levels of ATP and NADPH. Thus the dark reactions do not actually occur in the dark.

2.

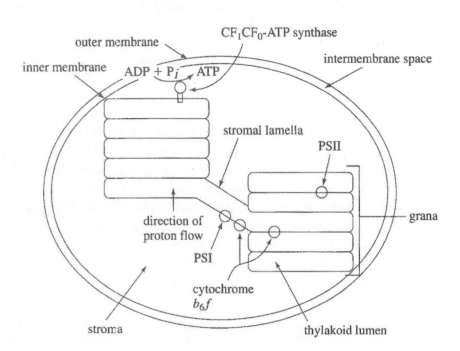

3.   Use Planck's law multiplied by Avogadro's number:

$$E = \frac{hc}{\lambda} N$$

$$= (6.626 \times 10^{-34} \text{ J·s}) (2.998 \times 10^8 \text{ m·s}^{-1}) (6.0221 \times 10^{23} \text{ mol}^{-1}) / \lambda$$
$$= (0.1196 \text{ J·m·mol}^{-1}) / \lambda$$

(a) $E = (0.1196 \text{ J·m·mol}^{-1}) / (4 \times 10^{-7} \text{ m}) = 3.0 \times 10^5 \text{ J·mol}^{-1} = 300 \text{ kJ·mol}^{-1}$
(b) $E = (0.1196 \text{ J·m·mol}^{-1}) / (5 \times 10^{-7} \text{ m}) = 2.4 \times 10^5 \text{ J·mol}^{-1} = 240 \text{ kJ·mol}^{-1}$
(c) $E = (0.1196 \text{ J·m·mol}^{-1}) / (6 \times 10^{-7} \text{ m}) = 2.0 \times 10^5 \text{ J·mol}^{-1} = 200 \text{ kJ·mol}^{-1}$
(d) $E = (0.1196 \text{ J·m·mol}^{-1}) / (7 \times 10^{-7} \text{ m}) = 1.7 \times 10^5 \text{ J·mol}^{-1} = 170 \text{ kJ·mol}^{-1}$

4.   The pigments in purple photosynthetic bacteria absorb radiation with longer wavelengths than visible light. This is the most intense radiation in the environments that they inhabit, the murky bottoms of stagnant ponds.

5.   The chlorophyll molecules in a reaction center have a slightly lower-energy excited state than that of the antenna chlorophyll molecules. This allows excitation energy to be transferred from the antennae molecules to the reaction center, where photooxidation occurs.

6.   (a) The electron ejected from the excited special pair must be transferred away so that it cannot return immediately to the special pair, which would allow the excitation energy to be released as heat (or possibly in a way that would damage the reaction center).
(b) After its reduction by an electron, the quinone (now a semiquinone) must await a second excitation event and electron transfer to become fully reduced to a quinol so that it can transfer both electrons to the membrane-bound quinol pool.

7.   Electrons from the bacterial reaction center flow to ubiquinone and then to cytochrome bc_1 and cytochrome c_2 before returning to the special pair. This cyclic electron flow does not yield reduced $NADP^+$, but it translocates four protons per electron pair to the periplasmic space.
        In the chloroplast, electrons flow in a linear fashion (the Z-scheme) from water to $NADP^+$, so that 2 NADPH are produced for every 2 H_2O oxidized to O_2. The 4 electrons pass from PSII to cytochrome b_6f and then to PSI before they reach ferredoxin–$NADP^+$ reductase. This results in the transmembrane movement of 12 protons. In cyclic electron flow, electrons cycle from PSI back to cytochrome b_6f and hence no NADPH is produced. Instead, this increases the number of protons translocated without affecting the stoichiometry of the $H_2O \rightarrow O_2$ reaction.

8.   The standard reduction potential for the reaction $O_2 + 4\,e^- + 4\,H^+ \rightarrow 2\,H_2O$ is 0.815 V. The reduction potential of the P870 bacterial reaction center is ~0.500 V (Figure 19-10), which is not sufficient to oxidize water (electrons spontaneously flow to centers with more positive reduction potentials). The two-reaction center Z-scheme (Figure 19-12) spans a redox range that allows both the oxidation of water and the reduction of $NADP^+$.

9. The longer wavelength activates only PSI (which contains P700), whereas the shorter wavelength activates both PSI and PSII (which contains P680). Since PSII feeds electrons to PSI, both photosystems must operate together for the redox reactions to proceed most efficiently. When only 700 nm light is available, PSII is unable to extract the electrons from H_2O necessary to form O_2. However, in the presence of 700 nm and 500 nm light, PSII can supply electrons to PSI, which can energize them with 700 nm light, thereby driving O_2 production.

10. DCMU blocks electron flow between PSII and PSI. This would cause the excitation energy of PSII to be dissipated by a mechanism other than photooxidation (since the electrons have nowhere to go). Some of the absorbed energy is released as fluorescence. Plants in which electron flow was blocked by DCMU can be detected by their fluorescence.

11. The S states in the OEC are chemical intermediates in the five-step reaction in which H_2O is oxidized to O_2. Dark-adapted chloroplasts are in the S_1 state, since O_2 is generated on the third flash of light (Figure 19-16).

12. The cytochrome b_6f complex, via the Q cycle, pumps the majority of the protons that make up the proton gradient required for ATP synthesis. Eight protons are translocated for every 2 H_2O oxidized; more during cyclic electron flow. The membrane-bound cytochrome b_6f complex resembles mitochondrial Complex III (cytochrome bc_1).

13. The Cu(II)/Cu(I) half-reaction normally has a standard redox potential of 0.158 V, but plastocyanin's standard redox potential is 0.370 V. In the protein, the Cu(II) atom is strained toward the tetrahedral coordination geometry of Cu(I), which promotes its reduction.

14. Both photosystems are membrane-bound protein complexes, contain special pairs of chlorophyll where photooxidation occurs, and have near-symmetrical arrangements of pigment molecules at the reaction center.

    The photosystems differ in overall structure, their redox groups, and the pathway of electron transfer.

15. The reduction potential for P680 should be greater than that of the O_2/H_2O half-reaction (0.815 V; Table 14-5) since electrons flow from H_2O to P680. The standard reduction potential of ferredoxin should be less than that for the $NADP^+/NADPH$ half-reaction (–0.320 V) since electrons flow from ferredoxin to $NADP^+$.

16. Under bright sun (high proportion of short-wavelength light), PSII is more active than PSI. As a result, reduced plastoquinone accumulates. This activates a protein kinase to phosphorylate the LHCs, which then move to the stromal lamellae, where they associate with and funnel more light energy to PSI.

    Under shady light (high proportion of long-wavelength light), PSI is more active than PSII. Oxidized plastoquinone therefore accumulates. This leads to dephosphorylation of the LHCs, which move to the grana to associate with and funnel light energy to PSII.

17. 3-Phosphoglycerate is the first stable sugar that incorporates the $^{14}CO_2$. This suggests that $^{14}CO_2$ is added to a 2-carbon compound. Ribulose-5-phosphate levels increase after the removal of $^{14}CO_2$, which instead suggests that ribulose-5-phosphate is the $^{14}CO_2$ acceptor.

18. Shown below are the 13 reactions of the Calvin cycle:

    (1) $3 \text{ Ru5P} + 3 \text{ ATP} \rightarrow 3 \text{ RuBP} + 3 \text{ ADP}$
    (2) $3 \text{ RuBP} + 3 \text{ CO}_2 \rightarrow 6 \text{ 3PG}$
    (3) $6 \text{ 3PG} + 6 \text{ ATP} \rightarrow 6 \text{ BPG} + 6 \text{ ADP}$
    (4) $6 \text{ BPG} + 6 \text{ NADPH} \rightarrow 6 \text{ GAP} + 6 \text{ P}_i + 6 \text{ NADP}^+$
    (5) $2 \text{ GAP} \rightleftharpoons 2 \text{ DHAP}$
    (6) $\text{GAP} + \text{DHAP} \rightleftharpoons \text{FBP}$
    (7) $\text{FBP} \rightarrow \text{F6P} + \text{P}_i$
    (8) $\text{F6P} + \text{GAP} \rightleftharpoons \text{Xu5P} + \text{E4P}$
    (9) $\text{E4P} + \text{DHAP} \rightleftharpoons \text{SBP}$
    (10) $\text{SBP} \rightarrow \text{S7P} + \text{P}_i$
    (11) $\text{S7P} + \text{GAP} \rightleftharpoons \text{Xu5P} + \text{R5P}$
    (12) $\text{R5P} \rightleftharpoons \text{Ru5P}$
    (13) $2 \text{ Xu5P} \rightleftharpoons 2 \text{ Ru5P}$

    Stage I (Reactions 1–4) is the production phase or energy-requiring phase and consists of carboxylation, phosphorylation, and reduction steps. Stage II (Reactions 5–13) is the rearrangement or recovery phase, which regenerates ribulose-5-phosphate.

    Stage I:
    $3 \text{ Ru5P} + 3 \text{ CO}_2 + 9 \text{ ATP} + 6 \text{ NADPH} \rightarrow 6 \text{ GAP} + 9 \text{ ADP} + 6 \text{ P}_i + 6 \text{ NADP}^+$
    Stage II:
    $5 \text{ GAP} \rightarrow 3 \text{ Ru5P} + 2 \text{ P}_i$

19. The interconversion of fructose-1,6-bisphosphate (FBP) and fructose-6-phosphate (F6P) is a potential futile cycle. The relevant enzymes are phosphofructokinase (PFK, for glycolysis) and fructose bisphosphatase (FBPase, for the Calvin cycle). To control these enzymes, a redox-sensing protein, thioredoxin, activates FBPase and deactivates PFK in the light, and activates PFK and deactivates FBPase in the dark.

20. Photorespiration is a side reaction of ribulose bisphosphate carboxylase in which O_2 (which competes with CO_2 for binding to the active site) reacts with RuBP to form 3PG and 2-phosphoglycolate. The 2-phosphoglycolate is eventually converted back to 3PG by a multistep pathway that consumes NADH and ATP and yields CO_2. Photorespiration therefore wastes some of the free energy captured in the photosynthetic light reactions as well as "unfixing" some of the CO_2 fixed by photosynthesis.

21. C_4 plants have an advantage when CO_2 is relatively scarce, since they concentrate CO_2 in mesophyll cells. However, this process consumes 2 ATP equivalents. The more energy-efficient C_3 plants therefore have an advantage when CO_2 is not limiting. Thus, a higher atmospheric concentration of CO_2 would favor C_3 plants over C_4 plants.

20

Lipid Metabolism

As we saw in Chapter 9, cells contain a diverse array of lipids. The major functions of lipids include forming a barrier to the extracellular environment, maintaining membrane fluidity, and serving as an important energy store, principally in the form of triacylglycerols. This chapter focuses on the diverse metabolic pathways of cellular lipids. First, the text considers the means by which ingested dietary lipids are degraded, absorbed, and packaged in lipoproteins for transport between tissues. The chapter then presents the oxidative pathways by which long acyl chains are converted to successive two-carbon acetyl-CoA fragments, which are either further oxidized via the citric acid cycle and the electron-transport system, converted to ketone bodies, or used in biosynthesis. Next, mechanisms of fatty acid biosynthesis by successive condensations of two-carbon fragments are presented. The steps leading to the biosynthesis of membrane phospholipids and glycolipids are outlined, followed by the pathway of cholesterol biosynthesis and its regulation. The reader should contemplate the varied metabolic reactions that result in the formation or degradation of lipids in the context of the nutritive, structural, and regulatory roles that lipids fulfill in the cell.

Essential Concepts

Lipid Digestion, Absorption, and Transport

1. The enzymatic digestion of triacylglycerols occurs at the lipid–water interface and is aided by the presence of bile acids, which help solubilize the lipids. Bile acids are cholesterol derivatives that are synthesized in the liver as taurine or glycine derivatives, stored in the gall bladder, and released into the small intestine.

2. Pancreatic lipase hydrolyzes triacylglycerols first to diacylglycerols and then to 2-monoacylglycerols plus free fatty acids. Binding of the enzyme to triacylglycerol at the lipid–water interface requires colipase. The interaction of this protein with lipase produces a hydrophobic surface which promotes binding of the protein complex to the lipid. Phospholipase A_2 also degrades phospholipids to lysophospholipids and fatty acids at a lipid–water interface.

3. Micelles containing bile acids promote the absorption of triacylglycerol and phospholipid hydrolysis products by the intestinal mucosa. Inside intestinal cells, fatty acids are complexed to a fatty acid–binding protein that, in effect, increases the aqueous solubility of these hydrophobic compounds.

4. Many lipids are transported between tissues through the circulation as components of lipoproteins. Lipoproteins are globular structures that are composed of a hydrophobic interior containing triacylglycerols and cholesteryl esters encased in an amphiphilic outer layer of protein, phospholipid, and cholesterol. There are five classes of lipoproteins: (a)

chylomicrons; (b) very low density lipoproteins (VLDL); (c) intermediate density lipoproteins (IDL); (d) low density lipoproteins (LDL); and (e) high density lipoproteins (HDL). The densities of lipoprotein classes increase as the quantity of lipid contained in the core decreases.

5. At least nine apolipoproteins comprise the protein components of human lipoproteins. Most of these have a high content of α helices whose nonpolar and polar residues are on opposite sides of the α helix. The nonpolar faces of the apolipoproteins interact with the phospholipid hydrophobic tails, and the polar faces interact with the phospholipid polar head groups.

6. Absorbed fatty acids are used for triacylglycerol synthesis in intestinal cells. These triacylglycerols are then incorporated into chylomicrons, which enter the bloodstream via the lymphatic system and eventually provide triacylglycerols to peripheral tissues, chiefly skeletal muscle and adipose tissue. The delivery process converts chylomicrons to much smaller chylomicron remnants, which are taken up by the liver.

7. The liver synthesizes other lipoproteins, including very low density lipoproteins (VLDL), which transport triacylglycerols and cholesterol from the liver to skeletal muscle and adipose tissue. In the capillaries, triacylglycerols are degraded by lipoprotein lipase to yield fatty acids (which the cells can either oxidize or reincorporate into triacylglycerols) and glycerol (which can be transformed to the glycolytic intermediate dihydroxyacetone phosphate). As they lose their triacylglycerol component, VLDL are converted first to intermediate density lipoproteins (IDL) and then to low density lipoproteins (LDL). Cells take up the cholesterol-rich LDL by receptor-mediated endocytosis.

8. Cholesterol is removed from cell surface membranes and carried as cholesteryl esters through the bloodstream to the liver by high density lipoproteins (HDL). There, the cholesteryl esters are transferred to VLDL.

Fatty Acid Oxidation

9. When energy needs dictate, triacylglycerols stored in adipose tissue are broken down (mobilized) by hormone-sensitive lipase. Released free fatty acids are transported in complex with serum albumin to the liver and other tissues.

10. Before being degraded by oxidation, fatty acids are first activated by the formation of an acyl-CoA in an ATP-dependent reaction catalyzed by thiokinase.

11. Since β oxidation takes place in the mitochondrial matrix, the acyl groups must cross the inner mitochondrial membrane, which is impermeable to fatty acyl-CoA derivatives. Therefore, the acyl group is transferred to carnitine by carnitine palmitoyl transferase I. The resulting acyl-carnitine readily crosses the membrane via a carrier protein. Once in the mitochondrial matrix, the acyl group is transferred back to a CoA molecule by carnitine palmitoyl transferase II, and the liberated carnitine crosses the membrane back to the cytosol.

12. Fatty acyl groups are degraded by a process called β oxidation, in which successive two-carbon fragments are removed as acetyl-CoA units.
    (a) Acyl-CoA dehydrogenase catalyzes formation of a *trans*-2,3 (α,β) double bond. The enzyme's bound FAD is thereby reduced to $FADH_2$.
    (b) Enoyl-CoA hydratase catalyzes the hydration of the double bond to produce a 3-L-hydroxyacyl-CoA.
    (c) 3- L-Hydroxyacyl-CoA dehydrogenase catalyzes the formation of a β-ketoacyl-CoA with the reduction of NAD^+ to NADH.
    (d) Thiolase catalyzes the thiolysis of the C2—C3 bond, releasing acetyl-CoA and forming a new acyl-CoA which is two carbons shorter than the starting substrate.
    This sequence of reactions is repeated until the acyl-CoA has been converted entirely to acetyl-CoA. For oxidation of palmitoyl-CoA, the sequence occurs 7 times to yield 8 acetyl-CoA.

13. The oxidation of fatty acids is highly exergonic. For example, palmitate's 8 acetyl-CoA can enter the citric acid cycle, and the $FADH_2$ and NADH generated by β oxidation and the citric acid cycle can be reoxidized by the electron-transport chain, which yields a total of 106 ATP.

14. The oxidation of unsaturated fatty acids requires additional enzymatic reactions to accommodate the double bond at C9 in monounsaturated fatty acids (e.g., oleic acid) and at three-carbon intervals in polyunsaturated fatty acids (e.g., linoleic acid). When a *cis*-3,4 (β,γ) double bond is encountered after several rounds of β oxidation, enoyl-CoA isomerase converts it to a *trans*-2,3 double bond. Oxidation of a polyunsaturated fatty acid also requires a reaction catalyzed by 2,4-dienoyl-CoA reductase, which removes a double bond at the expense of NADPH.

15. The final round of β oxidation of odd-chain fatty acids yields propionyl-CoA. This three-carbon compound is converted to succinyl-CoA, a citric acid cycle intermediate, by three reactions:
    (a) Propionyl-CoA carboxylase catalyzes an ATP-dependent carboxylation reaction that requires the coenzyme biotin and produces (*S*)-methylmalonyl-CoA.
    (b) Methylmalonyl-CoA racemase converts (*S*)-methylmalonyl-CoA to (*R*)-methylmalonyl-CoA.
    (c) Methylmalonyl-CoA mutase transforms (*R*)-methylmalonyl-CoA into succinyl-CoA in a reaction that requires the coenzyme 5′-deoxyadenosylcobalamin, which is derived from cobalamin (vitamin B_{12}).

16. The methylmalonyl-CoA mutase reaction rearranges the substrate's carbon skeleton. The reaction mechanism features homolytic cleavage of the C—Co bond in the coenzyme so that the C and Co atoms each retain one electron. Such homolytic cleavage is rare in biological systems; in biochemical reactions, bonds are usually broken by heterolytic cleavage in which one of the atoms acquires both electrons. The Co ion therefore functions as a generator of free radicals, which are essential for the reaction.

17. The peroxisome also carries out β oxidation. In animals, peroxisomes oxidize very long acyl chains (>22 carbons). These are shortened in the peroxisome, and β oxidation is then completed in the mitochondrion. Peroxisomal oxidation differs from oxidation in mitochondria in two ways:
    (a) No carnitine is required.
    (b) The first step of acyl-CoA oxidation by acyl-CoA oxidase involves transfer of electrons to O_2 with formation of hydrogen peroxide (H_2O_2) and a *trans*-2,3-enoyl-CoA. Thus, peroxisomal oxidation of fatty acids yields less energy than mitochondrial oxidation.

Ketone Bodies

18. Acetyl-CoA, in addition to undergoing oxidation via the citric acid cycle, can also undergo ketogenesis to form acetoacetate, D-β-hydroxybutyrate, and acetone. These water-soluble compounds are collectively called ketone bodies. Acetoacetate and D-β-hydroxybutyrate are important sources of metabolic energy under certain circumstances.

19. Ketogenesis, the formation of ketone bodies from acetyl-CoA, occurs as follows:
    (a) Two acetyl-CoA units condense to form acetoacetyl-CoA in a reversal of the thiolase reaction.
    (b) Hydroxymethylglutaryl-CoA synthase catalyzes the condensation of acetoacetyl-CoA with a third acetyl-CoA unit to form β-hydroxy-β-methylglutaryl-CoA (HMG-CoA).
    (c) HMG-CoA lyase cleaves HMG-CoA to form acetyl-CoA and acetoacetate.
    (d) β-Hydroxybutyrate dehydrogenase can reduce acetoacetate to β-hydroxybutyrate in an NADH-dependent reaction.
    (e) Acetoacetate also undergoes spontaneous decarboxylation to form acetone.

20. Acetoacetate and β-hydroxybutyrate formed in the liver are transported in the bloodstream to peripheral tissues and there are converted into two acetyl-CoA units. Succinyl-CoA supplies the CoA for this process and is therefore not utilized in the citric acid cycle to form GTP and succinate.

Fatty Acid Biosynthesis

21. Although fatty acid biosynthesis involves the condensation of successive acetyl-CoA units, this metabolic pathway is distinct from β oxidation in several respects:
    (a) It is a reductive process.
    (b) It takes place in the cytosol.
    (c) It utilizes NADPH as hydrogen donor.
    (d) It uses the C_3 dicarboxylic acid derivative, malonyl-CoA, as its C_2 donor.
    (e) The growing acyl chain is attached to acyl-carrier protein (ACP) rather than to CoA.
    (f) It employs entirely different enzymes and is independently regulated.

22. In order for fatty acid biosynthesis to proceed, sufficient amounts of both acetyl-CoA and NADPH must be available in the cytosol. Acetyl-CoA, which is generated by pyruvate dehydrogenase in the mitochondrion, cannot cross the inner mitochondrial membrane to reach the cytosol. Instead, transport occurs by means of the tricarboxylate transport system

in which acetyl-CoA reacts with oxaloacetate to form citrate, which readily crosses the membrane via a transporter. Once in the cytosol, citrate is converted to pyruvate in a series of reactions that liberates acetyl-CoA and also generates NADPH in a 1:1 ratio. Pyruvate then enters the mitochondrion and is converted to oxaloacetate.

23. The first committed step in fatty acid biosynthesis is catalyzed by acetyl-CoA carboxylase, a biotin-dependent enzyme, which converts acetyl-CoA to malonyl-CoA. In the first reaction step, bicarbonate becomes covalently attached to the biotin prosthetic group. This "activated" CO_2 is then transferred from biotin to acetyl-CoA, forming malonyl-CoA.

24. In eukaryotic cells, acetyl-CoA carboxylase is regulated both allosterically and by covalent modification. Citrate allosterically increase the V_{max} of the enzyme, whereas long-chain acyl-CoAs inhibit the reaction. AMP-dependent kinase phosphorylates Ser 79, thereby inactivating the enzyme. Glucagon acts through a cAMP-dependent pathway to inhibit the enzyme, possibly by preventing its dephosphorylation. In contrast, insulin enhances enzyme activity by promoting its dephosphorylation.

25. Fatty acid synthesis from acetyl-CoA and malonyl-CoA takes place by successive cycles of six enzymatic reactions. In *E. coli,* these reactions are catalyzed by separate enzymes. In eukaryotic cells, the process occurs within fatty acid synthase, a protein whose sequence contains all the required enzyme activities and an ACP domain. Fatty acid synthase catalyzes the following reactions:
    (a) In the first two reactions, the synthase is primed by transferring an acetyl group from acetyl-CoA to a Cys SH group and by transferring a malonyl group from malonyl-CoA to the terminal SH of the phosphopantetheine prosthetic group of ACP.
    (b) In the third reaction, malonyl-ACP is decarboxylated and condenses with the acetyl group to form a β-ketoacyl-ACP.
    (c) In reactions 4 to 6, the β-keto group is reduced by NADPH to form a hydroxyl group, the hydroxyl group is eliminated in a dehydration reaction to form a double bond, and a second reduction by NADPH reduces the double bond to produce an alkyl group.
    (d) The four-carbon acyl chain is then transferred from the ACP to the enzyme Cys SH. The cycle starts anew by the transfer of another malonyl group to the now vacant ACP site in the synthase.
    Seven cycles are required to synthesize a C_{16} acyl chain (palmitate), a process that consumes 8 acetyl-CoA and 14 NADPH (which are provided by glucose oxidation via the pentose phosphate pathway). The completed saturated acyl chain is released by thioester cleavage catalyzed by the seventh enzyme activity associated with the synthase.

26. Mammalian fatty acid synthase consists of two identical monomers arranged such that the Cys-linked acyl group in one polypeptide is located close to the phosphopantetheine moiety in the other. This allows the synthase dimer to simultaneously synthesize two acyl chains.

27. Palmitate may be lengthened and desaturated by elongases and desaturases, respectively. However, because animals, unlike plants, cannot introduce a double bond beyond C9 in an acyl chain, linoleic acid (9,12-*cis*-octadecadienoic acid) must be obtained from the diet. It is therefore called an essential fatty acid.

28. The synthesis of triacylglycerols begins with successive acylations of glycerol-3-phosphate to yield lysophosphatidic acid and then phosphatidic acid. Dephosphorylation produces 1,2-diacylglycerol, which then accepts another fatty acyl group. An alternative pathway involves acylation of dihydroxyacetone phosphate, followed by an NADPH-dependent reduction to produce lysophosphatidic acid.

Regulation of Fatty Acid Metabolism

29. Triacylglycerol metabolism, like that of glycogen, is important for the well-being of the whole organism and is regulated by hormones. Fatty acid oxidation is controlled by the rate of triacylglycerol hydrolysis in adipose tissue by hormone-sensitive triacylglycerol lipase. The lipase is stimulated by glucagon through cAMP-dependent phosphorylation, which also inhibits acetyl-CoA carboxylase, so fatty acid oxidation is enhanced and fatty acid synthesis is inhibited.

30. Insulin opposes the effects of glucagon by reducing cAMP levels, thereby inactivating triacylglycerol lipase and stimulating fatty acid synthesis. The ratio of glucagon to insulin therefore controls the status of fatty acid metabolism.

31. These mechanisms of short-term regulation are complemented by long-term hormonal regulation of fatty acid metabolism, which alters the levels of key enzymes such as acetyl-CoA carboxylase and triacylglycerol lipase.

Synthesis of Other Lipids

32. The formation of choline- and ethanolamine-containing glycerophospholipids involves three enzymatic steps:
    (a) The phosphorylation of the nitrogen-containing base.
    (b) Activation of the phosphocholine or phosphoethanolamine by CTP to form CDP–choline or CDP–ethanolamine.
    (c) Transfer of the activated base to 1,2-diacylglycerol to form the phospholipid.

33. The synthesis of phosphatidylglycerol and phosphatidylinositol involves activation of phosphatidic acid by reaction with CTP to form CDP–diacylglycerol. The activated lipid is then transferred to glycerol-3-phosphate or inositol. Cardiolipin is synthesized from two phosphatidylglycerol.

34. Enzymes that acylate glycerol-3-phosphate show a preference for introducing a saturated fatty acid in position 1 and an unsaturated fatty acid in position 2. However, there must be additional reactions, catalyzed by phospholipases and acyltransferases, which result in the exchange of acyl groups, to account for the fatty acid compositions of all membrane phospholipids.

35. In sphingolipid synthesis, ceramide (*N*-acylsphingosine) is formed in four reactions from palmitoyl-CoA and serine. Phosphatidylcholine donates its phosphocholine group to ceramide to produce sphingomyelin. Ceramide can also be glycosylated, with UDP–glucose or UDP–galactose serving as the sugar donor, to form cerebrosides. Additional

glycosylation of cerebrosides generates more complex sphingoglycolipids such as gangliosides.

36. Prostaglandin synthesis begins with the formation of a cyclopentane ring within the linear fatty acid. This reaction is catalyzed by prostaglandin H_2 synthase, which contains two catalytic activities, a cyclooxygenase and a peroxidase. Prostaglandin H_2 synthase has three isoforms: COX-1, COX-2, and the recently discovered COX-3.

37. Some common anti-inflammatory drugs interact with the various isoforms of prostaglandin H_2 synthase:
    (a) Aspirin acetylates a specific serine residue of the enzyme;
    (b) Ibuprofen and acetaminophen interact noncovalently and block the active site of prostaglandin H_2 synthase;
    (c) Vioxx and Celebrex specifically inhibit COX-2 but not COX-1.
    (d) Aspirin and ibuprofen are relatively nonspecific, but acetaminophen inhibits only COX-3.

Cholesterol Metabolism

38. All 27 carbon atoms of cholesterol are derived from acetate. The major stages in cholesterol formation are:
    (a) Acetate is converted to hydroxymethylglutaryl-CoA (HMG-CoA) and then via mevalonate to an isoprene unit, isopentenyl pyrophosphate.
    (b) Condensation of six isoprene units forms squalene, a linear 30-carbon compound.
    (c) Squalene is oxidized and cyclized to form lanosterol.
    (d) Further modification and removal of three carbons yields cholesterol.

39. After its synthesis in the liver, cholesterol may be transformed into bile acids or converted to cholesteryl esters for transport via lipoproteins. Cholesterol is also the precursor of steroid hormones.

40. Cholesterol is essential for cell membrane integrity, yet excess cholesterol may be harmful to the organism. Thus, its biosynthesis, utilization, and cellular distribution are carefully controlled. Cholesterol metabolism is regulated in three ways:
    (a) Cholesterol biosynthesis is controlled by HMG-CoA reductase, which catalyzes the rate-limiting conversion of HMG-CoA to mevalonate. In the short term, this enzyme is inactivated by phosphorylation (catalyzed by AMP-dependent kinase, the same enzyme that inactivates acetyl-CoA carboxylase) and activated by dephosphorylation. Long-term regulation involves changes in the level of enzymes in inverse proportion to the concentration of cholesterol.
    (b) The statins are a group of drugs that inhibit HMG-CoA reductase activity via complementary interactions in the active site of the enzyme.
    (c) Cholesterol transport and removal from blood is governed largely by the activity of LDL receptors on the liver cell surface, which in turn depends on the number of LDL receptors and hence on the rate of receptor synthesis.

41. The primary regulatory mechanism for HMG-CoA reductase activity is long-term control of cellular enzyme concentration, which occurs at the level of transcription of the HMG-CoA reductase gene. The control of this and 20 other genes participating in cholesterol biosynthesis and transport involves the binding of a specific regulatory protein, or transcription factor (see Chapter 28), at a DNA sequence called the sterol regulatory element (SRE). A SRE binding protein (SREBP) binds to the SRE to stimulate transcription.

42. Inactive SREBP, a transmembrane protein of the ER, is noncovalently bound to SREBP cleavage-activating protein (SCAP), another transmembrane protein, which remains localized to the ER when bound to cholesterol (as occurs when the cellular cholesterol concentration is high). As the cholesterol concentration falls, cholesterol dissociation from SCAP results in a conformational change in SCAP that promotes the migration of the SCAP–SREBP complex into the Golgi apparatus. There, two proteases cleave SREBP, releasing it into the cytosol and allowing its N-terminal fragment to migrate to the nucleus, where it interacts with available SRE sites on DNA.

43. Elevated levels of blood cholesterol, in the form of LDL, are strongly correlated with atherosclerosis, a risk factor for cardiovascular disease. High blood cholesterol results from the genetic disease familial hypercholesterolemia (which is characterized by an absence of LDL receptors) or from high dietary cholesterol intake (which tends to repress LDL receptor synthesis). In Tangier disease, a defective transport protein prevents cholesterol efflux from cells, which also contributes to atherosclerosis.

Questions

1. From a chemical perspective, why is the energy content of fats so much greater than that of carbohydrates or proteins?

Lipid Digestion, Absorption, and Transport

2. Why do individuals who have their gall bladders removed encounter difficulties in digesting large amounts of fats?

3. How do pancreatic lipase and phospholipase A_2 differ in the mechanism by which they promote hydrolysis of glycerolipids at interfaces?

4. Match each term on the left with its description on the right.

____ Bile acid	A.	Helps bind lipase to the lipid–water interface
____ Intestinal fatty acid	B.	Hydrolyzes phospholipids to yield binding protein lysophospholipids and free fatty acids

_____ Albumin

C.   Forms micelles that take up nonpolar lipid degradation products and helps transport them through the intestinal wall

_____ Phospholipase A$_2$

D.   Transports lipid digestion products through the lymphatic system and then the bloodstream to the tissues

_____ Colipase

E.   Transports through the bloodstream free fatty acids released from adipose tissue stores

_____ Chylomicrons

F.   Forms complexes with free fatty acids to shield intestinal cells from their detergent-like effects and increase their effective solubility in the blood.

5.   Determine the order of the following events involving lipoprotein-mediated transport of dietary triacylglycerols and cholesterol.

_____ Triacylglycerols are removed from circulating VLDL by lipoprotein lipase.
_____ Chylomicrons are transported through the lymphatic system and enter the bloodstream.
_____ Chylomicrons are formed in the intestinal mucosa.
_____ Cells take up cholesterol via receptor-mediated LDL endocytosis.
_____ Chylomicrons are degraded by lipoprotein lipase to chylomicron remnants.
_____ LDL components are rapidly degraded by lysosomal enzymes.
_____ VLDL are synthesized in liver.

6.   Why are cholesteryl esters located in the interior of a lipoprotein while cholesterol is located on the exterior?

7.   If the LDL receptor underwent a mutation that increased its affinity for apoB-100, what would be the effect on serum cholesterol levels?

Fatty Acid Oxidation

8.   What products would be isolated from urine when dogs are fed (a) phenylheptanoic acid and (b) phenyloctanoic acid? How does this experiment, originally performed by Knoop, shed light on the process of fatty acid oxidation?

9.   A patient develops an enlarged fatty liver and low blood glucose. These symptoms can be partially overcome only when massive amounts of carnitine are included in the diet. What enzyme defect might be responsible for this problem?

10. Which of the following statements is (are) correct? In the β oxidation of fatty acids,
    (a) The activation of fatty acids by acyl-CoA synthetase is driven by the hydrolysis of pyrophosphate.
    (b) The reaction catalyzed by acyl-CoA dehydrogenase uses FAD as electron acceptor.
    (c) The reaction catalyzed by enoyl-CoA hydratase produces a 3-D-hydroxyacyl-CoA.

11. Calculate the yield of ATP when one mole of stearic acid is completely oxidized to CO_2 and H_2O.

12. In the β oxidation of oleic acid, the presence of the double bond at the _____ position necessitates modification of the β oxidation cycle. The modification occurs after the _____ round of β oxidation because the _____ enoyl-CoA is not a substrate for _____. The problem is overcome by the action of the enzyme _____ which converts the _____ bond to a _____ bond so that β oxidation can continue.

13. In the β oxidation of linoleic acid, a second difficulty presents itself because of the additional double bond in the _____ position. In this case, after the fifth round of oxidation, a_____ CoA is formed, which is a poor substrate for _____. In mammals, two reactions are necessary for β oxidation to resume. In the first, NADPH-dependent _____ reductase reduces one double bond to form a _____ , and then the action of a _____ isomerase yields a _____ CoA that can participate in β oxidation.

14. Describe the role of cobalamin (vitamin B_{12}) in the methylmalonyl-CoA mutase reaction.

15. Explain why succinyl-CoA arising from the oxidation of odd-chain fatty acids cannot be directly oxidized by the citric acid cycle. How can it be further degraded?

16. Why does the β oxidation of fatty acids in peroxisomes yield less ATP than the corresponding process in mitochondria?

Ketone Bodies

17. Infants have high levels of ketone bodies in their blood and abundant 3-ketoacyl-CoA transferase in their tissues (except in liver) prior to weaning. What nutritional advantage does this confer?

18. Calculate the ATP that is produced when linoleic acid (9,12-octadecadienoic acid; 18:2) is (a) oxidized to CO_2 and H_2O or (b) converted to the ketone body acetoacetate in the liver and then oxidized to CO_2 and H_2O in the peripheral tissues.

19. Write equations for ketone body synthesis and degradation. What is the net result of combined synthesis and degradation?

Fatty Acid Biosynthesis

20. A rat liver cytosol preparation is incubated with acetate and all cofactors necessary for the biosynthesis of palmitate. Which carbon atoms of palmitate will be isotopically labeled when the preparation contains (a) $H^{14}CO_3^-$ or (b) $^{14}CH_3COO^-$?

21. What are the advantages of the multifunctional dimeric structure of animal fatty acid synthase in the formation of fatty acids?

22. Which of the following statements is(are) correct? The transport of acetyl-CoA from the mitochondrial matrix to the cytosol for fatty acid biosynthesis:
    (a) is necessary because the inner mitochondrial membrane is impermeable to acetyl-CoA.
    (b) uses the tricarboxylate transport system, which operates in both directions.
    (c) can generate equal numbers of acetyl-CoA and NADPH molecules in the cytosol .

Regulation of Fatty Acid Metabolism

23. Match each term on the left with its function in the short-term regulation of fatty acid metabolism.

    ____ Citrate   A. Activates acetyl-CoA carboxylase
    ____ cAMP-dependent   B. Inhibits carnitine palmitoyl transferase phosphorylation
    ____ Palmitate   C. Activates hormone-sensitive lipase
    ____ Malonyl-CoA   D. Inhibits acetyl-CoA carboxylase

Synthesis of Other Lipids

24. Match each term on the left with its metabolic role on the right.

    ____ CDP–diacylglycerol   A. Precursor of cardiolipin
    ____ Ceramide   B. Precursor of phosphatidylcholine
    ____ Phosphatidylglycerol   C. Contains a vinyl ether linkage
    ____ Plasmalogen   D. Precursor of sphingomyelin
    ____ CDP–Choline   E. Precursor of phosphatidylinositol

25. When tissues are incubated *in vitro* with ^{32}P-labeled P_i, the incorporation of radioactivity into phospholipids can be readily demonstrated. High concentrations of the drug propranolol dramatically alter the pattern of phospholipid labeling, such that radioactivity in phosphatidylcholine and phosphatidylethanolamine markedly declines while that in phosphatidylinositol and phosphatidylglycerol is greatly elevated. From this information and your knowledge of phospholipid biosynthetic pathways, determine which enzymatic reaction is principally affected by propranolol and explain how the drug alters phospholipid biosynthesis.

26. A metabolic disease may result from an enzyme deficiency that prevents the synthesis of an essential metabolite or that prevents the normal breakdown of a metabolite. Which type of defect is exhibited in Tay–Sachs disease and what enzyme is affected?

27. In lipid metabolism, the production of _____ is blocked by aspirin, which inhibits the reaction catalyzed by _____.

Cholesterol Metabolism

28. What are the two principal ways in which the cholesterol needs of many tissues are met?

29. Place the following intermediates in cholesterol biosynthesis in the correct order: squalene, farnesyl pyrophosphate, 2,3-oxidosqualene, hydroxymethylglutaryl-CoA (HMG-CoA), lanosterol, mevalonate, geranyl pyrophosphate.

30. Explain how the regulation of HMG-CoA reductase, the principal control site for cholesterol synthesis, can conserve cellular ATP.

31. Which of the following statements is(are) correct? Inhibitors of HMG-CoA reductase are used to decrease serum cholesterol levels. Such inhibitors would also:
    (a) reduce the intracellular level of mevalonate.
    (b) reduce the synthesis of LDL receptors.
    (c) reduce the synthesis of ubiquinone.

32. Match the following biochemical role or event with the substance listed on the right.

    ____ Cholesterol   A. Exposes proteolytic site in the cytosolic domain of SREBP

    ____ Site-1 protease   B. Interacts with SRE sequences
    ____ Site-2 protease   C. Binds to SCAP
    ____ WD repeat   D. Required for transport to the nucleus
    ____ bHLH   E. Domain of SCAP that binds to SREBP

Answers to Questions

1. The energy content of fats is greater because the carbon atoms of fatty acids are more reduced and therefore release more free energy upon oxidation than the carbons of carbohydrates or proteins. In addition, fats can be stored in large amounts in anhydrous form. In contrast, proteins cannot be stored in large amounts, and carbohydrates require a large volume for hydration.

2. The gall bladder stores the bile acids secreted by the liver and releases them on demand. Bile acids are essential for emulsifying triacylglycerols and thereby promoting their hydrolysis. In addition, bile acids and released fatty acids are components of mixed micelles that are absorbed across the intestinal wall. The liver secretes bile acids more or less continuously so that if large amount of fats are rapidly ingested, insufficient bile acids will be available to emulsify them.

3. Both pancreatic lipase and phospholipase A_2 hydrolyze their substrates at the lipid–water interface, where the enzymes undergo interfacial activation. Pancreatic lipase binds to the

interface only when complexed with pancreatic colipase. In the presence of a lipid micelle, the lipase undergoes a structural change that exposes the active site, forms an oxyanion hole, and, in conjunction with colipase, creates a large hydrophobic surface near the active site that helps to bind the lipase–colipase complex to the lipid.

Phospholipase A_2 does not alter its conformation but instead has a hydrophobic channel that enables a micellar phospholipid to gain access to the enzyme active site without having to pass through the aqueous phase.

4.

   C   Bile acid
   F   Intestinal fatty acid–binding protein
   E   Albumin
   B   Phospholipase A_2
   A   Colipase
   D   Chylomicrons

5.

   5   Triacylglycerols are removed from circulating VLDL by lipoprotein lipase.
   2   Chylomicrons are transported through the lymphatic system and enter the bloodstream.
   1   Chylomicrons are formed in the intestinal mucosa.
   6   Cells take up cholesterol via receptor-mediated LDL endocytosis.
   3   Chylomicrons are degraded by lipoprotein lipase to chylomicron remnants.
   7   LDL components are rapidly degraded by lysosomal enzymes.
   4   VLDL are synthesized in liver.

6. Cholesteryl esters are highly hydrophobic and therefore occupy the nonaqueous interior of the lipoprotein. Cholesterol, with an OH group, is weakly polar and therefore can interact with water molecules at the surface of the lipoprotein.

7. ApoB-100 is the LDL protein component that is recognized by the LDL receptor. Increased binding of LDL by its receptor would increase the cellular uptake of cholesterol-rich LDL. Consequently, the level of cholesterol in the serum would be lower than normal.

8. (a) Hippuric acid (a benzoic acid derivative). (b) Phenylaceturic acid (a phenylacetic acid derivative). This experiment suggested that fatty acids are degraded by oxidation through removal of successive two-carbon fragments.

9. The patient probably has a carnitine palmitoyl transferase I deficiency. As a result, acyl-CoA transport across the mitochondrial membrane is inadequate. Fatty acids released from adipose tissue stores would accumulate in the liver. Glucose levels would fall because it would be the primary metabolic fuel in the absence of oxidizable fatty acids. Treatment with high doses of carnitine would elevate tissue carnitine levels and hence enable some acyl-carnitine to form and enter the mitochondrion so that the acyl group could be oxidized by β oxidation.

10. (a) and (b) are correct. (c) is incorrect because the product is 3-L-hydroxyacyl-CoA.

11. 120 ATP. Stearic acid (an 18:0 fatty acid) is first activated by conversion to stearoyl-CoA, with the consumption of 2 equivalents of ATP. Stearoyl-CoA is then degraded by eight rounds of β oxidation to form 9 acetyl-CoA, 8 $FADH_2$, and 8 NADH. Oxidation of each acetyl-CoA by the citric acid cycle yields GTP, NADH, and $FADH_2$. Reoxidation of the $FADH_2$ and NADH, with the transfer of electrons to O_2 to form H_2O, yields ATP by oxidative phosphorylation. The stoichiometry of ATP production can be summarized as follows:

Process	Reduced coenzymes formed	ATPs formed
Fatty acid activation		–2
8 rounds of β oxidation	8 $FADH_2$	12
	8 NADH	20
9 rounds of citric acid cycle		9 (GTP)
	27 NADH	67.5
	9 $FADH_2$	13.5
Total		120

12. The missing terms are: cis Δ^9; third; cis Δ^3; enoyl-CoA hydratase; enoyl-CoA isomerase; cis Δ^3; trans Δ^2.

13. The missing terms are: cis Δ^{12}; 2,4-dienoyl-; enoyl-CoA hydratase; 2,4-dienoyl-CoA; *trans*-Δ^3-enoyl-CoA; 3,2-enoyl-CoA; *trans*-Δ^2-enoyl.

14. Cobalamin is essential for converting (*R*)-methylmalonyl-CoA to succinyl-CoA. The cobalt alternates between the Co(III) and Co(II) states and therefore functions as a free radical generator. The weak carbon—cobalt(III) bond of the coenzyme undergoes homolytic cleavage such that the carbon and cobalt each acquire one electron. This yields a deoxyadenosyl radical that can abstract a hydrogen atom from methylmalonyl-CoA, which then rearranges to form succinyl-CoA.

15. Since citric acid cycle intermediates function as catalysts and are continuously regenerated, the net oxidation of succinyl-CoA can occur only if it is converted to another compound that is oxidized by the operation of the cycle, such as pyruvate or acetyl-CoA. This is accomplished by transforming succinyl-CoA to malate and then transporting the malate to the cytosol, where it is oxidatively decarboxylated to pyruvate and CO_2 in a reaction catalyzed by malic enzyme. Pyruvate can then be transported into the mitochondrion and oxidized via pyruvate dehydrogenase and the citric acid cycle.

16. The first step in β oxidation in peroxisomes is catalyzed by acyl-CoA oxidase rather than acyl-CoA dehydrogenase as occurs in mitochondria. In the acyl-CoA oxidase reaction, electrons from acyl-CoA are transferred directly to O_2 to form H_2O_2 and therefore do not pass through the electron-transport chain with concomitant formation of ATP. For this reason, peroxisomal β oxidation yields less ATP than mitochondrial β oxidation.

17.   The large intake of milk means that fat supplies a major portion of the calories consumed by infants prior to weaning. The oxidation of fatty acids produces abundant acetyl-CoA which is partly converted to ketone bodies in the liver. The ketone bodies, like glucose, are water-soluble and easily transported. They are therefore readily available as metabolic fuels to support the rapid growth and development of the infant.

18.

(a)   The net production of ATP from the complete oxidation of linoleic acid can be calculated as in Problem 11. The total yield is 116.5 ATP. This is 3.5 less than for the oxidation of stearate because the first double bond of linoleate does not need an FAD to reduce it (FADH$_2$ is equivalent to 1.5 ATP) and the reduction of its second double bond consumes an NADPH (which is equivalent to ~2.5 ATP).

(b)   If the 9 acetyl-CoA generated by the β oxidation of linoleic acid were instead converted to ketone bodies, 4.5 molecules of acetoacetate would be formed. The transformation of acetoacetate back into acetyl-CoA does not directly consume ATP. However, it involves the consumption of 4.5 succinyl-CoA, which would otherwise provide the energy for a substrate level phosphorylation. Therefore, the net yield of ATP for this metabolic route can be considered to be 116.4 – 4.5 = 112 ATP.

19.   Ketogenesis:
   2 Acetyl-CoA + H$_2$O → acetoacetate + 2 CoASH

Ketone body breakdown:
   Acetoacetate + succinyl-CoA + CoASH → 2 acetyl-CoA + succinate

Combined synthesis and breakdown:
   Succinyl-CoA + H$_2$O → succinate + CoASH

20.

(a)   H^{14}CO$_3^-$ is incorporated into malonyl-CoA by the action of acetyl-CoA carboxylase, but the labeled carboxyl group is subsequently eliminated as ^{14}CO$_2$ during condensation of malonyl-ACP with the growing fatty acyl chain. Therefore, the biosynthesized palmitate will be unlabeled.

(b)   The radioactive methyl carbon in acetate will label each even-numbered carbon atom in the completed fatty acid.

21.   The proximity of multiple enzyme activities involved in fatty acid synthesis on a single polypeptide chain may enhance the efficiency of the process. Moreover, groups that are anchored to the ACP on one subunit are mainly processed by the enzymatic activities on the opposite subunit.

22.   (a) and (c) are correct. (b) is incorrect because the tricarboxylate transport system is irreversible, that is, unidirectional. This is because both the ATP-citrate lyase reaction (in the cytosol) and the pyruvate carboxylase reaction (in the matrix) involve ATP hydrolysis. Thus, the net result of one turn of the cycle is the hydrolysis of 2 ATP to 2 ADP + 2 P$_i$ and the transport of acetyl-CoA from the matrix to the cytosol (assuming that the NADH and NADPH in the malate dehydrogenase and malic enzyme reactions are equivalent).

23.

    __A__   Citrate
    __C__   cAMP-dependent phosphorylation
    __D__   Palmitate
    __B__   Malonyl-CoA

24.

    __E__   CDP–diacylglycerol
    __D__   Ceramide
    __A__   Phosphatidylglycerol
    __C__   Plasmalogen
    __B__   CDP–Choline

25.   Propranolol primarily inhibits phosphatidic acid phosphatase, which converts phosphatidic acid to 1,2-diacylglycerol. Since diacylglycerol is an intermediate in the synthesis of phosphatidylcholine and phosphatidylethanolamine, propranolol decreases the production of these lipids. Inhibition of phosphatidic acid phosphatase also increases the concentration of phosphatidic acid, which is then converted in greater amounts to phosphatidylinositol and phosphatidylglycerol.

26.   Tay–Sachs disease results from a defect in the enzyme hexosaminidase A, which breaks down ganglioside G_{M2} to ganglioside G_{M3}.

27.   In lipid metabolism, the production of <u>prostaglandins</u> is blocked by aspirin, which inhibits the reaction catalyzed by <u>prostaglandin H_2 synthase (COX)</u>.

28.   Cholesterol can be obtained either by synthesis *de novo* from acetyl-CoA or from circulating lipoproteins, primarily LDL, that enter cells by receptor-mediated endocytosis.

29.   The order is: HMG-CoA, mevalonate, geranyl pyrophosphate, farnesyl pyrophosphate, squalene, 2,3-oxidosqualene, lanosterol.

30.   HMG-CoA reductase catalyzes the first unique step in the synthesis of cholesterol, the conversion of HMG-CoA to mevalonate. This reaction is followed by three reactions that consume ATP. By regulating the activity of HMG-CoA reductase so that it operates only when cholesterol is needed, the cell avoids wasting metabolic energy on the production of cholesterol precursors.

31.   (a) and (c) are correct. (b) is incorrect because inhibition of cholesterol biosynthesis tends to increase the synthesis of LDL receptors.

32.

    __C__ Cholesterol
    __A__ Site-1 protease
    __D__ Site-2 protease
    __E__ WD repeat
    __B__ bHLH

21

Amino Acid Metabolism

This chapter surveys nitrogen metabolism, with its key focus on amino acid metabolism. The chapter begins with a brief discussion of protein turnover and pathways of protein degradation, including the lysosomal and ubiquitin-dependent pathways. The following sections describe the mechanisms by which amino acids are catabolized, beginning with amino acid deamination by transamination and oxidative deamination. Next is the urea cycle, a pathway that transfers an ammonium ion and the amino group of aspartate through a series of intermediates to arginine, which is cleaved to generate urea and regenerate the amino acid ornithine. The chapter then discusses the degradation of amino acids, which yield common intermediates that can be used for gluconeogenesis (from glucogenic amino acids) or fatty acid synthesis (from ketogenic amino acids). Many of these degradative pathways involve enzymatic reactions mediated by the coenzyme pyridoxal-5′-phosphate (PLP), which catalyzes transamination reactions, decarboxylation reactions, and reactions that break the C_α—C_β covalent bond of amino acids. This coenzyme, like thiamine pyrophosphate, is a resonance-stabilized electron sink that stabilizes carbanions. Some defects in amino acid metabolism are discussed in Boxes 21-1 and 21-2.

The sections in the final part of the chapter include the anabolic reactions of amino acid metabolism. Amino acid biosynthetic reactions are divided into two groups, those synthesizing the nonessential (for mammals) and essential amino acids. Within each group, the derivation of amino acids from common precursors is emphasized. The chapter then examines the synthesis of heme, which is assembled in a modular fashion beginning with glycine and succinyl-CoA. Disorders of heme metabolism are discussed in the text and in Box 21-3. The following section describes the conversion of select amino acids to hormones and neurotransmitters. This chapter concludes with a discussion of how microorganisms fix N_2 into NH_3 by an energetically costly process.

Essential Concepts

Protein Degradation

1. Proteins are continuously synthesized and degraded into amino acids. This serves three functions:
   (a) Proteins serve as a form of long-term energy storage that is utilized to provide gluconeogenic precursors and ketone bodies.
   (b) Degradation eliminates abnormal and damaged proteins, whose accumulation is potentially hazardous to cell function.
   (c) The regulation of metabolic activity under changing physiological conditions requires the degradation of one set of regulatory proteins and its replacement by a new set.

2. Proteins have highly variable half-lives, ranging from a few minutes to several days. The half-life of a protein varies with the identity of its N-terminal residue, a phenomenon which is referred to as the N-end rule. Other signals are contained in the protein's amino acid sequence. For example, proteins with segments rich in Pro (P), Glu (E), Ser (S), and Thr (T), which are called PEST sequences, have very short half-lives.

3. Intracellular proteins are degraded either in lysosomes or, after ubiquitination, in proteasomes.
   (a) Lysosomes degrade proteins (and other biological molecules) that are taken up by endocytosis, and they recycle cellular proteins that are enclosed in vacuoles. In well-nourished cells, lysosomal protein degradation is nonselective, but in starved cells, this pathway is more selective.
   (b) Intracellular proteins may be tagged for degradation by covalently linking them to a small protein called ubiquitin. Ubiquitination requires three enzymes: ubiquitin-activating enzyme (E1), ubiquitin-conjugating enzyme (E2), and ubiquitin-protein ligase (E3). E1 forms a thioester bond with ubiquitin in a reaction that consumes one ATP. E2 transfers this activated ubiquitin from E1 to itself, forming another thioester bond. E3 transfers activated ubiquitin to the Lys ε-amino group of a previously selected target protein.

4. Ubiquitinated proteins are degraded in an ATP-dependent manner inside a large multiprotein complex called the 26S proteasome. The complex has a barrel-like core (20S subunit) that degrades proteins into ~8 residue fragments, which diffuse out of the proteasome and are subsequently hydrolyzed by cytosolic peptidases. The 19S cap recognizes ubiquitinated proteins and unfolds them for degradation.

Amino Acid Deamination

5. The degradation of an amino acid usually begins with its transamination by a pyridoxal-5′-phosphate (PLP)-containing enzyme, in which the amino group is transferred to an α-keto acid (usually α-ketoglutarate). Transamination utilizing α-ketoglutarate generates glutamate, which can be subsequently oxidatively deaminated by glutamate dehydrogenase to regenerate α-ketoglutarate and release a free ammonium ion. High concentrations of ammonia are toxic, so the steady-state level of ammonia is kept low.

6. PLP functions as an electron sink in the resonance stabilization of a C_α carbanion derived from an amino acid. This carbanion is an intermediate in amino acid transamination, decarboxylation, and the breakage of the C_α—C_β bond.

7. Glutamate dehydrogenase catalyzes the oxidative deamination of glutamate to yield α-ketoglutarate and ammonium ion.

The Urea Cycle

8. Living organisms excrete excess nitrogen either as free ammonia (aquatic organisms only), urea (most terrestrial vertebrates), or uric acid (birds and terrestrial reptiles). The overall reaction of the urea cycle is

$$NH_3 + HCO_3^- + Aspartate$$

—3 ATP

►3 ADP + 2 P$_i$ + AMP + PP$_i$

Urea + Fumarate

9. The urea cycle takes place in the liver of adult vertebrates. Of the five reactions in the urea cycle, two are mitochondrial and three are cytosolic.
   (a) Ammonia (a product of the glutamate dehydrogenase reaction) reacts with HCO$_3^-$ and two ATP to form the "activated" product carbamoyl phosphate in a reaction catalyzed by mitochondrial carbamoyl phosphate synthetase.
   (b) Transfer of the carbamoyl group to ornithine results in the formation of citrulline, which is transported out of the mitochondrion to the cytosol.
   (c) Condensation of aspartate with citrulline in an ATP-dependent reaction forms argininosuccinate.
   (d) Argininosuccinate is cleaved into fumarate and arginine.
   (e) In the final reaction, hydrolytic cleavage of arginine yields urea and regenerates ornithine, which is transported back into the mitochondrion.

10. The urea cycle is regulated by the activity of carbamoyl phosphate synthetase I, which is allosterically activated by N-acetylglutamate. Increasing concentrations of glutamate, resulting from transamination during protein degradation, stimulate the synthesis of N-acetylglutamate, thereby boosting urea cycle activity to increase the excretion of amino groups as urea.

Breakdown of Amino Acids

11. The carbon skeletons of the 20 "standard" amino acids can be completely oxidized to CO$_2$ and H$_2$O. They can also be used to synthesize glucose or fatty acids:
   (a) Glucogenic amino acids are broken down to pyruvate, α-ketoglutarate, succinyl-CoA, fumarate, and oxaloacetate, which can be used as glucose precursors.
   (b) Ketogenic amino acids can be broken down to the ketone body acetoacetate or to acetyl-CoA, which can be used for fatty acid synthesis.

12. The amino acids can be grouped according to their common degradative products.
   (a) Alanine, cysteine, glycine, serine, and threonine are degraded to pyruvate.
   (b) Asparagine and aspartate are degraded to oxaloacetate.
   (c) Arginine, glutamate, glutamine, histidine, and proline are degraded to α-ketoglutarate.
   (d) Isoleucine, methionine, and valine are degraded to succinyl-CoA
   (e) Leucine and lysine are degraded to acetoacetate and acetyl-CoA.
   (f) Tryptophan is degraded to alanine and acetoacetate.
   (g) Phenylalanine and tyrosine are degraded to fumarate and acetoacetate.

13. Important cofactors involved in the degradation of amino acids are PLP, tetrahydrofolate (THF), biopterin, and *S*-adenosylmethionine (SAM).
    (a) PLP mediates the catalytic breakdown of the C_α—C_β covalent bond in threonine to generate acetaldehyde and glycine.
    (b) THF is composed of the following functional groups: 2-amino-4-oxo-6-methylpterin, *p*-aminobenzoic acid, and several glutamates. THF is derived from the vitamin folic acid, which must be reduced to form THF. THF serves as a one-carbon carrier in several reactions involved in amino acid catabolism. The carbon unit may exist in one of several oxidation states ranging from a relatively oxidized formyl group to a methyl group.
    (c) Biopterin is a redox cofactor involved in the oxidation of phenylalanine to tyrosine. Biopterin contains a pteridine ring that is similar to the isoalloxazine ring of the flavin coenzymes.
    (d) The reaction of ATP with methionine yields SAM, a potent methylating reagent in several reactions in which the transfer of a methyl group from THF is not favored.

14. The catabolism of the branched-chain amino acids (isoleucine, leucine, and valine), lysine, and tryptophan employs enzymatic conversions resembling those in the β oxidation of fatty acids and in the oxidative decarboxylation of α-keto acids such as pyruvate and α-ketoglutarate.

Amino Acid Biosynthesis

15. Amino acids that cannot be synthesized by mammals and that must therefore be obtained from their diet are called essential amino acids. Nonessential amino acids can be synthesized from common intermediates:
    (a) Alanine, aspartate, and glutamate are formed by one-step transamination reactions. Asparagine and glutamine are formed by amidation of aspartate and glutamate. The activation of glutamate by glutamine synthetase, which occurs prior to its amidation, is a key regulatory point in bacterial nitrogen metabolism.
    (b) Glutamate gives rise to proline and arginine (which is also considered to be an essential amino acid because, when synthesized, it is largely degraded to urea).
    (c) 3-Phosphoglycerate is the precursor of serine, which can be converted to cysteine and glycine. Glycine can also be produced from CO_2, NH_4^+, and N^5,N^{10}-methylene-THF.

16. The essential amino acids can be categorized into four groups based on their synthetic pathways:
    (a) The aspartate family. Aspartate serves as the precursor for the synthesis of lysine, threonine, and methionine. This pathway also produces homoserine and homocysteine.
    (b) The pyruvate family. Pyruvate serves as a precursor for the synthesis of valine, leucine, and isoleucine.
    (c) Aromatic amino acids. Phosphoenolpyruvate and erythrose-4-phosphate serve as the precursors for tyrosine, phenylalanine, and tryptophan. Serine is also required for the synthesis of tryptophan.
    (d) Histidine. 5-Phosphoribosyl-α-pyrophosphate is a precursor for the synthesis of histidine.

Other Products of Amino Acid Metabolism

17.   Heme synthesis occurs in a modular fashion similar to the synthesis of cholesterol. Heme synthesis begins in the mitochondrion with the condensation of glycine and succinyl-CoA. The product, δ-aminolevulinic acid (ALA), diffuses into the cytosol, where two ALA condense to form porphobilinogen (PBG). Four PBG then condense to form the porphyrin uroporphyrinogen III. Decarboxylation reactions form coproporphyrinogen III, which diffuses into the mitochondrion and is converted to heme.

18.   Heme synthesis occurs in erythroid cells and in the liver. In the liver, heme acts by feedback inhibition to regulate ALA synthase activity. In reticulocytes (immature red blood cells), heme stimulates the synthesis of globin and possibly itself. Once heme synthesis is "switched on," it appears to occur at its maximal rate.

19.   Red blood cells survive in the bloodstream for about 120 days and then are removed by the spleen and destroyed. Heme catabolism results in the formation of bilirubin, which is transported to the gall bladder for secretion into the intestinal tract. Bilirubin is converted by microbes in the colon into urobilinogen then into stercobilin, which gives the feces their red-brown pigment. Some urobilinogen is reabsorbed and converted in the kidneys to urobilin, which gives urine its characteristic yellow color.

20.   Several hormones and neurotransmitters are derived from amino acids. PLP-mediated decarboxylation forms histamine from histidine and γ-aminobutyric acid (GABA) from glutamate. Decarboxylation and hydroxylation of tryptophan yield serotonin. Hydroxylated tyrosine is the precursor of L-DOPA and melanin. PLP-dependent decarboxylation of L-DOPA yields dopamine, which is hydroxylated to form norepinephrine. Donation of a methyl group from SAM to norepinephrine yields epinephrine.

21.   Nitric oxide (NO) is produced by a five-electron oxidation of arginine. NO, a relatively stable gaseous free radical, is synthesized by vascular endothelial cells and induces vasodilation. A variety of cells, including neuronal cells and leukocytes, synthesize NO, which acts as a signaling molecule and is involved in the generation of antimicrobial hydroxyl radicals. Sustained NO release is implicated in endotoxic shock and neuronal damage.

Nitrogen Fixation

22. N_2, an extremely stable gaseous molecule, can be reduced, or fixed, by species of bacteria called diazotrophs, such as those of the genus *Rhizobium,* which live symbiotically in the root nodules of leguminous plants. Diazotrophs convert N_2 to NH_3 by the following net reaction

$$N_2 + 8\,H^+ + 8\,e^- + 16\,ATP + 16\,H_2O \rightarrow 2\,NH_3 + H_2 + 16\,ADP + 16\,P_i$$

The ammonia formed is added to α-ketoglutarate or glutamate by glutamate dehydrogenase or glutamine synthetase, respectively.

23. Nitrogenase, which catalyzes the reduction of N_2 to NH_3, is a two-subunit enzyme. Its Fe-protein contains one [4Fe–4S] cluster and two ATP-binding sites, and its MoFe-protein contains a P-cluster (a [4Fe–4S] cluster linked to a [4Fe–3S] cluster) and a FeMo cofactor (a [4Fe–3S] cluster and a [Mo–3Fe–3S] cluster bridged by 3 sulfide ions). The electrons to reduce N_2 come from oxidative or photosynthetic reactions, depending on the species. Although electron transfer must occur at least 8 times per N_2 molecule fixed (requiring a total of 16 ATP), the actual physiological cost of N_2 fixation is as high as 20–30 ATP.

24. Several of the enzymes involved in amino acid metabolism, including *E. coli* carbamoyl phosphate synthetase I, tryptophan synthase, and glutamate synthetase, exhibit channeling, in which reaction intermediates travel some distance from one active site to the next without dissociating from the enzyme. Channeling increases the rate of a metabolic pathway by generating high local concentrations of intermediates and by preventing their loss or degradation.

Questions

Protein Degradation

1. Indicate whether the following statements are true:
   (a) Proteins with the sequence Lys-Phe-Glu-Arg-Gln are selectively degraded by proteasomes.
   (b) Proteins containing sequences rich in Pro, Glu, Ser, and Thr often have short half-lives.
   (c) The addition of ubiquitin protects segments of a protein from proteolysis.
   (d) Lysosomal proteases degrade only extracellular proteins that enter the cell by endocytosis.
   (e) The ubiquitin-transfer reactions catalyzed by E2 and E3 do not require the input of free energy in the form of ATP.

Amino Acid Deamination

2. Draw the chemical structures for the products of transamination reactions involving α-ketoglutarate and (a) threonine, (b) isoleucine, and (c) glycine.

3. In what form is PLP found in aminotransferases prior to reacting with an amino acid? In what form is this coenzyme found after the release of the α-keto acid?

4. Aminotransferase reactions occur via a Ping Pong mechanism (see p. 376). Use Cleland notation to describe the reaction shown in Figure 21-8, including substrates, intermediates, and products.

5. In which direction would you expect flux through the glutamate dehydrogenase reaction in starved individuals? Why?

The Urea Cycle

6.   Write the overall equation for the formation of urea. What is the total free energy cost (in terms of "high-energy" phosphoanhydride bonds) per mole of urea synthesized?

7.   Which reactions of the urea cycle take place in the mitochondrion? Which step is the first committed step?

8.   Complete the diagram below, which shows the flux of alanine's amino group from its entry into the liver to its exit as urea.

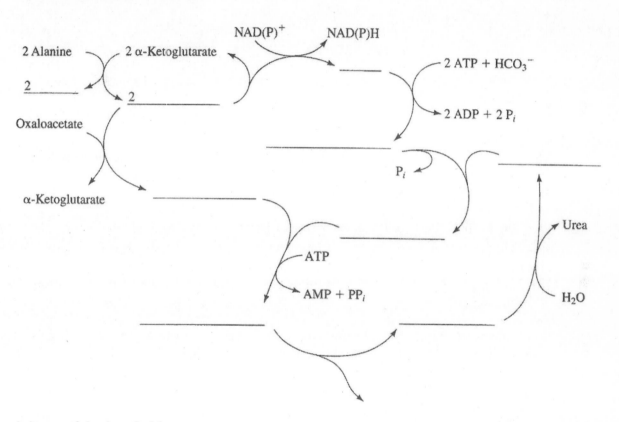

Breakdown of Amino Acids

9.   Which amino acids yield citric acid cycle intermediates upon transamination?

10. Reactions that involve PLP can be classified according to which bond to the alpha carbon is broken. For the amino acid–PLP adduct shown below, indicate what kind of reaction involves cleavage of bonds *a*, *b*, and *c* and provide an example of each.

Amino acid–PLP Schiff base (aldimine)

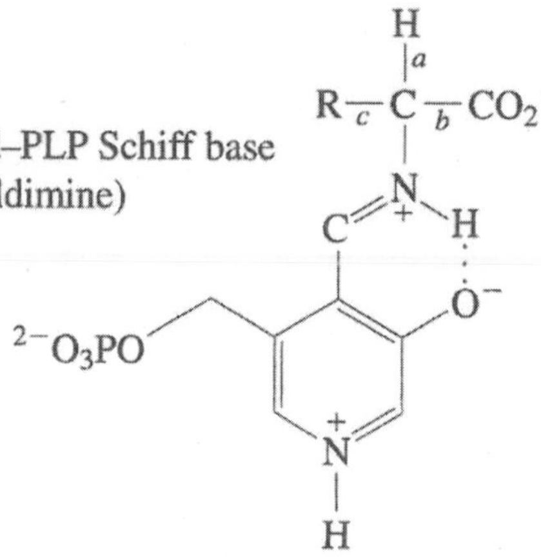

11. Many glucogenic amino acids are broken down to a citric acid cycle intermediate, which merely increases the catalytic activity of the cycle. How does the cell obtain a net yield of glucose from such amino acids?

12. Examine Figure 21-21 for the degradation of isoleucine, valine, and leucine.
    (a) Which reaction is analogous to the reaction catalyzed by the pyruvate dehydrogenase complex?
    (b) Which reaction is analogous to the reaction catalyzed by acyl-CoA dehydrogenase in fatty acid oxidation?

13. Some forms of maple syrup urine disease are amenable to treatment with large doses of dietary thiamine. What metabolic defect is being treated in these cases?

14. Which coenzymes mediate one-carbon transfer reactions?

15. How do sulfonamides selectively inhibit bacterial growth? Why aren't animals also adversely affected by these antibiotics?

16. *S*-adenosylmethionine (SAM) is a key methylating agent in several physiological processes, but its regeneration depends on the presence of another methylating agent. How is SAM regenerated?

17. Calculate the free energy yield for the complete oxidation of the carbon skeletons of (a) aspartate, (b) glutamine, and (c) lysine. Express your answer in ATP equivalents and assume that NADPH is equivalent to NADH.

Amino Acid Biosynthesis

18. Rationalize the significance of α-ketoglutarate as an activator of glutamine synthetase.

19.   The four intermediates of the urea cycle are α-amino acids.
      (a) Which one of these is considered to be an essential amino acid in children?
      (b) Outline a pathway by which adults can synthesize this amino acid from glucose.

20.   Chorismate is the seventh intermediate in the synthesis of tryptophan from
      phosphoenolpyruvate and erythrose-4-phosphate (Figure 21-34). Why is it considered a
      branch point in amino acid biosynthesis?

21.   What is the advantage of channeling? Why is it especially important for indole in the
      tryptophan synthase reaction?

Other Products of Amino Acid Metabolism

22.   Porphyrin is synthesized in a modular fashion, much like cholesterol. How many molecules
      of succinyl-CoA are required?

23.   How many carbons of heme would be labeled with ^{14}C if the starting material included (a)
      glycine with a ^{14}C-labeled carboxyl group and (b) succinyl-CoA with a ^{14}C-labeled
      carboxyl group?

24.   Heme oxygenase converts heme to the linear _____ and releases
      _____ and _____.

25.   Bilirubin serves as an antioxidant (i.e., it undergoes oxidation to protect other cell
      components from oxidation). How is bilirubin regenerated after its oxidation?

Nitrogen Fixation

26.   In mitochondrial respiration, electrons transferred from NADH and $FADH_2$ reduce O_2 to
      H_2O. Why can't these same coenzymes participate in the reduction of N_2 to NH_3?

Answers to Questions

1.   (a) False, (b) true, (c) false, (d) false, (e) true.

2. (a)

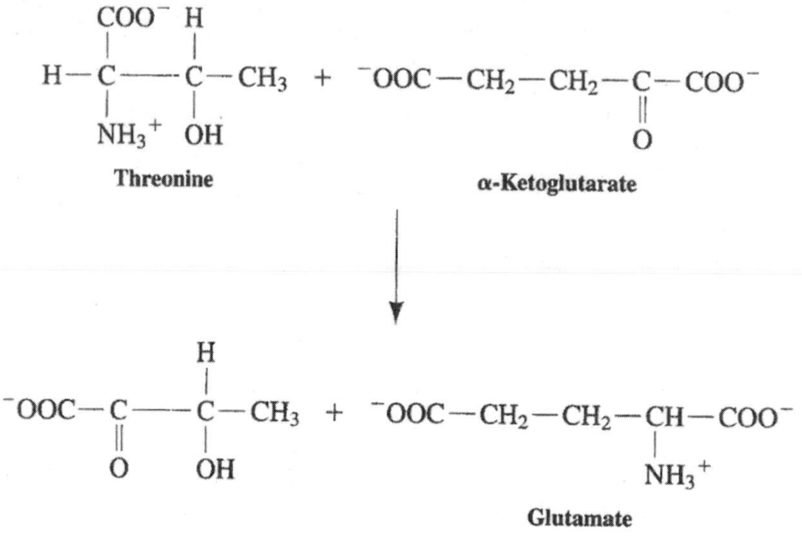

$$
\begin{array}{c}
\underset{\text{Threonine}}{
\overset{\displaystyle \text{COO}^-\ \text{H}}{
H-\underset{NH_3^+}{C}-\underset{OH}{C}-CH_3}}
\quad + \quad
\underset{\alpha\text{-Ketoglutarate}}{
{}^-OOC-CH_2-CH_2-\underset{O}{\overset{\|}{C}}-COO^-}
\end{array}
$$

$$\downarrow$$

$$
\begin{array}{c}
{}^-OOC-\underset{O}{\overset{\|}{C}}-\underset{OH}{\overset{H}{C}}-CH_3
\quad + \quad
\underset{\text{Glutamate}}{
{}^-OOC-CH_2-CH_2-\underset{NH_3^+}{CH}-COO^-}
\end{array}
$$

(b)

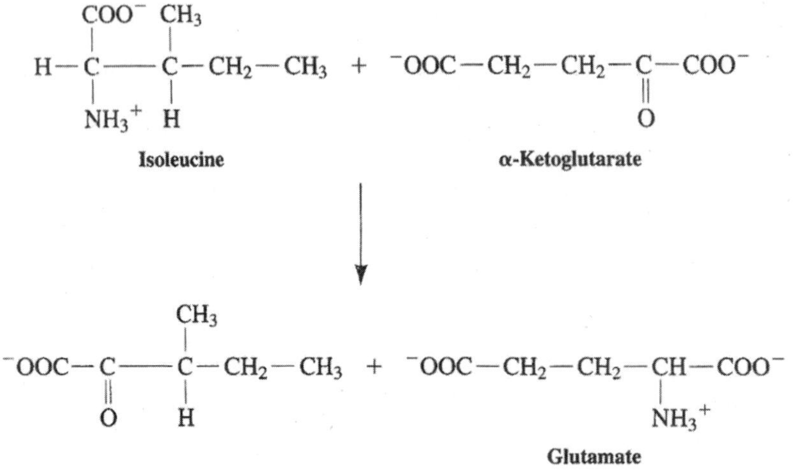

$$
\begin{array}{c}
\underset{\text{Isoleucine}}{
\overset{\displaystyle \text{COO}^-\ \text{CH}_3}{
H-\underset{NH_3^+}{C}-\underset{H}{C}-CH_2-CH_3}}
\quad + \quad
\underset{\alpha\text{-Ketoglutarate}}{
{}^-OOC-CH_2-CH_2-\underset{O}{\overset{\|}{C}}-COO^-}
\end{array}
$$

$$\downarrow$$

$$
\begin{array}{c}
{}^-OOC-\underset{O}{\overset{\|}{C}}-\underset{H}{\overset{CH_3}{C}}-CH_2-CH_3
\quad + \quad
\underset{\text{Glutamate}}{
{}^-OOC-CH_2-CH_2-\underset{NH_3^+}{CH}-COO^-}
\end{array}
$$

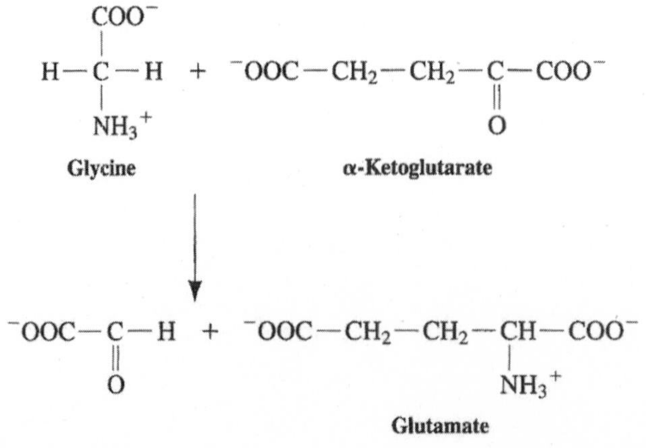

(c)

$$
\begin{array}{c}
\underset{\text{Glycine}}{
\overset{\displaystyle \text{COO}^-}{
H-\underset{NH_3^+}{C}-H}}
\quad + \quad
\underset{\alpha\text{-Ketoglutarate}}{
{}^-OOC-CH_2-CH_2-\underset{O}{\overset{\|}{C}}-COO^-}
\end{array}
$$

$$\downarrow$$

$$
\begin{array}{c}
{}^-OOC-\underset{O}{\overset{\|}{C}}-H
\quad + \quad
\underset{\text{Glutamate}}{
{}^-OOC-CH_2-CH_2-\underset{NH_3^+}{CH}-COO^-}
\end{array}
$$

3.   Prior to reacting with an amino acid, PLP is covalently bound to an enzyme Lys side chain via an imine linkage (Schiff base). With the release of the α-keto acid, PLP is converted to PMP, which is no longer covalently attached to the enzyme.

4.

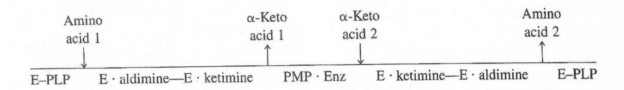

5.   Flux in this reaction would favor the production of α-ketoglutarate. Starved individuals break down protein (mostly muscle protein) and oxidize the resultant amino acids for energy. The α-ketoglutarate would be needed as an amino-group acceptor in the transamination of amino acids to be used as fuel.

6. NH_3 + HCO_3^- + aspartate + 3 ATP + H_2O $\rightarrow$
     Urea + fumarate + 2 ADP + AMP + 4 P_i

The total free energy cost is four ATP equivalents, since the PP_i produced in Step 3 is hydrolyzed by inorganic pyrophosphatase.

7.   Carbamoyl phosphate synthetase and ornithine transcarbamoylase are mitochondrial enzymes. The reaction catalyzed by carbamoyl phosphate synthetase is the first committed step.

8.

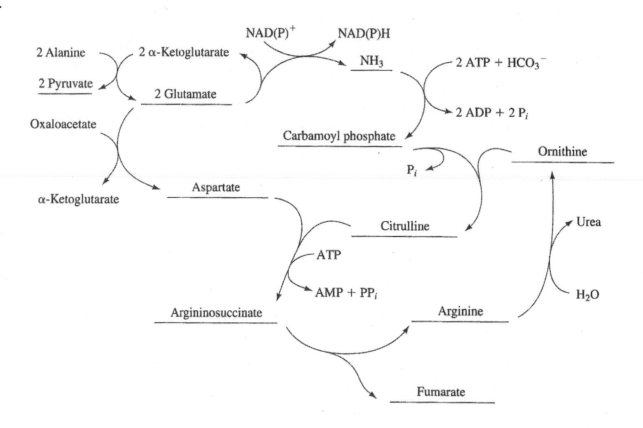

9. Aspartate and glutamate yield oxaloacetate and α-ketoglutarate, respectively.

10. Bond *a*: Transamination reactions such as occurs in the transfer of glutamic acid's amino group to oxaloacetate to yield α-ketoglutarate and aspartate.
    Bond *b*: The decarboxylation of α-amino acids such as occurs in the synthesis of physiologically active amines.
    Bond *c*: A C—C bond cleavage such as occurs in the serine hydroxymethyltransferase reaction.

11. All citric acid cycle intermediates can form oxaloacetate, which can then be decarboxylated and phosphorylated by PEP carboxykinase. The resulting PEP can then be converted to glucose by gluconeogenesis.

12.
    (a) Reaction 2
    (b) Reaction 3

13. In such cases, the defect is poor binding of thiamine pyrophosphate (TPP) to the E1 subunit of branched-chain α-keto acid dehydrogenase. The administration of large doses of the vitamin precursor of TPP improves the catalytic efficiency of the defective enzyme, which breaks down isoleucine, valine, and leucine.

14. Biotin, *S*-adenosylmethionine, and tetrahydrofolate.

15. Sulfonamides are structural analogs of the *p*-aminobenzoate group of folate and thereby interfere with folate synthesis. Unlike bacteria and plants, most animals cannot synthesize folate and hence are unaffected by these drugs.

16. After SAM donates its methyl group, the resulting *S*-adenosylhomocysteine is hydrolyzed to adenosine and homocysteine. Homocysteine is then methylated by N^5-methyltetrahydrofolate to form methionine, which is then adenylated by ATP to regenerate SAM (see Figure 21-18).

17. (a) 12.5 ATP. Aspartate is converted to oxaloacetate by transamination. Oxaloacetate is decarboxylated by PEP carboxykinase and the resulting PEP is converted to pyruvate and then to acetyl-CoA for entry into the citric acid cycle.

Reaction	Reduced coenzymes produced	ATP equivalents
PEP carboxykinase		−1 (GTP)
Pyruvate kinase		1
Pyruvate dehydrogenase	1 NADH	2.5
Citric acid cycle		1 (GTP)
	3 NADH	7.5
	1 FADH$_2$	1.5
Total		12.5

(b) 22.5 ATP. Glutamine is converted to glutamate and then oxidized to α-ketoglutarate (Figure 21-17), which is then converted to oxaloacetate. Oxidation of oxaloacetate proceeds as outlined in part *a*. (Alternatively, α-ketoglutarate can be converted to malate by the citric acid cycle. Malic enzyme converts malate to pyruvate in a reaction that produces NADPH. The net ATP yield is the same.)

Reaction	Reduced coenzymes produced	ATP equivalents
Glutamate dehydrogenase	1 NADPH	2.5
α-Ketoglutarate dehydrogenase	1 NADH	2.5
Succinyl-CoA synthetase		1 (GTP)
Succinate dehydrogenase	1 FADH$_2$	1.5
Malate dehydrogenase	1 NADH	2.5
PEP carboxykinase		−1 (GTP)
Pyruvate kinase		1
Pyruvate dehydrogenase	1 NADH	2.5
Citric acid cycle		1 (GTP)
	3 NADH	7.5
	1 FADH$_2$	1.5
Total		22.5

(c) 28 ATP. Lysine is converted to acetoacetate in 11 reactions (Figure 21-22), several of which yield reduced coenzymes. The acetoacetate is converted to 2 acetyl-CoA at the expense of 1 GTP (since a CoA group is donated by succinyl-CoA).

Reaction	Reduced coenzymes produced	ATP equivalents
Lysine degradation		
Reaction 1	–1 NADPH	–2.5
Reaction 2	1 NADH	2.5
Reaction 3	1 NAD(P)H	2.5
Reaction 5	1 NADH	2.5
Reaction 6	1 FADH$_2$	1.5
Reaction 9	1 NADH	2.5
3-Ketoacyl-CoA transferase		–1 (succinyl-CoA)
2 Citric acid cycle		2 (GTP)
	6 NADH	15
	2 FADH$_2$	3
Total		28

18. The oxidative deamination of glutamate by glutamate dehydrogenase yields ammonia and α-ketoglutarate. Hence rising levels of α-ketoglutarate signal rising levels of free ammonia. The activation of glutamine synthetase by α-ketoglutarate increases the production of glutamine to help prevent the accumulation of ammonia, which is toxic in high concentrations.

19.

(a)) Arginine
(b) Glucose is degraded to 2 pyruvate by glycolysis. One pyruvate is converted to oxaloacetate by the pyruvate carboxylase reaction. The other pyruvate, converted to acetyl-CoA by pyruvate dehydrogenase, combines with the oxaloacetate to form citrate, then isocitrate and then α-ketoglutarate (the first three reactions of the citric acid cycle). Glutamate dehydrogenase, which operates close to equilibrium, converts α-ketoglutarate and NH$_3$ to glutamate. Three additional reactions convert glutamate to ornithine (Figure 21-30). Since ornithine is a urea cycle intermediate, the reactions of the urea cycle convert it to the other cycle intermediates, including arginine.

20. Chorismate is also the precursor of phenylalanine and tyrosine, whose biosynthesis branches from the tryptophan pathway at chorismate.

21. Channeling, the transfer of an intermediate directly from one active site to another, increases the efficiency of the overall reaction by preventing the loss of the intermediate, which could then diffuse away or be degraded. Channeling is important for indole, a nonpolar molecule that might otherwise be lost from the cell.

22. Eight succinyl-CoA units are required to synthesize the porphyrin ring:

8 Succinyl-CoA + 8 glycine → 8 ALA → 4 BPG → uroporphyrinogen III

23.
 (a) None. The carboxyl group of glycine is lost as carbon dioxide during condensation of glycine with succinyl-CoA.
 (b) Two. Uroporphyrinogen III contains all eight succinyl-derived carboxyl groups. However, subsequent steps remove six of these (Figure 21-37).

24. Heme oxygenase converts heme to the linear <u>biliverdin</u> and releases Fe^{3+} and <u>CO</u>.

25. Oxidized bilirubin is biliverdin, which is reduced by NADPH to regenerate bilirubin.

26. The reduction potentials of NADH (–0.315 V) and $FADH_2$ (–0.219 V) are not sufficiently negative for these coenzymes to reduce N_2 ($\mathscr{E}°' = -0.340$ V).

22

Mammalian Fuel Metabolism: Integration and Regulation

This chapter summarizes and integrates the major pathways involved in fuel metabolism. It first examines the variations of fuel metabolism in the brain, muscle, adipose tissue, liver, and kidney. Each of these organs exhibits specialized needs and functions, which influence the regulation of fuel metabolism in that organ. The discussion in this chapter then turns to the ways in which fuel metabolism is interconnected between the liver and skeletal muscle via the Cori cycle and the glucose–alanine cycle. Fuel metabolism is regulated principally by the action of hormones, which bind to specific receptor proteins. Three signal transduction pathways regulate fuel metabolism: the adenylate cyclase signaling system and the phosphoinositide pathway activated by G protein–coupled receptors, and signaling pathways activated by receptor tyrosine kinases. This chapter concludes with a discussion of disturbances in fuel metabolism, in particular, starvation, diabetes, and obesity, with a brief diversion into the discovery of insulin (Box 22-1).

Essential Concepts

Organ Specialization

1. The biochemical pathways for the synthesis and breakdown of the three major metabolic fuels (carbohydrates, fatty acids, and protein) converge on a small number of metabolites: pyruvate, acetyl-CoA, and citric acid cycle intermediates. In animals, flux through these pathways is tissue specific. In mammals, the liver is one of the few organs that can carry out most of the biochemical pathways involved in energy metabolism (see Figure 22-1).

2. Some organs do not synthesize fuel molecules but only catabolize these molecules. Muscle and brain are two such organs. The brain's exclusive fuel source is normally glucose, and because the brain stores no glycogen, it relies on glucose present in the bloodstream. Prolonged fasting results in a gradual switch from glucose catabolism to ketone body catabolism in the brain. Skeletal and heart muscles rely on fatty acids when resting and use glucose from stored glycogen or from the blood upon exertion.

3. Adipose tissue is largely involved in the long-term storage of fatty acids (in the form of triacylglycerols). It releases fatty acids into the bloodstream in response to hormones (e.g., epinephrine). The synthesis of triacylglycerols requires the catabolism of glucose to glycerol-3-phosphate, to which fatty acids are esterified.

4. The liver maintains homeostatic levels of circulating fuel molecules for use by all tissues. The liver metabolizes excess glucose by converting it to glucose-6-phosphate (G6P). This reaction is catalyzed by glucokinase (an isozyme of hexokinase) whose relatively high K_M (~5 mM) makes it sensitive to changes in glucose concentration over the physiological

range (~5 mM). In contrast, hexokinase (K_M < 0.1 mM) is saturated with glucose at these glucose levels.

5. In the liver, G6P has several fates:
    (a) Polymerization into glycogen.
    (b) Conversion to glucose, which then equilibrates with the glucose in the bloodstream.
    (c) Catabolism into acetyl-CoA, which can be further catabolized for the production of ATP, or diverted to anabolic routes for the synthesis of fatty acids.
    (d) Conversion via the pentose phosphate pathway to ribose-5-phosphate, an important precursor for nucleotide synthesis, and NADPH for use in biosynthetic reactions.

6. The liver can synthesize and degrade triacylglycerols based on the body's metabolic needs. It also degrades amino acids to a variety of intermediates, which are then used for fatty acid synthesis (from the ketogenic amino acids) and gluconeogenesis (from the glucogenic amino acids). The kidney also carries out gluconeogenesis from the α-ketoglutarate generated when ammonia is released from glutamine and glutamate.

7. Some metabolic pathways, called interorgan pathways, are segregated between tissues that exchange metabolites (if such a pathway occurred entirely within a single organ, it would constitute a futile cycle). The Cori cycle is defined by the exchange of glucose and lactate between muscle and liver. The lactate generated by anaerobic glycolysis in muscle permits the muscle to anaerobically generate ATP required for contraction, while postponing the aerobic production of ATP in the liver necessary to resynthesize glucose.

8. Another interorgan pathway is the glucose–alanine cycle. Glycolytically produced pyruvate serves as an amino-group acceptor in transamination reactions in the muscle during protein degradation. The resulting alanine is transported to the liver, where it is reconverted to pyruvate for gluconeogenesis with the released amino group appearing in urea.

Hormonal Control of Fuel Metabolism

9. Hormones are chemical signals that reach their target cells via the bloodstream. Each hormone binds to a specific receptor, and its binding results in the transduction of the hormone signal into a biochemical response inside the cell. Many factors may modulate the response of a cell to a specific hormone, such as the number of receptors and the activities of other receptors.

10. Energy metabolism is regulated largely by pancreatic and adrenal hormones. The pancreas secretes the polypeptide hormones glucagon and insulin, which act antagonistically to each other to regulate blood glucose concentration. The adrenal medulla secretes the catecholamines epinephrine and norepinephrine, which bind to α- and β-adrenoreceptors. These receptors often respond to the same ligand in opposite ways in different tissues.

11. A variety of signal transduction pathways regulate metabolism (see Figure 22-10). Glucagon and the aforementioned catecholamines exert their regulation via adenylate cyclase and/or the phosphoinositide pathways, while regulation by insulin occurs via the

receptor tyrosine kinase–Ras signaling cascade (though it can activate the phosphoinositide pathway as well (see Figure 13-31).

Metabolic Homeostasis: Regulation of Energy Metabolism, Appetite, and Body Weight

12. AMP-dependent protein kinase (AMPK) serves as a major sensor for the regulation of metabolic homeostasis. Allosteric control of the phosphorylated form of AMPK by AMP and ATP results in AMPK-mediated activation of metabolic breakdown pathways and inhibition of biosynthetic pathways in order to conserve ATP during metabolic stress. Adiponection is a key polypeptide hormone secreted by adipocytes that leads to the activation of AMPK.

13. Five peptide hormones that regulate appetite include leptin, neuropeptide Y, insulin, ghrelin, and PYY_{3-36}. Insulin, leptin, and PYY_{3-36} suppress appetite, while ghrelin and neuropeptide Y stimulate appetite. Diminished response to leptin or the secretion of ghrelin by the stomach results in the secretion of neuropeptide Y by the hypothalamus. Conversely, both leptin and insulin inhibit the secretion of neuropeptide Y. While ghrelin appears largely to control short-term appetite, leptin may control both long-term and short-term appetite. The gastrointestinal tract secretes the appetite-suppressing peptide hormone PYY_{3-36}.

Disturbances in Fuel Metabolism

14. Starvation requires the mobilization of stored fuel molecules. Early in a fast, the liver accelerates glycogenolysis and increases gluconeogenesis from available materials. This makes glucose available continuously to the central nervous system. Skeletal muscle protein becomes a source of glucose via glucogenic amino acids released upon proteolysis. After several days, the liver shifts its metabolic machinery toward the breakdown of fatty acids in order to produce ketone bodies. The brain eventually adapts to use ketone bodies rather than glucose as a primary metabolic fuel.

15. Diabetes mellitus is a pathological condition that results from insufficient secretion of insulin or inefficient stimulation of its target cells. There are two major forms of diabetes:
    (a) Insulin-dependent or juvenile-onset diabetes mellitus (type I) results from an autoimmune response that destroys pancreatic β cells, which secrete insulin.
    (b) Non-insulin-dependent or maturity-onset diabetes mellitus (type II) may be caused by a decrease in the number of insulin receptors or their ability to bind insulin. This leads to the clinical condition called insulin resistance, in which insulin levels are far higher than normal but target tissues exhibit little or no response to the hormone.

16. While obesity can result from the loss of proper leptin production or responsiveness, most cases of obesity are caused by high-calorie diets that result in a new set point for metabolic control. Obesity and long-term high-calorie diets appear to be major contributors to a phenomenon referred to as metabolic syndrome. This syndrome is characterized by insulin resistance, systemic inflammation, and an increased likelihood of developing atherosclerosis, hypertension, and type 2 diabetes, which can lead to coronary heart disease.

Questions

Organ Specialization

1. List the possible products of acetyl-CoA metabolism in animals. Include the pathway involved and, where appropriate, the key regulatory enzyme that commits acetyl-CoA to that pathway.

Fate	Pathway	Key Enzyme
1.		
2.		
3.		
4.		
5.		

2. List the metabolic products of pyruvate metabolism in animals. Include the pathway involved and, where appropriate, the key regulatory enzyme that commits pyruvate to that pathway.

Fate	Pathway	Key Enzyme
1.		
2.		
3.		
4.		

3. List the metabolic products of glucose-6-phosphate metabolism. Include the pathway involved and the key regulatory enzymes.

Fate	Pathway	Key Enzyme
1.		
2.		
3.		
4.		

4. Match the metabolic pathways below with the cellular compartments in which they occur.
   A. Cytosol
   B. Inner mitochondrial membrane
   C. Mitochondrial matrix
   D. Mitochondrial intermembrane space
   E. Peroxisome
   F. Endoplasmic reticulum

   ____ Glycolysis
   ____ Lactic fermentation
   ____ Pyruvate dehydrogenation
   ____ Citric acid cycle
   ____ Oxidative phosphorylation of ADP
   ____ Pentose phosphate pathway
   ____ Fatty acid elongation and desaturation

   ____ Palmitoyl-CoA oxidation
   ____ Palmitate synthesis
   ____ Lignoceric (24:0) acid synthesis
   ____ Amino acid degradation
   ____ Urea cycle
   ____ Gluconeogenesis
   ____ Cholesterol synthesis

5.   Match the metabolic function below with the appropriate pathway(s).

A.   Glycolysis
B.   Pentose phosphate pathway
C.   Gluconeogenesis
D.   Fatty acid oxidation
E.   Glycogen synthesis
F.   Amino acid degradation

(a) Provides reducing equivalents for biosynthesis.
(b) Provides glucose during the early phase of a fast, after glycogen stores are depleted.
(c) Provides energy for heart muscle.
(d) Stores metabolic fuel in the liver.
(e) Provides ATP during a rapid burst of skeletal muscle activity.
(f) Generates the nucleotide precursor ribose-5-phosphate.
(g) Provides energy for gluconeogenesis during a fast
(h) Provides energy immediately after a meal

6.   What is the significance of high levels of hexokinase activity in the brain?

7.   What similarities are shared by the Cori cycle and the glucose–alanine cycle? What distinguishes them?

8.   Muscle phosphofructokinase is allosterically stimulated by NH_4^+. What is the physiological function of this stimulation?

9.   How does an increase in blood glucose affect triacylglycerol metabolism in adipocytes?

Hormonal Control of Fuel Metabolism

10. Without referring to the text, fill in the following table by writing increase, decrease or no effect:

Tissue	Glucagon	Insulin	Epinephrine
Muscle			
Glucose uptake			
Glycogen synthesis			
Adipose			
Glucose uptake			
Lipolysis			
Lipogenesis			
Liver			
Glycogen synthesis			
Lipogenesis			
Gluconeogenesis			

11. Which glucose transporter (GLUT2 or GLUT4) would you expect to find in the insulin-secreting β-islet cells of the pancreas? Provide a rationale.

Metabolic Homeostasis: Regulation of Energy Metabolism, Appetite, and Body Weight

12. Why does AMPK respond to AMP and ATP instead of ADP and ATP?

13. Fill in the following table with the effects of AMPK in cardiac muscle, skeletal muscle, liver, and adipose tissue. What general metabolic flux is favored by the activities of AMPK in these tissues?

Tissue	*AMPK activity*
Cardiac Muscle	
Skeletal Muscle	
Liver	
Adipose Tissue	

14. Which hormones regulate the production of the appetite-stimulating hormone neuropeptide Y? Which acts as a long-term regulator? Which is a short-term regulator?

Disturbances in Fuel Metabolism

15. List the order in which the following energy sources are used by skeletal muscle to produce ATP: protein, phosphocreatine, fatty acids, and glycogen.

16. List the order in which the liver uses the following substances to provide the body with metabolic fuel during starvation: glycogen, fatty acids, muscle protein, and nonmuscle protein. Explain your answer.

17. You obtain a liver homogenate from a fasting mouse and "spike" it with a high concentration of glucose. In what form would you expect glycogen phosphorylase?

18. During a fast to lose weight, is it important to be physically active or might it be better to remain sedentary? Explain.

Answers to Questions

1.   Note that animals cannot undertake the net conversion of acetyl units to glucose.

Fate	Pathway	Key Enzyme
1. CO_2 and H_2O	Citric acid cycle	Citrate synthase
2. Acetoacetate and/or β-hydroxybutyrate	Ketogenesis	Mitochondrial HMG-CoA synthase
3. Fatty acids	Fatty acid synthesis	Acetyl-CoA carboxylase
4. Cholesterol	Cholesterol synthesis	Cytosolic HMG-CoA synthase
5. Amino acids	Pathways utilizing citric acid cycle intermediates	Varies; often catalyzes an exergonic reaction coupled to ATP hydrolysis, e.g., glutamine synthetase

2.   Note that the conversion of lactate to pyruvate is reversible.

Fate	Pathway	Key Enzyme
1. CO_2 and H_2O	Pyruvate dehydrogenation/ citric acid cycle	Pyruvate dehydrogenase
2. Oxaloacetate for citric acid cycle	Anaplerotic reactions	Pyruvate carboxylase
3. Glucose	Gluconeogenesis	Phosphofructokinase-2, which alters the levels of F2,6P, a key allosteric regulator of PFK-1 and FBPase-1
4. Amino acids	Transamination and pathways utilizing citric acid cycle intermediates	Varies; the transamination of pyruvate to alanine is a reversible transamination

3.

Fate	Pathway	Key Enzyme
1. Glucose	Glucose transport (liver)	Glucose-6-phosphatase
2. Glycogen	Glycogen synthesis	Glycogen synthase
3. Acetyl-CoA	Glycolysis/pyruvate dehydrogenation	PFK-1 and pyruvate dehydrogenase
4. Ribose-5-phosphate and NADPH	Pentose phosphate pathway	G6P dehydrogenase

4.

  A   Glycolysis   A   Palmitoyl-CoA oxidation
  A   Lactic fermentation   A   Palmitate synthesis
  C   Pyruvate dehydrogenation   E   Lignoceric (24:0) acid synthesis
  C   Citric acid cycle   A,C Amino acid degradation
  B   Oxidative phosphorylation of ADP   A,C Urea cycle
  A   Pentose phosphate pathway   A,C Gluconeogenesis
  F   Fatty acid elongation and desaturation   A   Cholesterol synthesis

5.   (a) B; (b) C, F; (c) D; (d) E; (e) A; (f) B; (g) D; (h) A, F

6.   The brain (and the rest of the central nervous system) has a high energy demand, and under normal conditions, its sole source of metabolic fuel is blood-delivered glucose. Hexokinase has a high affinity for glucose, so that glucose that enters a brain cell via facilitated diffusion can be rapidly converted to glucose-6-phosphate, thereby allowing additional glucose to enter the cell.

7.   Both cycles operate under conditions of metabolic stress. In both cycles, metabolites arriving from skeletal muscle are converted by the liver to glucose, which is returned to the bloodstream. The Cori cycle involves the synthesis of glucose from lactate produced by lactic fermentation in muscle during sustained vigorous exercise. The glucose–alanine cycle involves the synthesis of glucose from alanine produced in the muscle. It is thereby a mechanism of transporting nitrogen from muscle to liver. However, under conditions of starvation, the cycle is broken and the muscle-derived glucose is supplied to other tissues.

8.   The degradation of proteins yields amino acids whose further breakdown, in many cases [e.g., glycine, cysteine, serine, and threonine (Figure 21-14); and glutamine (Figure 21-17)], yields ammonia. The ammonia is then transferred to α-ketoglutarate via the glutamate dehydrogenase reaction to yield glutamate, and the resulting α-amino group is transferred to pyruvate by a transaminase to yield alanine. The alanine is then transported to the liver where, via transaminases, glutamate dehydrogenase, and urea cycle, its α-amino group is transferred to urea for excretion by the kidneys.

The elevation of the NH_4^+ level in muscle is indicative of increased protein breakdown (e.g., due to starvation). The activation by NH_4^+ of muscle phosphofructokinase (PFK) stimulates glycolysis (PFK is the major flux-determining enzyme) and thereby increases the rate of pyruvate production to accommodate the increased need to synthesize alanine.

9.   Increasing levels of blood glucose favor triacylglycerol synthesis as the rate of glucose uptake by adipocytes increases. The three-carbon backbone of triacylglycerols is glycerol-3-phosphate, which is obtained by the reduction of dihydroxyacetone phosphate. A high level of glycolytic activity in adipocytes results in a sufficiently large pool of glycerol-3-phosphate to favor fatty acid esterification.

10.   See Table 22-1 on p. 801.

11.   The pancreas is the sensor organ for blood glucose concentration, while the liver serves as the primary effector organ that regulates blood glucose concentration. For the pancreas, it makes more biological sense to have a transporter that is constitutively expressed, such as the GLUT2 transporter, which is in fact the case.

12.   Recall from Section 15-4A that ATP, ADP, and AMP concentrations are rapidly equilibrated by adenylate kinase (2 ADP $\leftrightarrows$ ATP + AMP). Small (<10%) decreases in [ATP] consequently result in rapid, large increases in [ADP] and at least 4-fold increases in [AMP].

13.   The activity of AMPK promotes metabolic pathways that rapidly produce ATP while inhibiting energy-storing biosynthetic pathways.

Tissue	AMPK activity
Cardiac Muscle	Activation of glycolysis
Skeletal Muscle	Activation of fatty acid oxidation and glucose uptake
Liver	Inhibition of lipogenesis and gluconeogenesis
Adipose Tissue	Inhibition of lipolysis

14.   Decreases in leptin act as a long-term regulator of neuropeptide Y production, while ghrelin and PYY_{3-26} appear to act as short-term regulators of neuropeptide Y production.

15.   Phosphocreatine, glycogen, fatty acids, protein.

16. The order of usage of these metabolic fuels is glycogen, muscle protein, fatty acids, and nonmuscle protein. Glucose is the primary metabolic fuel of muscle and the central nervous system. After the first day of a fast, the glycogen stores of the liver (and muscle) are exhausted. In order to maintain circulating glucose levels, muscle protein is broken down into amino acids, many of which are glucogenic. Extended fasting alters the metabolic machinery of the liver to favor fatty acid degradation into ketone bodies, thereby conserving critical muscle mass. Severe, extended starvation eventually leads to loss of nonmuscle protein, which can severely compromise organ function and can be life-threatening.

17. During a fast, glucagon stimulates glycogen degradation via PKA, which phosphorylates and activates glycogen phosphorylase (glycogen phosphorylase a). The addition of a large amount of glucose shifts the $T \rightleftharpoons R$ equilibrium toward the T state. Stabilization of the T state by glucose binding promotes dephosphorylation of Ser 14, which converts the enzyme to the less active glycogen phosphorylase b.

18. A certain level of activity facilitates the transition from metabolism of glucose to metabolism of fatty acids. In a sedentary fasting individual, inactive muscle would provide more metabolic fuel via protein degradation than active muscle; hence, a completely sedentary individual might experience significant loss of muscle mass without a major loss of adipose tissue mass, the primary target for weight loss. Therefore, it is important to be active rather than sedentary while dieting.

23
Nucleotide Metabolism

This chapter presents nucleotide metabolism, including the synthesis of purine and pyrimidine ribonucleotides, the conversion of ribonucleotides to deoxyribonucleotides, and the degradation of nucleotides. As you saw in Chapter 3, nucleotides are composed of a purine or pyrimidine which is linked to C1′ of either ribose or deoxyribose. The pentose in turn is esterified through C5′ to one or more phosphate groups. Nucleoside triphosphates are precursors of nucleic acids and also, through the hydrolysis of one or both of their phosphoanhydride bonds, provide the free energy that drives many biochemical reactions. This chapter begins with a detailed discussion of the synthesis of purine nucleotides, describing the source of each of the carbons and nitrogens of the purine ring. The chapter then discusses the intricate system of negative and positive feedback regulation that balances the synthesis of ATP and GTP. Cells recycle the purines resulting from the turnover of nucleic acids via salvage pathways. The following section describes the synthesis of pyrimidines, in which negative feedback mechanisms balance the amounts of UTP and CTP in the cell. Next, the chapter discusses the formation of deoxyribonucleotides by the action of ribonucleotide reductase. This enzyme has several unique features, including its mechanism of action and its intricate allosteric regulatory mechanisms. The formation of deoxyUTP provides the precursor for deoxythymidylate. The key enzyme involved here is thymidylate synthase, which is a target of cancer chemotherapy, as described in Box 23-1. Finally, the chapter considers the degradation of nucleotides, which is in some respects a continuation of nitrogen metabolism discussed in Chapter 21. The catabolism of purines leads to the formation of uric acid, and the catabolism of pyrimidines leads to the formation of β-alanine and β-aminoisobutyrate. The fate of uric acid varies among animal groups depending on their strategies for excreting excess nitrogen.

Essential Concepts

Synthesis of Purine Ribonucleotides

1. Early investigations of the nucleotide biosynthetic pathway identified the precursors that supply the carbon and nitrogen atoms of the purine ring: N1 from aspartate; C2 and C8 from formate; N3 and N9 from glutamine; C4, C5, and N7 from glycine; and C6 from bicarbonate.

2. The purines are synthesized as ribonucleotides rather than as free bases. The purine nucleotide biosynthetic pathway produces inosine 5′-monophosphate (IMP) in 11 steps:
    (a) Ribose is activated by reaction with ATP to form 5-phosphoribosyl-α-pyrophosphate (PRPP).
    (b) The amide nitrogen of glutamine (which becomes N9) displaces the pyrophosphate group of PRPP, inverting the configuration at C1′ to form β-5-phosphoribosylamine.

(c) A molecule of glycine is added, providing C4, C5, and N7 and forming glycinamide ribotide (GAR).

(d) The free amino group of GAR reacts with the one-carbon carrier N10-formyltetrahydrofolate to add C8 and give formylglycinamide ribotide (FGAR).

(e) The amide nitrogen of a second glutamine is added to FGAR in an ATP-dependent reaction, which provides N3 and yields formylglycinamidine ribotide (FGAM).

(f) The purine imidazole ring is closed in a reaction requiring ATP and producing 5-aminoimidazole ribotide (AIR).

(g) Purine C6 is incorporated into AIR in a carboxylation reaction to give carboxyaminoimidazole ribotide (CAIR).

(h) Aspartate contributes N1 by forming an amide bond with C6 to yield 5-aminoimidazole-4-(N-succinylocarboxamide) ribotide (SACAIR).

(i) Fumarate is cleaved from SACAIR to produce 5-aminoimidazole-4-carboxamide ribotide (AICAR). Note that reactions (h) and (i) resemble reactions of the urea cycle.

(j) The last purine ring atom, C2, is added through formylation by N10-formyltetrahydrofolate to form 5-formaminoimidazole-4-carboxamide ribotide (FAICAR).

(k) In the final step, the larger ring is closed to form IMP.

3. IMP is rapidly transformed into AMP and GMP. To form AMP, the amino group of aspartate displaces the carbonyl oxygen at C6 of the purine base, followed by removal of fumarate. To form GMP, IMP is oxidized in an NAD^+-dependent reaction to yield xanthosine monophosphate (XMP), which then accepts a glutamine amide nitrogen to produce the guanine nucleotide. Note that AMP synthesis from IMP requires cleavage of GTP to GDP + P_i, whereas GMP formation requires ATP cleavage to AMP + PP_i.

4. Nucleoside monophosphates are converted to di- and triphosphate derivatives in two steps. First, a specific nucleoside monophosphate kinase catalyzes phosphorylation of a nucleoside monophosphate by a nucleoside triphosphate to yield two nucleoside diphosphates. Second, a nucleoside diphosphate is phosphorylated to the corresponding triphosphate by the action of nucleoside diphosphate kinase.

5. Purine nucleotide biosynthesis is carefully regulated. IMP production is controlled at its first two steps. Ribose phosphate pyrophosphokinase, which generates PRPP, is inhibited through negative feedback by ADP and GDP. Amidophosphoribosyl transferase, which produces phosphoribosylamine, is inhibited by adenine nucleotides and by guanine nucleotides, which bind at separate inhibitory sites. PRPP also allosterically stimulates amidophosphoribosyl transferase.

6. Additional regulation occurs immediately beyond the IMP branch point, because AMP and GMP inhibit their own synthesis. Moreover, a balance in the synthesis of the two purine nucleotides is achieved because GTP is necessary for AMP synthesis and ATP is required for GMP formation (see Figure 23-4).

7. Purines released from the degradation of nucleic acids are re-utilized via metabolic salvage pathways. In mammals, distinct enzymes catalyze the reactions of adenine and guanine with PRPP to form AMP and GMP, respectively. Hypoxanthine–guanine

phosphoribosyltransferase (HGPRT) uses guanine as a substrate but also transforms the purine base hypoxanthine into IMP. An inherited deficiency in HGPRT results in Lesch–Nyhan syndrome, a disease characterized by bizarre behavioral abnormalities.

Synthesis of Pyrimidine Ribonucleotides

8. Pyrimidine biosynthesis utilizes only three precursors. Aspartate contributes N1, C4, C5, and C6 of the pyrimidine ring, whereas C2 is derived from bicarbonate and N3 is donated by glutamine.

9. In pyrimidine nucleotide formation, the ring is completed before being linked to ribose-5-phosphate. The six steps of the UMP biosynthetic pathway are as follows:

   (a) Bicarbonate and the amide nitrogen of glutamine combine to form carbamoyl phosphate in a reaction that requires 2 ATP and is catalyzed by carbamoyl synthetase II. This enzyme is distinct from carbamoyl synthetase I, which forms carbamoyl phosphate during the urea cycle, a process in which ammonia is the nitrogen donor.

   (b) Carbamoyl phosphate reacts with aspartate to yield carbamoyl aspartate. The reaction is catalyzed by aspartate transcarbamoylase, a key enzyme for pyrimidine biosynthesis regulation in *E. coli* but not in animals.

   (c) Carbamoyl aspartate is converted to the ring compound dihydroorotate by elimination of water.

   (d) Dihydroorotate is dehydrogenated to form orotate, with a mitochondrial quinone serving as the ultimate electron acceptor.

   (e) Orotate reacts with PRPP to produce orotidine-5'-monophosphate (OMP). The same enzyme can also salvage uracil and cytosine by catalyzing their conversion to UMP and CMP, respectively.

   (f) OMP undergoes decarboxylation to form UMP.

10. In microorganisms, the six reactions of UMP biosynthesis are carried out by separate enzymes. However, in animals, several of the reactions are catalyzed by polypeptides that possess more than one enzyme activity. This is also true for certain steps in purine formation. In these instances, pathway intermediates are not liberated from the multifunctional proteins but are channeled directly to the next active site.

11. As in purine nucleotide biosynthesis, UTP is formed from UMP by the successive actions of a nucleoside monophosphate kinase and nucleoside diphosphate kinase.

12. CTP is synthesized through amination of UTP by glutamine (in animals) or ammonia (in bacteria) in an ATP-dependent reaction.

13. The regulation of pyrimidine biosynthesis in bacteria is accomplished by the allosteric modulation of aspartate transcarbamoylase, such that the binding of ATP to the enzyme stimulates activity and the binding of CTP or UTP inhibits it. Pyrimidine formation in animals is regulated by changes in carbamoyl synthetase II activity. This enzyme is inhibited by UDP and UTP and stimulated by ATP and PRPP. In addition, decarboxylation of OMP is diminished by elevated UMP levels.

Formation of Deoxyribonucleotides

14. Deoxyribonucleotides are synthesized by the reduction of the hydroxyl group at C2′ of the corresponding ribonucleotide. Enzymes that catalyze these reactions are called ribonucleotide reductases.

15. Ribonucleotide reductase in eukaryotes and some prokaryotes reduces a ribonucleoside diphosphate (NDP) to a deoxyribonucleoside diphosphate (dNDP). The enzyme is a heterotetramer that can be dissociated into inactive homodimers, R1 and R2. R1 contains a substrate-binding site and three independent allosteric sites that control enzymatic activity and substrate specificity. Each subunit of R2 has an Fe(III) prosthetic group that interacts with Tyr 122 to form a tyrosyl free radical, which plays a critical role in the reaction mechanism. Five cysteine residues in each subunit of R1 also participate, some acting as electron donors to reduce the substrate and generate a disulfide bond.

16. Following NDP reduction, the ribonucleotide reductase disulfide bond is reduced via disulfide interchange with the small protein thioredoxin, which regenerates the active enzyme. Oxidized thioredoxin is subsequently reduced by thioredoxin reductase, which obtains electrons from NADPH. This coenzyme is thus the ultimate reducing agent in the conversion of an NDP to a dNDP.

17. The production of balanced amounts of deoxyribonucleotides needed for normal DNA synthesis is achieved by elaborate allosteric control of ribonucleotide reductase. Binding of nucleotides to the specificity site, activity site, and hexamerization site govern the enzyme's substrate specificity, catalytic activity, and oligomerization (which also affects catalytic activity).

18. Nucleoside diphosphate kinase catalyzes the ATP-dependent phosphorylation of dNDPs to produce the dNTPs needed for DNA synthesis.

19. In the synthesis of thymine deoxyribonucleotides, dUTP is first hydrolyzed to dUMP and then methylated by thymidylate synthase to form dTMP. The methyl donor in this reaction is N^5,N^{10}-methylenetetrahydrofolate, which is simultaneously oxidized to dihydrofolate. Dihydrofolate is reduced to tetrahydrofolate with NADPH by dihydrofolate reductase. Tetrahydrofolate then accepts a hydroxymethyl group from serine to regenerate the methyl donor. Finally, dTMP is phosphorylated to dTTP. Drug-induced inhibition of dTMP synthesis is an important tool in cancer therapy (see Box 23-1).

Nucleotide Degradation

20. The breakdown of purine nucleotides leads to production of uric acid. Degradation involves the dephosphorylation of the nucleotides to nucleosides, followed by the hydrolysis or phosphorolysis of the nucleosides to yield the free purine base and ribose or ribose-1-phosphate. The bases may be reincorporated into purine nucleotides via the salvage reactions. Adenosine is deaminated to form inosine prior to further degradation.

21. Both hypoxanthine, which is derived by phosphorolysis of inosine, and guanine are oxidized sequentially to xanthine and then uric acid. Xanthine oxidase catalyzes the two reactions that convert hypoxanthine to uric acid, via a complex series of electron-transfer events.

22. Uric acid is the final product of purine degradation for primates, birds, terrestrial reptiles, and some insects. Excretion of uric acid is advantageous in that it requires little accompanying water. Other organisms oxidize uric acid to allantoin, allantoic acid, urea, or ammonia prior to excretion.

23. A variety of factors may lead to abnormally high levels of uric acid, which can result in gout.

24. Pyrimidine nucleotide catabolism occurs through reactions similar to those for purine nucleotides to yield the free bases uracil and thymine. These are further broken down to the amino acids β-alanine and β-aminoisobutyrate, which can be further metabolized to furnish energy-yielding compounds.

Questions

Synthesis of Purine Ribonucleotides

1. Which of the following statements is (are) true? The enzyme ribose phosphate pyrophosphokinase:
   (a) Catalyzes the transfer of pyrophosphate from 5-phosphoribosyl-α-pyrophosphate (PRPP) to activate 5-phosphoribosylamine.
   (b) Catalyzes the formation PRPP from ribose-5-phosphate and ATP.
   (c) Transfers a pyrophosphoryl group from ATP to the C1 of ribose.
   (d) Produces pyrophosphate as a reaction product.

2. Which of the following compounds contribute atoms in the synthesis of the purine ring?

NH_4^+	N^5,N^{10}-methylenetetrahydrofolate
S-adenosylmethionine	Aspartate
Glutamine	Fumarate
N^{10}-formyltetrahydrofolate	CO_2
Serine	Glycine

3. In the synthesis of AMP from inosine monophosphate, the amino group that replaces the 6-keto group is obtained from_____ with release of _____, whereas GMP formation involves an oxidation reaction in which the electron acceptor is_____, followed by transfer of an amino group from _____.

4. Which of the following statements is (are) true? Purine nucleotide biosynthesis is regulated in part by:
   (a) Feedback inhibition of amidophosphoribosyl transferase by CTP and UMP.
   (b) Feedback inhibition of amidophosphoribosyl transferase by PRPP.
   (c) Feedback inhibition of amidophosphoribosyl transferase by ADP and GDP.
   (d) Feedback inhibition of ribose phosphate pyrophosphokinase by ADP and GDP.

5. In most tissues, most purine bases have been salvaged rather than synthesized *de novo*. What is the advantage of this metabolic strategy?

Synthesis of Pyrimidine Ribonucleotides

6. Identify the compound shown below.

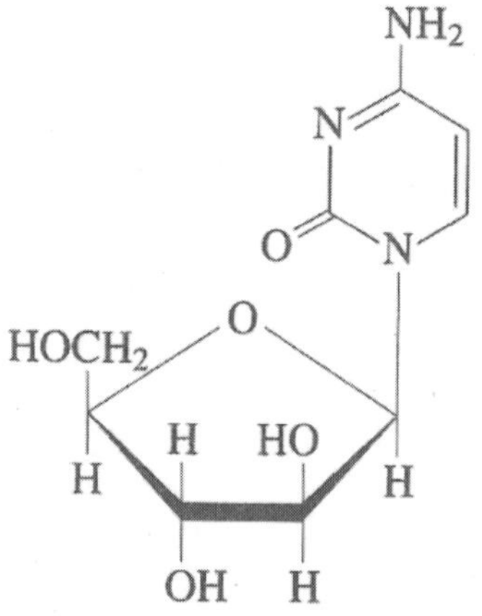

7. Determine which of the following compounds participate in CTP biosynthesis and place them in the correct order:

Orotic acid	Aspartic acid
Glutamine	N^{10}-formyltetrahydrofolate
Uracil	HCO_3^-
Cytidine	Uridylic acid
Dihydroorotate	PRPP

8. In what way is UMP synthesis similar to fatty acid synthesis in mammals?

9. Why is the administration of uridine an effective treatment for human orotic aciduria? Why is it preferable to use uridine instead of UMP?

Formation of Deoxyribonucleotides

10. The pentose phosphate pathway converts glucose to the ribose-5-phosphate needed for DNA synthesis. What other pentose phosphate pathway product is required for DNA synthesis?

11. Match each component of the ribonucleotide reductase reaction on the left with its function on the right.

    _____ dATP   A.   Location of tyrosyl radical
    _____ R1 subunit   B.   Stimulates GDP reduction and inhibits CDP and UDP reduction
    _____ R2 subunit   C.   Physiological reducing agent of the enzyme
    _____ dTTP   D.   Inhibits reduction of all NDPs
    _____ ATP   E.   Location of substrate specificity site
    _____ thioredoxin   F.   Stimulates CDP and UDP reduction

12. Why is 5-fluorouracil a powerful antitumor agent? Why is it considered a suicide substrate?

13. What is the rationale for administering a high dose of methotrexate followed by a massive amount of N^5-formyltetrahydrofolate and thymidine to a cancer patient?

Nucleotide Degradation

14. Adenosine deaminase (ADA) catalyzes the first step in the catabolism of adenine nucleotides. Why does an absence of ADA seriously compromise the immune system?

15. Summarize the metabolic pathway by which ^{14}C initially present at position 5 of uracil could appear in long-chain fatty acids.

16. Some terrestrial organisms secrete excess nitrogen as uric acid rather than urea. Why is this an advantage?

Answers to Questions

1. (b) and (c) are correct

2. Glutamine, N^{10}-formyltetrahydrofolate, aspartate, CO_2, and glycine.

3. Aspartate; fumarate; NAD^+; glutamine

4. (c) and (d) are correct

5. The synthesis of the purine ring is energetically expensive (7–9 ATP equivalents). In contrast, the salvage pathways in which a pre-existing purine base reacts with PRPP to reform a nucleotide requires only the energy of the activated PRPP.

6.   This molecule is cytosine arabinoside, an analog of cytidine in which arabinose replaces ribose.

7. HCO_3^- and glutamine; aspartic acid; dihydroorotate; orotic acid; PRPP; uridylic acid (UMP), glutamine (second use).

8.   In both pathways in mammals, multiple enzyme activities are located on a single polypeptide chain. This increases the rate of the multistep process and allows intermediates to be channeled from one active site to the next without being lost to the surrounding medium.

9.   Orotic aciduria is caused by a deficiency of the bifunctional enzyme that converts orotic acid to UMP. Uridine can be phosphorylated to UMP, thereby bypassing the metabolic block and acting as precursor for both uridine and cytidine nucleotides. Uridine is preferable to UMP because, being an uncharged molecule, it can readily cross cell membranes.

10.   The pentose phosphate pathway also produces NADPH, which is the ultimate electron donor for the reduction of NDPs to dNDPs via thioredoxin reductase, thioredoxin, and ribonucleotide reductase.

11.
D	dATP
E	1 subunit
A	2 subunit
B	dTTP
F	ATP
C	thioredoxin

12.   5-Fluorouracil is readily taken up by cancer cells and converted to 5-fluorodeoxyuridylate (FdUMP), which is an inactivator of thymidylate synthase. The rapid proliferation of tumor cells requires a constant supply of dTTP for DNA synthesis. Most slow-growing normal cells are relatively insensitive to 5-fluorouracil treatment. FdUMP inactivates thymidylate synthase only after going partially through the catalytic cycle. Since the enzyme actually generates the species that inactivates it, FdUMP is termed a mechanism-based inhibitor or a suicide substrate.

13.   Methotrexate is a dihydrofolate (DHF) analog that binds avidly to dihydrofolate reductase and consequently prevents the reduction of dihydrofolate to tetrahydrofolate (THF). Rapidly growing cancer cells, which require THF for the synthesis of dTTP, are particularly sensitive to methotrexate. A high dose of methotrexate is intended to rapidly kill cancer cells. Subsequent administration of a tetrahydrofolate derivative and thymidine supplies sufficient amounts of folate and a dTTP precursor to enable the more slowly growing normal cells to survive.

14. In the absence of ADA activity, deoxyadenosine accumulates and is phosphorylated. The elevated concentration of dATP inhibits ribonucleotide reductase. Inadequate production of deoxyribonucleotides inhibits DNA synthesis and hence cell proliferation. Lymphocytes exhibit especially active phosphorylation of deoxyadenosine and are therefore particularly vulnerable to ADA deficiency.

15. In the catabolism of pyrimidines, uracil is degraded in several steps to malonyl-CoA (Figure 23-24), which is a substrate for fatty acid synthesis (Section 20-4C). C5 of the pyrimidine ring becomes the methylene carbon of malonyl-CoA and can therefore be incorporated into fatty acyl chains.

16. Uric acid excretion requires little water since it is eliminated as insoluble crystals. Urea is water-soluble and requires large volumes of water to be excreted. Thus uric acid excretion is advantageous over urea excretion in environments where water is scarce.

24
Nucleic Acid Structure

This chapter takes a closer look at the structures of nucleic acids, which were first introduced in Chapter 3. This chapter begins by examining the structure of DNA in some detail, focusing on the different helical conformations and the limited flexibility of DNA that results from restrictions on rotation of various covalent bonds. Next, you are introduced to supercoiling, a structural feature of DNA that has important consequences for the biological activity of DNA, and topoisomerases, the enzymes that alter supercoiling. The forces that stabilize double helical structures are considered next. Knowledge of these forces is central to understanding the techniques used to isolate, analyze, and manipulate nucleic acids.

The chapter then turns to the interactions of DNA with proteins, giving as examples the restriction endonucleases and well-characterized prokaryotic and eukaryotic sequence-specific regulators of transcription. These transcription factors include the repressor from bacteriophage 434, the E. coli trp repressor, and the E. coli met repressor. The eukaryotic transcription factors that are discussed are proteins that contain zinc finger DNA-binding motifs and the leucine zipper dimerization motif (including those containing the basic helix–loop–helix motif near the DNA-binding region). In all cases, your attention is drawn to the common molecular interactions found in all of these proteins.

Finally, the structure of eukaryotic chromosomes is considered. This section explores how the histone proteins interact with DNA to generate higher-order structures that condense DNA over 50,000-fold. First, the structure of the nucleosome is discussed. This is followed by descriptions of the 30-nm filament of chromatin and the organization of highly condensed metaphase chromosomes.

Essential Concepts

1. DNA and RNA are the two kinds of nucleic acids that store genetic information and make it available to the cell. These molecules must therefore have the following properties:
   (a) Genetic information must be stored in a form that is stable and manageable in size.
   (b) Genetic information must be decoded by transcription and translation, which together convert nucleic acid sequences into protein sequences.
   (c) The information in DNA or RNA must be accessible to proteins and other nucleic acids.
   (d) Replication, in the case of DNA, must be template-driven so that each daughter cell receives the same genetic information.

The DNA Helix

2. DNA occurs in three major structural forms called A-DNA, B-DNA, and Z-DNA. B-DNA is the most common form and has the structural form first described by Watson and Crick.

3.   The key structural features of B-DNA include:
     (a) The two antiparallel strands wind in a right-handed manner around a common axis, producing a double helix that is about 20 Å in diameter.
     (b) Pairs of nucleotide bases are nearly perpendicular to the axis of the helix. The base pairs are in the interior of the double helix, while the sugar–phosphate backbone is on the outside, thus giving the appearance of a spiral staircase.
     (c) The "ideal" B-DNA double helix has ten base pairs (bp) per turn with a pitch of about 34 Å.
     (d) The double helix contains a wide and deep major groove and a narrow and deep minor groove.

4.   A-DNA is a wider and flatter right-handed helix than B-DNA. The planes of its base pairs are tilted about 20° to the helix axis. Its major groove is deep and narrow, while its minor groove is shallow and wide. A-DNA forms under dehydrating conditions.

5.   Z-DNA is a left-handed DNA helix that contains a deep minor groove but no discernible major groove. It forms *in vitro* in nucleic acids containing alternating purines and pyrimidines. The existence of Z-DNA-binding proteins suggests that Z-DNA forms *in vivo*.

6.   RNA is single-stranded but can form regions of double helix by folding back on itself. 2'-OH groups preclude B-form structures; instead, double helical regions assume conformations resembling A-DNA. DNA–RNA hybrids also show A-form conformations.

7.   Segments of B-DNA deviate from the ideal conformation, often in a sequence-dependent manner, which may be important for sequence-specific recognition of DNA by proteins involved in regulating specific gene activity.

8.   The conformational flexibility of DNA is limited by steric hindrance at the glycosidic bond between the nitrogenous base and the ribose moiety. Steric hindrance also induces a specific pucker in the ribose and restricts rotation in the bonds of the sugar–phosphate backbone.

9.   The twisting of a double helix that is often observed in covalently closed, circular, double-stranded DNA is called supercoiling or superhelicity. A key topological property of a closed circular double helix is that the number of supercoils cannot be altered without first cutting at least one strand of DNA. In the mathematical relationship

$$L = T + W$$

L is the linking number (the number of times that one strand of the double helix winds around the other), T is the twist (the number of complete revolutions that one strand makes around the axis of the double helix), and W is the writhing number (the number of turns the duplex axis makes around the superhelix axis, a measure of the supercoiling of the circular DNA). When $W < 0$, the DNA is negatively supercoiled; when $W > 0$, the DNA is positively supercoiled. Naturally occurring DNA is negatively supercoiled, which promotes strand separation for processes such as DNA replication and transcription.

10. Topoisomerases alter DNA supercoiling by making transient single-strand breaks (Type I) or double-strand breaks (Type II).

    (a) Type IA topoisomerases relax only negatively supercoiled DNA by changing the linking number in increments of one; while type IB topoisomerases can relax negative and positive supercoils by a controlled rotation mechanism. In both cases, the energy of the broken phosphodiester bond is conserved in the form of a phospho-Tyr diester linkage.

    (b) Type II topoisomerases of both prokaryotes and eukaryotes relax negative and positive supercoils in an ATP-dependent manner. Only the prokaryotic version (also called DNA gyrase) can introduce negative supercoils. Negative supercoils in eukaryotes result primarily from packaging DNA into nucleosomes.

Forces Stabilizing Nucleic Acid Structures

11. Like the denaturation of proteins, the denaturation of DNA is cooperative; i.e., the unwinding of one part of DNA destabilizes the remaining double helix. This can be measured by observing the increase in ultraviolet light absorbance in going from double-stranded to single-stranded DNA. The midpoint of this "melting" curve is called the melting temperature, T_m.

12. The T_m of double-stranded DNA depends on the solvent, the kind and concentration of ions in solution, the pH, and the mole fraction of $G \cdot C$ base pairs.

13. While base pairing provides specificity to the structure of DNA, it contributes little to the stability of DNA. Hydrophobic interactions (which tend to cause free base pairs to aggregate) and van der Waals interactions between base pairs (called stacking interactions) contribute the most to the stability of double-stranded nucleic acids. Stacking interactions between $G \cdot C$ base pairs are stronger than those between $A \cdot T$ base pairs. Hence, the greater stability of GC-rich DNA is a reflection not of the greater number of hydrogen bonds but of greater stacking energies.

14. Most RNA is single-stranded and adopts a much wider variety of shapes than does DNA. These varied shapes are due to various sections of RNA forming double-stranded regions via specific base pairing. For example, 5S ribosomal RNA (rRNA) contains helices, hairpin loops, internal loops, and bulges. The compact shape of transfer RNA (tRNA) is a result of base pairing and extensive stacking interactions.

15. RNA also has catalytic activity. Catalytic RNAs are similar to protein enzymes in their rate enhancement, substrate specificity, ability to use metal cofactors or imidazole groups, and regulation by small allosteric effectors. An RNA found in certain plant viruses, called the hammerhead ribozyme, catalyzes the cleavage of a specific RNA during posttranscriptional processing. Naturally occurring and synthetic versions of ribozymes can catalyze many other reactions, including some required for RNA synthesis.

Fractionation of Nucleic Acids

16. DNA and RNA in cells are invariably associated with proteins, so nucleic acid preparations must be deproteinized as part of their purification. Deproteinization can be accomplished by treating cell lysates with detergents, chaotropic agents, or proteases. The nucleic acid can then be recovered by precipitation with ethanol.

17. Double-stranded DNA can be separated from single-stranded nucleic acid, nucleotides, and other soluble molecules by chromatography using a column of hydroxyapatite, a form of calcium phosphate.

18. Eukaryotic messenger RNA (mRNA) can be separated from total RNA (of which it constitutes no more than 5% by mass), DNA, nucleotides, and other soluble molecules by affinity chromatography. Most eukaryotic mRNAs contain a poly(A) sequence at their 3′ ends, which is not found in rRNA or tRNA. Poly(U) or poly(T) covalently attached to a cellulose or agarose matrix can bind eukaryotic mRNA by complementary base pairing under high salt conditions; the mRNA is subsequently recovered by exposing the matrix to a low salt buffer.

19. Small nucleic acids (< 1000 nucleotides) can be easily resolved by polyacrylamide gel electrophoresis. Larger nucleic acids (up to 100,000 bp) can be separated using less cross-linked agarose gels. For extremely large DNAs (up to 10^7 bp), pulsed-field gel electrophoresis is used. Nucleic acids can be visualized in these gels by adding planar aromatic cations such as ethidium bromide, acridine orange, or proflavin. These dyes are called intercalation agents because they intercalate between the stacked bases, where they exhibit much greater fluorescence than do the free dyes. Hybridization with a labeled oligonucleotide probe is the basis for Southern blotting (to detect DNA) and Northern blotting (to detect RNA).

DNA–Protein Interactions

20. Many proteins bind DNA nonspecifically, primarily through interactions between the protein's functional groups and the sugar–phosphate backbone of DNA. Histones and certain DNA replication proteins are examples of such proteins.

21. Sequence-specific DNA-binding proteins usually interact with base pairs in the major groove of DNA by hydrogen bonding, either directly or indirectly via intervening water molecules. Ionic interactions with the sugar–phosphate backbone also occur, probably to facilitate contact between these proteins and DNA, after which the DNA can be "scanned" for a specific binding site. Dissociation constants for sequence-specific DNA-binding proteins are 10^3 to 10^7 times lower than those for nonspecific DNA-binding proteins.

22. Prokaryotic sequence-specific DNA-binding proteins recognize base pairs directly or through indirect readout, in a manner that depends on the sequence-dependent conformation and/or flexibility of the DNA backbone. Many DNA-binding proteins recognize palindromic (or nearly palindromic) sequences.

23. A large class of eukaryotic sequence-specific DNA-binding proteins are transcription factors. Many of these proteins form dimers and promote transcription at specific sites.

24. Many eukaryotic transcription factors contain structural motifs called zinc fingers. Zinc fingers are compact ~30 residue modules containing one or two Zn^{2+} ions liganded by His and/or Cys residues.

25. Another type of eukaryotic transcription factor contains a structural motif called the leucine zipper. The leucine zipper is a seven-residue pseudorepeating unit (*a-b-c-d-e-f-g*)$_n$ in which a hydrophobic strip along one side of the α helix formed by these residues promotes dimerization. The DNA-binding domain of such proteins may consist of extensions of the zipper helices or may be separated from the zipper by a basic helix–loop–helix motif.

Eukaryotic Chromosome Structure

26. Prokaryotic genomes typically consist of a single circular DNA molecule ranging from several hundred thousand bp to several million bp. The 23 chromosomes of the 3-billion-bp human genome have a total extended length of nearly one meter. Individual human chromosomes, with extended lengths between 1.6 and 8.2 cm, are condensed to varying degrees in different cells at different times, down to 1.3 to 10 μm long during mitosis.

27. Eukaryotic DNA is packaged into units called nucleosomes. Each nucleosome contains an octamer of four different basic proteins called histones [(H2A)$_2$(H2B)$_2$(H3)$_2$(H4)$_2$] and 146 bp of DNA. Approximately 55 bp of DNA (called linker DNA) links two nucleosomes and is associated with a fifth histone, H1. Histones are basic proteins with a large proportion of Arg and Lys residues, which interact with the phosphate oxygens of the sugar–phosphate backbone via salt bridges, hydrogen bonds, and helix dipoles.

28. The winding of DNA in nucleosomes to form the beaded chain seen in electron micrographs reduces its length by sevenfold. Further reduction of length is achieved by coiling the beaded chain into a ~30-nm-diameter filament. Higher-order structures, requiring nonhistone proteins are less well characterized. The most condensed structure is the metaphase chromosome, which has a central scaffold of fibrous protein that anchors loops of DNA and has an overall diameter of ~1.0 μm.

29. Prokaryotic DNA is also packaged into nucleosome-like structures with highly basic proteins that functionally resemble histones.

Key Equation

$$L = T + W$$

Questions

The DNA Helix

1. What form of DNA might you expect to see in desiccated (but viable!) brine shrimp eggs? Why?

2. Double-stranded DNA is relatively stiff, whereas single-stranded DNA is a flexible coil. What factors influence the structure of single-stranded versus double-stranded DNA?

3. How does the ribose pucker affect DNA structure?

4. A sample of a circular plasmid is digested with Type IA topoisomerase and analyzed by agarose gel electrophoresis followed by staining with ethidium bromide. Fifteen bands of DNA are visible. What do these bands represent and how do their linking numbers differ?

Forces Stabilizing Nucleic Acid Structures

5. Why is DNA not susceptible to hydrolysis by NaOH?

6. High concentrations of denaturing agents such as urea or formamide ($HCONH_2$) tend to favor a rodlike conformation for single-stranded DNA, rather than a flexible coil. What molecular interactions promote this behavior?

7. Explain how the following affect the T_m of double-stranded DNA:
   (a) Increasing the monovalent salt concentration.
   (b) Decreasing the pH.
   (c) Increasing the pH.
   (d) Increasing the concentration of formamide.

8. Would the UV absorbance of a solution containing partially stacked poly(A) increase or decrease when $[Na^+]$ increases?

9. What forces stabilize the tertiary structure of tRNA?

Fractionation of Nucleic Acids

10. Ice-cold ethanol is used to precipitate DNA. What is the significance of the temperature of the ethanol? Is this more important for shorter or longer DNA strands?

11. Samples of a circular plasmid are incubated with increasing concentrations of ethidium bromide and then analyzed on an agarose gel. Which gel below shows the results of this procedure?

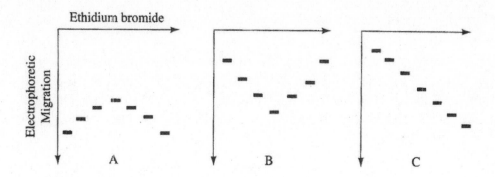

DNA–Protein Interactions

12. Specific DNA-binding proteins mainly contact DNA through hydrogen bonds in the major groove of B-DNA.
    (a) Why might sequence-specific binding be more common in the major groove than in the minor groove?
    (b) What hydrogen-bond contacts can proteins make with the bases in the major groove? Are these different from those in the minor groove?

13. In what ways do HTH proteins and the *met* repressor represent two general modes of DNA–protein interaction?

Eukaryotic Chromosome Structure

14. What insight into the structure of the eukaryotic chromosome was obtained from nuclease digestion studies?

15. Is histone H1 present in the electron micrograph shown in Figure 24-41? What is the relationship of H1 to the nucleosome structure?

16. Estimate the packing ratio of the DNA in human metaphase chromosomes, which have an aggregate length of 200 μm and contain 6×10^9 bp.

17. Limited digestion of chromatin with a bacterial nuclease yields a 166-bp DNA fragment, whereas limited digestion with pancreatic DNase I yields 10-bp fragments. Explain the difference in these results.

Answers to Questions

1. The A form of DNA, favored under dehydrating conditions, is the most likely form to occur in desiccated cysts or spores or in such things as brine shrimp eggs, which survive for years in a desiccated form.

2. Hydrogen bonding between base pairs and the limited rotation of the bases with respect to the sugar residues impose limitations on the rotation of the bonds in the ribose–phosphate backbone of double-stranded DNA. However, in single-stranded DNA, the ribose–phosphate bonds have greater freedom to rotate, allowing the polymer to take up a greater range of conformations.

3. The ribose pucker governs the relative orientations of the phosphate groups to each sugar residue. Residues in B-DNA have the C2'-*endo* conformation; in A-DNA, they have the C3'-*endo* conformation; and in Z-DNA, purine nucleotides are C3'-*endo* and pyrimidine nucleotides are C2'-*endo*.

4. The fifteen bands represent the native, negatively supercoiled plasmid (migrating the fastest because it is the most compact) and progressively less supercoiled DNA molecules (with the most relaxed DNA migrating the slowest because it is the least compact). Because Type IA topoisomerase relaxes DNA by making single-strand cuts, the linking numbers of the 15 DNA bands change by increments of one from bottom to top.

5. DNA does not contain a 2'-OH group that can be deprotonated and then serve as a nucleophile to attack the phosphate group at the 3' position.

6. Chaotropic agents such as urea and formamide tend to disrupt the structure of water. Hence, in their presence, water's ability to solvate DNA's anionic phosphate groups is reduced. Consequently, the phosphate groups repel one another more strongly than they do in the absence of the chaotropic agents, which induces the DNA to take up an extended rodlike conformation.

7.
    (a) Monovalent salts increase the T_m because they attenuate the repulsions between the negatively charged phosphate groups.
    (b) A large decrease in pH disrupts hydrogen bonding between base pairs by protonating some of the bases and therefore decreases Tm.
    (c) A large increase in pH disrupts hydrogen bonding between base pairs by deprotonating some of the base pairs and therefore decreases Tm.
    (d) Formamide disrupts water structure and thereby promotes denaturation, leading to a decrease in T_m.

8. An increase in ionic strength reduces the repulsions between phosphate groups and hence promotes base stacking. UV absorbance, which increases when the bases melt apart, would therefore decrease.

9. The tertiary structure of tRNA is stabilized by non-Watson–Crick hydrogen bonding and by stacking interactions.

10. Ethanol decreases the T_m of the DNA as well as decreasing its solubility, so it is important to keep the solution cold to keep the duplex intact. Shorter DNA molecules are more vulnerable to melting and hence strand separation, since T_m depends in part on the length of the DNA.

11. Gel A best represents the expected banding pattern. With increasing ethidium bromide, the plasmid unwinds, becoming progressively less supercoiled until it is a relaxed open circle. The relaxed open circle is less compact than the native plasmid; therefore, it migrates more slowly during electrophoresis. Further increases in ethidium bromide induce positive supercoils, so the plasmid again becomes more compact and migrates faster.

12.
    (a) The bases are more exposed in the major groove of B-DNA, which is wider than the minor groove, and are therefore more accessible to binding proteins. In addition, more base-specific hydrogen bonding donors and acceptors are exposed in the major groove.
    (b) In the major groove, the groups available for hydrogen bonding are N7 and N6 of adenine, N7 and O6 of guanine, N4 of cytosine, and O4 of thymine. In the minor groove, the groups available for hydrogen bonding are N3 of adenine, N3 and N2 of guanine, O2 of cytosine, and O2 of thymine.

13. Both kinds of protein exhibit a twofold symmetry that is reflected in the palindromic DNA sequence at their binding sites. The α helices of HTH proteins contact the major groove of DNA directly or indirectly via H_2O bridges. In *met* repressor-like proteins, β strands contact the major groove of DNA.

14. Studies using micrococcal nuclease digestion indicated that the nucleosome contains ~200 bp of DNA. Further digestion trims this DNA to ~146 bp, leaving the nucleosome core particle.

15. Histone H1 is probably absent from the preparation shown in Figure 24-41; compare with Figure 24-44*b*. H1 appears to compact the DNA by binding to the ends of the DNA entering and leaving the nucleosome core.

16. The uncondensed DNA would have a length of $(6 \times 10^9$ bp)(0.34 nm/bp) = 2.04 m. The packing ratio is therefore 2.04 m/200 μm = ~10,000.

17. The bacterial nuclease does not cleave the DNA of the nucleosome; however, DNase I appears to be able to cleave nucleosome-bound DNA, cutting it once per helical turn.

25

DNA Replication, Repair, and Recombination

This chapter introduces you to the remarkably complex biochemistry of DNA replication, repair, and recombination. The chapter begins with a review of the classic Meselson and Stahl experiment showing that DNA is replicated semiconservatively; that is, each daughter DNA contains one new strand hydrogen bonded to one old, or parental, strand. The chapter reviews the essential steps in the replication of DNA and then takes a more detailed look at the proteins and enzymes involved in this remarkable process. Our best information comes from the enzymology of prokaryotic DNA replication, so this is the focal point of the discussion. Unique features of eukaryotic DNA replication are covered, particularly the problems presented by replicating a linear DNA molecule. The chapter then looks at the genesis of point mutations and small insertions and deletions of nucleotides during DNA replication. This naturally leads to a discussion of the mechanisms cells use to repair damaged DNA or mistakes incurred during replication. Again, prokaryotic mechanisms are understood best, and hence are the focus of the discussion. The chapter then considers the mechanisms by which prokaryotes engage in DNA recombination. This chapter concludes with an introduction to mobile DNA elements called transposons (and retrotransposons in eukaryotes), which can move to new locations in the genome by recombination during their own replication.

Essential Concepts

Overview of DNA Replication

1. The structure of DNA provides a template-driven mechanism for its replication. Experiments by Meselson and Stahl showed that each polynucleotide strand serves as a template for a new daughter strand. On completion of replication, each daughter strand, which is hydrogen bonded to its template, or parental strand, segregates to one of the daughter cells. This mode of DNA replication is called semiconservative DNA replication.

2. DNA polymerase requires a template, all four deoxynucleoside triphosphates (NTPs), and a primer from which to extend the chain. The polymerization reaction involves the nucleophilic attack of the growing DNA chain's (or a primer's) 3′-OH group on the α-phosphoryl group of a free NTP that is hydrogen bonded to the template. The liberated PP_i is subsequently hydrolyzed, making the polymerization irreversible. Because the 3′ end of the chain grows, polymerization is said to proceed from 5′ to 3′.

3. DNA replication is bidirectional, meaning that DNA synthesis proceeds in both directions from the site of initiation of replication. Circular DNA that contains a bubble where the parental DNA has unwound to allow replication is said to be undergoing θ replication.

4. Since DNA replication can proceed only in the $5'\rightarrow3'$ direction, the daughter strand that extends $3'\rightarrow5'$ in the direction that the replication fork travels, the lagging strand, must be synthesized as a series of fragments, that is, semidiscontinuously. The fragments, which are are called Okazaki fragments (named after the scientist who first observed their synthesis), are eventually joined together by DNA ligase. In the daughter strand that extends $5'\rightarrow3'$ in the direction that the replication fork travels, the leading strand, DNA synthesis is continuous from the site of initiation.

5. DNA synthesis *in vivo* begins as the extension of an RNA primer synthesized by primase (or RNA polymerase) at the site of initiation of DNA replication. Primers are subsequently removed by the $5'\rightarrow3'$ exonuclease activity of a DNA polymerase, and the polymerase fills in the gap, which is then sealed by DNA ligase. Ligation is endergonic and requires the free energy of ATP or NAD^+ hydrolysis.

Prokaryotic DNA Replication

6. There are three DNA polymerases in prokaryotes:
    (a) DNA polymerase I, the enzyme discovered by Arthur Kornberg, removes RNA primers (via its $5'\rightarrow3'$ exonuclease activity) and subsequently fills in the gaps (via its $5'\rightarrow3'$ polymerase). These two reactions together are referred to as nick translation. Pol I also has $3'\rightarrow5'$ exonuclease activity to eliminate misincorporated nucleotides.
    (b) DNA polymerase II is involved in DNA repair.
    (c) DNA polymerase III is the primary DNA replicating enzyme. It is the largest of the three with at least 10 subunits. Pol III has $5'\rightarrow3'$ polymerase activity and a $3'\rightarrow5'$ exonuclease activity that eliminates misincorporated nucleotides.
    The polymerase active site forms sequence-independent hydrogen bonds with double-stranded DNA and can therefore detect mispairings. Catalysis requires two metal ions, usually Mg^{2+}.

7. The initiation of DNA replication in *E. coli* occurs at a specific sequence called the *oriC* locus, where the DNA is melted apart, unwound, and the leading strand primer is synthesized.
    (a) DNA melting, or local denaturation, requires ATP and is accomplished by a complex of DnaA protein.
    (b) The hexameric protein DnaB (a helicase) further unwinds DNA in an ATP-dependent reaction.
    (c) The tetrameric single-strand binding protein (SSB) prevents reannealing of the DNA. Numerous molecules of SSB coat single-stranded DNA.
    (d) A complex of proteins called the primosome, which contains DnaB, primase (DnaG protein), and five other proteins, carries out primer synthesis. The primosome displaces SSB in an ATP-dependent process.

8. DNA synthesis of both leading and lagging strands is carried out by the replisome, a complex containing two Pol III enzymes. In order for the replisome to move as a single unit in the $5'\rightarrow3'$ direction, the lagging strand template must loop around once so that the Okazaki fragment and the leading strand can be synthesized in the same direction.

(a) The β subunit of Pol III forms a sliding clamp that moves along DNA, allowing Pol III to replicate the DNA with a processivity of >5000 residues.

(b) On the lagging strand, the γ complex (or clamp loader) of Pol III reloads the enzyme back onto the DNA template every ~1000 bp (the size of an Okazaki fragment).

9. Specific DNA sequences (*Ter* sites) signal the termination of DNA synthesis. However, arrest of DNA replication requires the activity of Tus protein, which binds to a *Ter* site and prevents strand displacement by DnaB.

10. The fidelity of DNA replication in *E. coli* is very high: ~1 error per 1000 bacteria per generation! Four factors contribute to this high fidelity:
    (a) Cells maintain balanced levels of all four deoxynucleotides.
    (b) Polymerization by Pol I, II, or III occurs only after an incoming nucleotide is properly positioned in the catalytic site and on the template.
    (c) The 3′→5′ exonuclease activity of DNA polymerases detects and excises mismatches in the new DNA strand.
    (d) DNA repair systems detect and correct residual errors after DNA synthesis is complete.

Eukaryotic DNA Replication

11. Eukaryotes contain at least 13 DNA polymerases that are grouped, along with prokaryotic polymerases, in six families according to sequence homology.

12. The major polymerases of DNA replication in eukaryotes are polymerases α and δ.
    (a) DNA polymerase α, which lacks exonuclease activity and is only moderately processive, associates with a primase and is responsible for initiating DNA replication.
    (b) DNA polymerase δ is the main polymerase for the leading and lagging strands and has 3′→5′ exonuclease activity. It has unlimited processivity when it is in complex with a protein called proliferating cell nuclear antigen (PCNA), which functions as a sliding clamp.

13. DNA polymerase ε appears to have a regulatory function. DNA polymerase γ is found in mitochondria (chloroplasts contain a similar enzyme), where it presumably replicates mitochondrial DNA.

14. A viral RNA-directed DNA polymerase is a reverse transcriptase. It has no primase or 3′→5′ exonuclease activity but does contain ribonuclease (RNase H) activity. Reverse transcriptase uses a tRNA primer to replicate the viral genome. Reverse transcriptase is an essential tool for cloning mRNA sequences, because it can be used to synthesize a complementary DNA molecule (cDNA).

15. In yeast, DNA replication begins at DNA sequences called autonomously replicating sequences (ARS). In higher eukaryotes, no analogous sequences have been identified. Many initiation sites appear simultaneously in higher eukaryotes and presumably require some alteration of nucleosome structure. As replication of the genome ensues, the RNA primers are removed by RNase H1 and flap endonuclease-1 (FEN1) in a two-step process. There are no known termination sequences in eukaryotic DNA.

16. Eukaryotic chromosomes are linear, so the RNA primers at the 3′ ends of a chromosome cannot be replaced with DNA (DNA polymerase must extend a DNA strand from 5′ to 3′). This would lead to progressively shorter chromosomes, with the concomitant loss of genetic information, with each cell generation. However, eukaryotic chromosomes contain telomeres, repetitive sequences at the ends of chromosomes. These sequences are added by an enzyme called telomerase. Telomerase, a ribonucleoprotein, contains an RNA molecule that serves as a template for a reverse transcriptase activity that extends the telomere. Telomeric sequences tend to fold back on themselves and also serve as binding sites for certain proteins.

17. Telomerase activity is absent in somatic tissues of animals, which exhibit a finite replicative lifespan (about 60–80 divisions after birth in humans). Telomeres appear to shorten as cells age in culture. Hence, progressive loss of telomeres may have an impact on aging in at least some highly proliferative tissues. Many immortal cells (such as cancer cells) contain telomerase activity. Hence, telomerase is an attractive target for antitumor drug development.

DNA Damage

18. A mutation is a random, heritable alteration of genetic information and can arise in several ways:
    (a) Misincorporation of the wrong base during DNA relication.
    (b) Addition or deletion of one or more nucleotides.
    (c) Chemical modification of a base so that it pairs with a different base during DNA replication.
    (d) Translocation of a segment of DNA from one region of the genome to another.

19. Chemicals that induce mutations, called chemical mutagens, produce two classes of DNA damage:
    (a) Point mutations, in which one base pair replaces another. In a transition, a purine (or pyrimidine) is replaced by another. In a transversion, a purine is replaced by a pyrimidine, or vice versa.
    (b) Insertion/deletion mutations, in which one or more nucleotide pairs are inserted or deleted.

20. The bases in DNA may be chemically altered by deamination, oxidation, and alkylation, and the base itself may be lost by hydrolysis of the glycosidic bond. Intercalating agents can generate insertion/deletion mutations. Adenine and cytosine in cellular DNA are modified by methylation in a sequence-specific manner that regulates gene expression.

21.   Many mutagens are also carcinogens (chemicals that cause cancer). This property is used in the Ames test to assess the carcinogenic potential of chemicals. In the Ames test, a *his⁻* strain of bacteria is exposed to a potential mutagen on an agar plate containing no His. The mutagenicity of a substance is scored as the number of *his⁺* colonies that appear, minus the few spontaneous ones that appear in the absence of the putative mutagen.

DNA Repair

22.   DNA repair mechanisms range from simple one-enzyme systems (such as recognition of alkylated bases or pyrimidine dimers) to more elaborate multienzyme systems (such as nucleotide excision repair, recombination repair, and the SOS response).

23.   Simple single-enzyme systems correct specific alkylated bases: e.g., O^6-methylguanine residues are recognized by O^6-alkylguanine–DNA alkyltransferase, which transfers the aberrant methyl group to a Cys residue on the enzyme. Pyrimidine dimers, which form from the absorption of UV light by two stacked pyrimidine residues, are repaired by light-activated DNA photolyases.

24.   In base excision repair, the damaged base is removed by a DNA glycosylase, an endonuclease cleaves the backbone at the resulting abasic site, and an exonuclease removes several additional residues. The gap is then filled in by a DNA polymerase and the nick is sealed by DNA ligase.

25.   Nucleotide excision repair is a more sophisticated method to correct pyrimidine dimers or other DNA lesions involving several base pairs. Nucleotide excision repair in *E. coli* is carried out by several enzymes:
       (a)   UvrABC endonuclease excises an oligonucleotide containing the damaged bases.
       (b)   Pol I fills in the single-stranded region.
       (c)   DNA ligase catalyses the formation of a phosphodiester linkage to restore the intact DNA molecule.

26.   The mismatch repair (MMR) system corrects replication mispairings overlooked by the DNA polymerase proofreading activity and corrects insertions and deletions up to 4 base pairs in length. Defects in enzymes in this system have been implicated in high incidence of cancer, including hereditary nonpolyposis colorectal cancer syndrome. Three proteins carry out MMR in *E. coli*:
       (a)   the MutS homodimer recognizes and binds to mismatched base pairs;
       (b)   the MutL homodimer forms a tetramer with MutS to loop out the region of DNA containing the mismatched base pairs in a ATP-driven reaction;
       (c)   the MutS/MutL tetramer recruits MutH and activates MutH endonuclease activity. MutH makes a single-strand cut (nick) adjacent to an unmethylated GATC sequence. The UvrD helicase and an exonuclease remove the DNA, and the gap is filled in by DNA pol III.

27. Eukaryotes contain at least 5 DNA polymerases, called error-prone polymerases, that can replicate past various DNA lesions. These polymerases have no proofreading exonuclease activity and, in the case of DNA polymerase η, insert incorrect bases once every ~30 nucleotides.

28. Double-stranded DNA breaks are repaired by two mechanisms: recombination repair and nonhomologus end-joining (NHEJ) repair.

29. A complex DNA repair system in *E. coli* is the SOS response, which is activated when DNA damage prevents replication. SOS repair is error prone and therefore mutagenic. The SOS response involves RecA, which also participates in recombination and error-prone DNA polymerases.

Recombination

30. Homologous (general) recombination occurs between two sections of DNA with extensive homology. The aligned strands of DNA are cleaved, and the strands cross over to form a four-branched structure called a Holliday junction. The junction can dissociate into two new duplexes in two ways, both yielding new DNA molecules.

31. Recombination in *E. coli* is mediated by RecA, which polymerizes on single-stranded gaps in DNA duplexes. RecA partially unwinds the duplexes and exchanges two strands in an ATP-dependent reaction. All recombination events require additional proteins to further unwind DNA, produce nicks, maintain proper supercoiling, drive branch migration, and seal nicks.

32. In recombination repair, DNA that contains a nick or gap (and which therefore cannot be further replicated) undergoes recombination so that an intact homologous daughter strand can serve as a template in place of the damaged parental strand, allowing replication to proceed. Double-strand breaks can also be repaired by recombination with an intact homologous chromosome.

33. Transposons are DNA segments in prokaryotes and eukaryotes that can move from one site in the genome to another by a mechanism that involves their replication. The simplest transposons are insertion elements (IS), which contain a transposase gene that mediates the recombination reactions involved in transposition. This recombination occurs between short target sequences located at the ends of the transposon. More complex transposons contain genes required for transposition as well other genes, such as antibiotic resistance genes.

34. Transposons promote inversions, deletions, and rearrangements of host DNA and thereby affect gene expression and permit chromosome evolution. Most eukaryotic transposons resemble the DNA from retroviruses rather than bacterial transposons; hence they are called retrotransposons. Transposition of a retrotransposon begins with its transcription to RNA, followed by synthesis of DNA from the RNA template by reverse transcriptase; the DNA then inserts randomly into the host genome.

Questions

Overview of DNA Replication

1. What feature of DNA replication, as shown in Figure 25-4, is inconsistent with the known enzymatic properties of DNA polymerases? What observation reconciled this inconsistency?

2. What prevents the reverse reaction of DNA polymerization *in vivo*?

Prokaryotic DNA Replication

3. Describe the roles of the exonuclease activities of Pol I.

4. For each of the enzymes and proteins listed below, match the function or feature related to DNA replication in *E. coli*.

    A. DNA polymerase I B.   DNA polymerase II C. DNA polymerase III
    D. DNA ligase   E.   DnaB   F. Single-strand binding protein
    G. Primase   H.   Tus protein   I. DNA gyrase

    _____ Required for the initiation of DNA synthesis
    _____ Essential for the condensation of Okazaki fragments
    _____ A helicase required for unwinding the DNA duplex
    _____ Involved in the termination of DNA strand replication
    _____ The enzyme that is strictly involved in DNA replication
    _____ The enzyme critical for relieving buildup of positive supercoils
    _____ The enzyme that participates in DNA repair
    _____ Prevents the reannealing of DNA at the replication fork
    _____ Excision and replacement of RNA primers
    _____ Incapable of nick translation
    _____ Requires NAD^+ in *E. coli*
    _____ The most abundant DNA polymerase

5. Radioactive DNA probes can be made by nick translation of linear pieces of DNA. Why is a trace amount of DNase I necessary to do so?

6. From the gut of a slug in your garden you have identified a new circular, double-stranded DNA–containing bacteriophage. Electron micrographs of infected bacteria during production of phage progeny reveal the structures shown below (thick lines represent double-stranded DNA and thin lines represent single-stranded nucleic acids). Alkali treatment of these replicative DNA forms destroys their single-stranded "tails." Bacteria incubated with ^{15}N-nucleosides were infected with this bacteriophage. CsCl equilibrium density centrifugation of DNA (which separates DNA roughly by mass) reveals two bands of phage DNA. The heavier band contains 100 times more DNA than the lighter band.

(a) What is the composition of the single-stranded nucleic acid in the structures shown below?

(b) Suggest a mechanism for DNA replication in this bacteriophage.

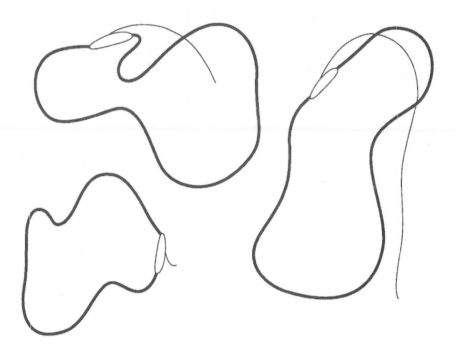

Eukaryotic DNA Replication

7. Which eukaryotic DNA polymerase is most likely responsible for the following activities: (a) synthesizing the leading strand, (b) synthesizing the lagging strand, (c) synthesizing mitochondrial DNA, (n) initiating DNA replication?

8. Why does reverse transcription of retroviral DNA require a specific host tRNA for DNA synthesis? What section of the tRNA is most likely to be involved?

9. When reverse transcriptase is used in the oligo(T)-primed synthesis of double-stranded DNA from a specific cellular mRNA, the aggregate length of the resulting double-stranded cDNA is sometimes twice as long as that of the mRNA template. Explain this phenomenon.

10. What mammalian cell type must retain its telomerase activity?

DNA Damage

11. In the pairs of DNA sequences below, the lower duplex represents a mutagenized daughter duplex. Identify the mutation as a transition, a transversion, an insertion, or a deletion.

  (a)   5′ GCCTAGAACCAGTAC 3′
        3′ CGGATCTTGGTCATG 5′

        5′ GCACTAGAACCAGTA 3′
        3′ CGTGATCTTGGTCAT 5′

  (b)   5′ CGCATAGCTACTGGAA 3′
        3′ GCGTATCGATGACCTT 5′

        5′ CGCAGAGCTACTGGAA 3′
        3′ GCGTCTCGATGACCTT 5′

12. What kinds of mutations does 5-bromouracil generate?

13. Match the compound on the left with the kind of mutation on the right.

  _____ Acridine orange   A. Transition
  _____ Nitrous acid   B. Transversion
  _____ Ethylnitrosourea   C. Insertion or deletion
  _____ Dimethyl sulfate
  _____ Ethidium bromide

DNA Repair

14. Match the proteins below with their roles in DNA repair.

  A. O^6-alkylguanine–DNA alkyltransferase   B. UvrABC endonuclease
  C. Uracil–DNA glycosylase   D. RecA

  _____ Removes 7-methyladenine residues from damaged DNA
  _____ Serves as a sink for methyl residues abstracted from O^6-methylguanine residues
  _____ Removes deaminated cytosine residues
  _____ Removes damaged DNA via recombination repair
  _____ Removes thymine dimers
  _____ Initiates the SOS response

Recombination

15. Why is transposition a misnomer in describing the operation of bacterial transposons?

16. You transformed a strain of *E. coli* with a plasmid containing a transposon that includes an ampicillin-resistance gene. To recover the plasmid, you pick a colony from a culture plate containing ampicillin. Much to your surprise, the plasmid you recover is twice the size you expected. You then pick 100 more colonies and discover that only one of them has plasmid of the expected size.
    (a) Why is the recovery of plasmid of expected size so low?
    (b) Why can you recover any plasmid of expected size at all?

Answers to Questions

1. The largely symmetric labeling of each replication fork suggests that both strands of DNA are synthesized simultaneously. However, this implies that one strand is synthesized from 5′ to 3′ and the other strand is synthesized from 3′ to 5′. A polymerase that acts in the 3′→5′ direction has never been found. Okazaki demonstrated that one of the new strands is synthesized continuously in the 5′→3′ direction and one strand is synthesized discontinuously in the 5′→ 3′ direction, after which the DNA fragments are ligated together.

2. The polymerase reaction is irreversible *in vivo* because the pyrophosphate released on incorporation of a nucleotide into a DNA strand is quickly cleaved to 2 P_i by inorganic pyrophosphatase.

3. The 3′→5′ exonuclease activity, located near the polymerase active site, recognizes and removes mismatched bases. The 5′→ 3′ exonuclease site of Pol I binds to DNA at single-strand breaks and initiates nick translation to replace RNA primers with DNA.

4.
G	Required for the initiation of DNA synthesis
D	Essential for the condensation of Okazaki fragments
E	A helicase required for unwinding the DNA duplex
H	Involved in the termination of DNA strand replication
C	The enzyme that is strictly involved in DNA replication
I	The enzyme critical for relieving buildup of positive supercoils
A,B	The enzyme that participates in DNA repair
F	Prevents the reannealing of DNA at the replication fork
A	Excision and replacement of RNA primers
B, C	Incapable of nick translation
D	Requires NAD^+ in *E. coli*
A	The most abundant DNA polymerase

5. A trace amount of DNase is necessary to produce single-strand nicks that are the starting point for the 5′→3′ exonuclease and polymerase activities of Pol I.

6.

(a) The single-stranded nucleic acid is likely to be composed entirely of RNA, suggesting that replication of the phage genome begins with the synthesis of a "genomic" RNA.

(b) The appearance of only two bands of phage DNA suggests that the "genomic" RNA serves as a template to generate new double-stranded DNA. In this scenario, ^{15}N nucleosides are incorporated into a DNA strand complementary to the RNA and then into a second DNA strand complementary to the first. Hence, both strands of newly synthesized DNA are labeled with ^{15}N. During infection, when progeny phage are produced, there is much more of the heavy new phage DNA than there is of the original light ^{14}N-containing DNA.

7.

(a) DNA polymerase δ
(b) DNA polymerase δ
(c) DNA polymerase γ
(c) DNA polymerase α

8. The host tRNA serves as an RNA primer for the synthesis of the DNA. The 3′ end is the most likely section of the molecule to serve as the primer, since it provides the 3′-OH group to which deoxynucleotides can be attached.

9. The 3′ end of the newly synthesized cDNA strand folds back to prime the synthesis of a complementary DNA strand, presumably on initiation of RNase H activity, which digests away the original mRNA template. This folded DNA then serves as a template for the synthesis of a complementary strand by DNA polymerase.

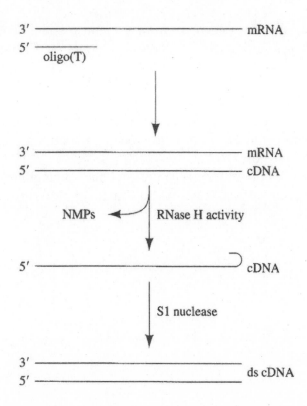

10. The cells that give rise to eggs and sperm must retain telomerase activity in order to produce immortal gametes.

11.
   (a) The mutation is an insertion of an A·T base pair in the third position.
   (b) The mutation is a transversion at the fifth position (T·A to G·C).

12. 5-Bromouracil (5BU) is a base analog that can substitute for thymine and base pair with adenine. However, its enol tautomer, which forms more readily than that of T, can also base pair with guanine, thereby generating a T·A → C·G transition. When 5BU substitutes for cytosine, it generates a C·G → T·A transition.

13.
   __C__ Acridine orange
   __A__ Nitrous acid
   __B__ Ethylnitrosourea
   __B__ Dimethyl sulfate
   __C__ Ethidium bromide

14.
   __B__ Removes 7-methyladenine residues from damaged DNA
   __A__ Serves as a sink for methyl residues abstracted from O^6-methylguanine residues
   __C__ Removes deaminated cytosine residues
   __D__ Removes damaged DNA via recombination repair
   __B__ Removes thymine dimers
   __D__ Initiates the SOS response

15. There is no transfer of a DNA segment from a donor to a recipient; instead, the transposon is replicated during insertion so that both the donor and recipient ultimately contain transposon sequences.

16.
   (a) The transposon may contain a defective resolvase so that it can integrate into the host DNA but cannot complete site-specific recombination that lets the original plasmid be recovered from the host DNA.
   (b) RecA (provided by the host cell) can resolve the cointegrate at a low level, so that the cointegrate is resolved in at least 1 in 100 cases.

26
Transcription and RNA Processing

This chapter focuses on the biochemistry of RNA synthesis, from the initiation of transcription through the subsequent modification of the transcription products into biologically functional RNAs. The chapter begins by discussing the structure and biochemical properties of prokaryotic RNA polymerase, comparing and contrasting them with those of DNA polymerase. The discussion then progresses to the three main activities of transcription: template binding and RNA chain initiation, chain elongation, and chain termination. The chapter then examines the remarkably more complex biochemistry of transcription in eukaryotic cells, which involves multiple DNA sequence elements (promoters, enhancers, and silencers) and many proteins (transcription factors). In addition, eukaryotic cells use three different isoforms of RNA polymerase to synthesize different classes of RNA. The chapter concludes with a discussion of posttranscriptional processing, focusing first on the extensive modifications of eukaryotic mRNAs. Here you are introduced to some of the minor forms of RNA involved in RNA processing (small nucleolar RNAs and guide RNAs). Finally, rRNA and tRNA processing in prokaryotes and eukaryotes is considered.

Essential Concepts

1. Cells contain several types of RNA, mainly ribosomal RNA (rRNA), transfer RNA (tRNA), and messenger RNA (mRNA). rRNA constitutes better than 80% of the total mass of cellular RNA and accounts for two-thirds of the mass of ribosomes. tRNAs are small RNAs that deliver amino acids to the ribosome. mRNAs contain the nucleotide sequences that encode the amino acid sequences of polypeptides. In addition to these three large classes, which account for over 95% of the mass of RNA in eukaryotic cells, there are small nuclear RNAs (snRNAs) involved in mRNA splicing, guide RNAs involved in RNA editing, and other small RNAs.

Prokaryotic RNA Transcription

2. The transcription of DNA into RNA is carried out by RNA polymerases, which contain multiple different subunits. Bacteria contain one RNA polymerase core enzyme, whereas eukaryotes have four or five isoforms. All RNA polymerases catalyze the reaction

$$(\text{RNA})_{n \text{ residues}} + \text{NTP} \rightleftharpoons (\text{RNA})_{n+1 \text{ residues}} + \text{PP}_i$$

The enzyme uses the substrates ATP, CTP, GTP, and UTP, synthesizing an RNA chain in the $5' \rightarrow 3'$ direction. Hydrolysis of the released PP_i makes polymerization irreversible.

3. The RNA polymerase holoenzyme in *E. coli* contains five types of subunits: $\alpha_2\beta\beta'\omega\sigma$. The σ subunit dissociates from the complex on the initiation of transcription, leaving the core enzyme to continue transcription.

4. Many prokaryotic genes exist in contiguous sets of functionally related genes called operons and are transcribed as polycistronic mRNAs. Most protein-encoding genes (structural genes) in eukaryotes, however, are transcribed individually as monocistronic mRNAs.

5. The prokaryotic RNA holoenzyme binds to DNA sequences called promoters, which are "upstream" from the coding sequence of a gene or operon. RNA polymerase initiates transcription about 10 nucleotides "downstream" from a consensus DNA sequence called the Pribnow box.

6. Different genes or operons have different promoter sequences, which allows for differential gene expression in response to environmental needs or developmental changes in the organism. Promoter selectivity is regulated by the σ subunit of RNA polymerase. The rate of transcription of a gene is directly proportional to the rate at which the holoenzyme forms a stable initiation complex.

7. A promoter region can be identified by a technique called footprinting. RNA polymerase and DNA containing the putative promoter are incubated with an alkylating reagent at DNA bases that are not protected by contact with the RNA polymerase. This results in cleavage of the DNA backbone at the alkylated residues. The products of this reaction can be analyzed by electrophoresis to determine the region of DNA that binds to RNA polymerase.

8. The RNA polymerase core enzyme binds DNA tightly ($K \approx 5 \times 10^{-12}$ M, with a half-life of ~60 min). The RNA polymerase holoenzyme, however, binds DNA relatively weakly ($K \approx 10^{-7}$ M) except at specific promoter sequences recognized by the σ subunit. Hence, the σ subunit allows the holoenzyme to rapidly scan many sites until it finds its promoter. RNA polymerase then melts ~11 bp of DNA to form an open complex. Once initiation of an RNA chain has occurred, the σ subunit is jettisoned, and the tight-binding core enzyme elongates the RNA chain with high processivity until a termination site is reached.

9. Transcriptional initiation requires the formation of unwound DNA, which creates a mechanical strain as RNAP maintains a firm grip on the promoter. This frequently results in the release of short transcripts (abortive initiation). The strain eventually provides the energy for RNAP to pull or "clear" the promoter past the enzyme, allowing for successful initiation. Subsequent elongation of the initiated transcript is highly processive and rapid.

10. Transcription termination is relatively imprecise. In prokaryotes, termination occurs in one of two major ways:
    (a) RNA polymerase halts when it encounters a hairpin-forming GC-rich region that is followed by a series of U residues on the DNA sense strand.

(b) Termination is mediated by Rho factor, a hexameric protein that recognizes a nascent RNA and unwinds the DNA–RNA duplex in the open complex, thereby releasing the RNA transcript at a particular site.

Transcription in Eukaryotes

11. Eukaryotes have three distinct isoforms of RNA polymerase (RNAP):
    (a) RNA polymerase I, located in nucleoli, synthesizes the precursors of most rRNAs.
    (b) RNA polymerase II, located in the nucleoplasm, synthesizes mRNA precursors.
    (c) RNA polymerase III, also located in the nucleoplasm, synthesizes the precursors of 5S rRNA, tRNA, and a variety of small nuclear and cytosolic RNAs.

12. Each RNA polymerase contains up to 15 subunits, several of which are homologous to the subunits of prokaryotic RNA polymerase. RNAP II has a positively charged DNA-binding cleft, two Mg^{2+} ions at the active site, and a clamp structure that enhances processivity. It accommodates the growing RNA chain as a hybrid DNA–RNA helix for about one turn.

13. In eukaryotic promoters, which are much more complex and diverse than prokaryotic promoters, the position and orientation may vary considerably relative to the transcription start site.

14. RNA polymerase I recognizes only one promoter sequence, which is species specific. In contrast, RNA polymerases II and III recognize many different promoters. Some RNA polymerase III promoters are located within the gene's transcribed sequence.

15. There are two classes of RNA polymerase II promoters, both analogous to prokaryotic promoter sequences:
    (a) Most constitutively expressed genes (housekeeping genes) contain CpG islands (see Box 25-4) near the transcription start site.
    (b) Many genes that are selectively expressed in specific cells at specific times contain an AT-rich region known as the TATA box or Inr (initiator) element (see Figure 26-14).
    In addition, there are other upstream sequences that are critical for transcriptional regulation.

16. DNA sequences upstream of the gene play crucial roles in the regulation of transcription. These sequences include enhancers (which stimulate transcription initiation rates) and silencers (which repress transcription initiation rates). Consequently, the regulatory proteins that bind to these sequences are called activators or repressors, respectively. All these regulatory proteins are considered to be transcription factors.

17. In eukaryotes, at least six proteins (called general transcription factors or GTFs) are required to recruit RNA polymerase to a promoter and initiate transcription.

18. The multiprotein complex that initiates transcription is called the preinitiation complex (PIC). The formation of the PIC begins with the binding of the TATA-binding protein (TBP) to the TATA box of a promoter. Several other GTFs bind in succession, followed by RNA polymerase II and the remaining GTFs. One of the PIC proteins (TFIIH) is an ATP-dependent helicase, which forms the open complex.

19. Shortly after transcription commences, RNA polymerase undergoes a conformational change that involves phosphorylation of its C-terminal domain (CTD). Some GTFs dissociate from the transcribing polymerase, and an Elongator complex binds. In eukaryotes, transcription termination is linked to RNA processing.

20. RNA polymerases I and III also use TBP but different sets of GTFs at their TATA-less promoters.

Posttranscriptional Processing

21. In prokaryotes, primary transcripts are translated without further modification; however, in eukaryotes most primary transcripts must be specifically modified by:
    (a) The addition of nucleotides to the 5′ and 3′ ends and, in rare cases, in the interior of the transcript.
    (b) The exo- and endonucleolytic removal of polynucleotide sequences.
    (c) Modification of specific nucleotide residues.

22. Eukaryotic mRNAs have a cap structure consisting of a 7-methylguanosine residue joined to the 5′ end of the mRNA via a 5′– 5′ triphosphate linkage. Cap formation requires a triphosphatase, capping enzyme, and one or more methyltransferases.

23. Most eukaryotic mRNAs also have a string of adenosine residues, the poly(A) tail, at their 3′ end. The primary transcript is cleaved past an AAUAAA sequence and the poly(A) tail is added by poly(A) polymerase. The poly(A) tail, which ranges from 20 to 250 nt, protects the mature mRNA from ribonuclease activity in cells.

24. Unlike prokaryotic genes, most eukaryotic structural genes give rise to primary transcripts, called heterogeneous RNA (hnRNA), that are much longer than predicted from the size of the polypeptides they encode. These transcripts are processed by splicing, which is the excision of intervening sequences (introns) and the ordered joining of the flanking expressed sequences (exons) to form a mature mRNA.

25. Each splicing reaction involves two transesterification reactions mediated by ribonucleoprotein complexes called spliceosomes, which contain 5 RNAs known as small nuclear RNAs (snRNAs) and at least 100 polypeptides. The snRNAs mediate proper alignment of splice junctions within the spliceosome. Splicing begins with the nucleophilic attack by the 2′-OH group of an intron adenosine residue on the intron's 5′-phosphate group. The second transesterification reaction splices the 3′-OH group of the 5′ exon with the 5′ end of the 3′ exon, releasing the intron in the form of a lariat.

26.   The DNA sequences of exon–intron junctions are conserved. In particular, the 5′ dinucleotide of an intron is invariably GU and the 3′ dinucleotide is invariably AG.

27.   In mammals, the number of introns per gene varies from none to 234 (in the titin gene), and the total length of introns can be 4–40 times longer than that of exons. The exons in a gene can be alternatively spliced to yield either isoforms of a class of proteins (e.g., tropomyosins) or proteins whose function has changed (e.g., *Drosophila* sex determination proteins). Approximately 15% of human genetic diseases are caused by point mutations that delete splice sites or result in altered gene splicing.

28.   In certain organisms, mRNA undergoes editing in which numerous U residues are inserted and removed to generate a translatable mRNA. Guide RNAs base pair with the immature mRNAs to direct these alterations. Substitutional RNA editing may result in significant protein diversity.

29.   In *E. coli,* primary rRNA transcripts contain the sequences for 5S rRNA, 16S rRNA, 23S rRNA, and up to four tRNAs. These RNAs are released by the action of several specific RNases that commence processing while the primary transcript is still being synthesized.

30.   Eukaryotic 45S rRNA primary transcripts contain the sequences for 5.8S rRNA, 18S rRNA, and 28S rRNA. Roughly 110 sites are methylated prior to the nucleolytic removal of spacer sequences that separate the three rRNAs. Small nucleolar RNAs (snoRNAs) guide the methyltransferases that modify the rRNAs.

31.   Some eukaryotic RNA introns are self-splicing, meaning that catalysis of splicing is completely mediated by the RNA itself. Self-splicing occurs via transesterification reactions with no free energy input. Group I introns are removed by a self-splicing reaction that requires a free guanine nucleotide. Group II introns react via a lariat intermediate and require no free nucleotide; hence, this self-splicing mechanism is nearly identical to that of spliceosome-mediated pre-mRNA splicing.

32.   Prokaryotic and eukaryotic transfer RNA is processed by trimming the ends of the primary transcript and modifying specific bases. Many eukaryotic pre-tRNAs also have an intron that is spliced out. In addition, tRNA nucleotidyltransferase adds two C's and an A to the 3′ end of the immature tRNA (prokaryotic pre-tRNA already contains this trinucleotide).

Questions

Prokaryotic RNA Transcription

1.   The prokaryotic RNA polymerase holoenzyme contains the core enzyme and a σ factor. What is the function of the core enzyme? What are its limitations in terms of enzymatic activity? What is the role of the σ factor?

2. In a binding assay, a bacterial RNA polymerase binds its 70-base promoter 100 times faster when the promoter is part of an 8 kb plasmid than when the promoter is part of a 200-bp linear DNA. What might account for this difference in binding behavior? Assume that the polymerase is incubated with an equal molar concentration of each type of DNA.

3. What evidence suggests that newly transcribed RNA is not wound around its template DNA?

4. You are interested in discerning the role of σ factors in prokaryotic transcription, using a system of purified core polymerase, a DNA template, a σ factor, and labeled nucleotides ($^{32}pppN$). The incorporation of ^{32}P into RNA at 2 different core enzyme concentrations (10^{-10} M and 10^{-11} M) and constant concentration of σ factor (10^{-12} M) is shown below.
   (a) What does the graph measure?
   (b) What can you conclude about the behavior of σ factor?
   (c) What can you conclude about the behavior of the core enzyme?

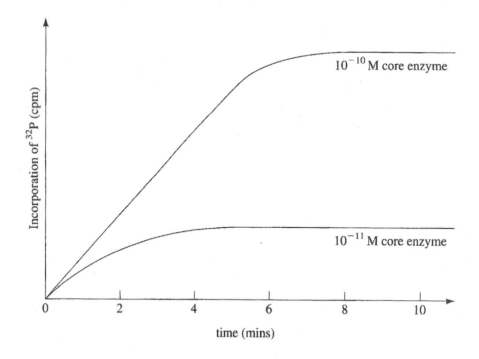

5. What evidence indicates that a GC-rich region of DNA followed by an AT-rich region is sufficient but not necessary to promote RNA chain termination?

Transcription in Eukaryotes

6. Distinguish between enhancers and promoters.

7. You have discovered a novel eukaryotic RNA polymerase. To study this enzyme, you have developed an *in vitro* assay with an inhibitor of RNA chain elongation, 3'-deoxy-5'[α-^{32}P] CTP. You are surprised by the fact that no ^{32}P is found in the transcribed RNA! What does this result tell you about the mechanism of the polymerase?

Posttranscriptional Processing

8. Indicate whether the posttranscriptional modifications listed below occur in prokaryotes or in eukaryotes.

   (a) 5' Cap
   (b) Polyadenylation
   (c) Methylation of nucleotide residues
   (d) Endonucleolytic cleavage
   (e) Splicing

9. Which of the following is contained in the rRNA cistron of *E. coli*?
   (a) 16S rRNA
   (b) 23S rRNA
   (c) a tRNA gene
   (d) 5S RNA

10. What RNA-processing enzyme complexes contain RNA? In which does RNA participate in catalysis?

11. What kinds of posttranscriptional processing occur in eukaryotic tRNA precursors?

12. Group I introns incubated with increasing concentrations of poly(C) show Michaelis–Menten kinetics for the cleavage of poly(C); that is, in the presence of increasing amounts of poly(C), the rate increases to a maximum. Is this proof that these introns are catalysts? Explain.

13. The mRNA capping reaction involves cleavage of a methylated guanosine triphosphate to release its pyrophosphate. You wish to measure RNA chain initiation rates in isolated nuclei. Which of the isotopes shown below would you use? Explain. What assumptions are necessary to use this experimental strategy?

    ^{32}pppA   p^{32}ppA   pp^{32}pA

14. What is the role of SXL and U2AF in generating alternatively spliced *tra* mRNA transcripts in the *Drosophila* sex-determination pathway?

15. Does GTP serve as a substrate, catalyst, cofactor, or allosteric effector in the self-splicing reaction of *Tetrahymena* pre-rRNA? Explain.

16. Describe the primer you would use to initiate reverse transcriptase–catalyzed synthesis of DNA that is complementary to a eukaryotic mRNA molecule.

Answers to Questions

1. The core enzyme carries out RNA synthesis but it cannot initiate transcription. The σ factor binds specifically to a promoter; this allows the DNA to melt apart so that RNA polymerase can begin transcribing.

2. The polymerase binds nonspecifically to the plasmid and finds its promoter in a one-dimensional search, which proceeds faster than when the enzyme must locate its promoter through a three-dimensional search.

3. *E. coli* topoisomerase mutants show a buildup of positive or negative supercoils during transcription, which would not occur if RNA polymerase simply followed the helical path of the DNA template. In fact, electron micrographs such as Figure 26-8 indicate that the RNA extends away from the DNA as transcription proceeds.

4.
    (a) Since radioactivity from ^{32}pppN is incorporated only into the 5′ nucleotide, the graph measures RNA chain initiations.
    (b) Since more initiations occur when more core enzyme is present despite constant σ factor concentration, the σ factor can dissociate from the core enzyme after initiation and bind to a new core enzyme to initiate a new transcript.
    (c) Since the curves flatten at both core enzyme concentrations, the core enzyme does not dissociate from the DNA template (another factor must facilitate dissociation of the core enzyme from the template). Consequently, once all the core enzyme binds to DNA and initiates transcription, no new RNA chain initiations are observed.

5. Termination resulting from the presence of GC- and AT-rich regions does not always occur. In some cases, particularly where these sequences are missing, Rho factor–dependent termination occurs.

6. Enhancers are eukaryotic transcriptional control elements where specific proteins bind and thereby promote RNA polymerase binding at the promoter. Enhancers are located at variable distances from the gene and in different orientations. Promoters, in contrast, are located near the transcription start site. A promoter is sufficient for transcription initiation and can operate with or without an enhancer. An enhancer cannot operate without a promoter.

7. Normal RNA chain elongation requires a 3′-OH group. Because the inhibitor lacks this functional group, it should normally be incorporated into the RNA but prevent further polymerization. The absence of ^{32}P in this experiment suggests that RNA chain elongation may proceed in the 3′→5′ direction. In this case, the RNA chain would grow through nucleophilic attack by the 3′-OH group of the incoming nucleotide, a reaction in which the 3′-deoxy compound cannot participate.

8.

(a) 5' Cap: eukaryotes

(b) Polyadenylation: eukaryotes

(c) Methylation of nucleotide residues: eukaryotes and prokaryotes

(d) Endonucleolytic cleavage: eukaryotes and prokaryotes

(e) Splicing: eukaryotes

9.   All of these elements are contained in the rRNA primary transcript.

10. RNA-processing enzyme complexes that contain RNA include RNase P, snRNPs, the mRNA-editing machinery, and the rRNA-methylating machinery. Only the RNase P RNA participates in catalysis.

11. Nucleolytic removal of nucleotides, splicing of two exons, addition of CCA residues at the 3' end, and chemical modification of certain nucleotides.

12. No. It must be demonstrated that the intron is regenerated in the course of the reaction. It is possible that the intron is irreversibly cleaved in the course of poly(C) cleavage.

13. The β-labeled compound $p^{32}ppA$ should be used because the ^{32}P will appear only in the first incorporated nucleotide (in subsequent residues, the β and γ phosphates are eliminated). If the α-labeled compound were used, all A residues in the mRNA would be labeled. During cap addition, two phosphates of the methylated GTP and one terminal mRNA phosphate are removed, so only a β label would remain. The assumptions in this experiment are that the initial residue is A rather than G and that the capping reaction is not rate limiting, so that the rate of appearance of β-labeled nucleotide indicates the rate of RNA chain initiation.

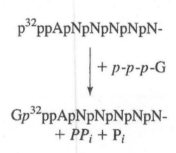

14. The role of U2AF is to bind to the splice junction of exon 2 in the *tra* pre-mRNA to orchestrate splicing of exon 1 with exon 2. In the absence of SXL (in males), U2AF binds to the proximal 3' splice site to yield an mRNA that contains an exon with a premature Stop codon, resulting in a truncated, nonfunctional TRA protein. In the presence of SXL (in females), this splice site is blocked by SXL; U2AF binds to a distal 3' splice site to join the 3' portion of exon 2, which no longer has the premature Stop codon and thereby produces a functional TRA protein.

15. GTP serves as a substrate since it becomes part of the excised intron and is not regenerated.

16. The primer would be a poly(T) sequence, because this sequence can pair with the poly(A) sequence at the 3′ end of the mRNA. The reverse transcriptase can then extend the poly(T) sequence to generate a cDNA.

27
Protein Synthesis

This chapter explores the mechanisms by which messenger RNA directs the synthesis of a polypeptide. The discussion begins with an examination of the genetic code, including the arguments and experiments that led to the conclusion that an mRNA nucleotide sequence is decoded as a nonoverlapping series of triplets. Hence, in this coding scheme, there are 61 triplet combinations, or codons, that encode the 20 amino acids and three Stop codons. Because most amino acids are specified by more than one codon, the genetic code is said to be degenerate but nonrandom. Whereas some amino acids are encoded by as many as six codons, others are encoded by only one; this phenomenon may reflect the way the genetic code evolved (Box 27-1). The discussion then turns to the molecules that "translate" the genetic code: transfer RNAs. The structures of the tRNAs are explored, from their modified bases to their unusual secondary and tertiary structures. Each tRNA can covalently attach an amino acid at its 3′ end to form an aminoacyl–tRNA. Two families of aminoacyl–tRNA synthetases (aaRS), which show surprisingly little sequence similarity, mediate this aminoacylation. Structural information on the mechanisms by which aaRSs recognize their cognate tRNAs is also presented. The chapter then discusses the "wobble hypothesis," which accounts for the degeneracy of the genetic code at the level of base-pairing interactions between the codons of mRNA and the anticodons of tRNAs. Some organisms have a "21st amino acid," selenocysteine, which is incorporated into certain proteins using a unique tRNA (Box 27-2). The chapter continues with a description of ribosome structure and the mechanisms of polypeptide chain initiation, elongation, and termination. Included in this section are discussions of the nature of the peptidyltransferase activity, the role of GTP hydrolysis in protein synthesis and translational accuracy, and the mechanisms by which antibiotics interfere with translation (Box 27-3). The chapter concludes with a discussion of posttranslational processing, including proteins involved in protein folding, covalent modification, and secretion.

Essential Concepts

1. Translation shares important features with DNA transcription and replication:
    (a) Translation occurs in a large multiprotein machine, which includes accessory factors that regulate the initiation, elongation, and termination of polypeptide synthesis.
    (b) Even though it is amino acids that are polymerized, translation depends on Watson–Crick base pairing between mRNA and tRNA so that amino acids are joined as genetically specified.
    (c) Accuracy is essential, so the translational machinery has proofreading capabilities.
    (d) The process is endergonic and requires the cleavage of "high-energy" phosphoanhydride bonds.

The Genetic Code

2. The DNA sequence encoding a protein is made up of a triplet of nucleotides called a codon. A triplet code allows more than one codon to specify an amino acid; hence, the code is said to be degenerate. For example, leucine is specified by six codons, serine by four, and glutamate by two. Codons that specify the same amino acid are termed synonyms.

3. Transfer RNA deciphers the codon. Each tRNA contains a trinucleotide sequence, called the anticodon, which is complementary to an mRNA codon specifying the amino acid attached to the tRNA.

4. The code is read sequentially and contains no internal punctuation. The sequence AUG (read 5′ to 3′ along the mRNA), or occasionally GUG, at the 5′ end of the mRNA marks the first codon to be translated. The codons UAA, UAG, and UGA are called Stop codons because they signal the termination of translation.

5. The arrangement of codons is nonrandom. Most synonyms share the first two nucleotides. Interestingly, codons with different nucleotides in the first position tend to specify chemically similar amino acids. Codons with second position pyrimidines encode mostly hydrophobic amino acids, and those with second position purines encode mostly polar amino acids. This suggests a nonrandom evolution of the genetic code that minimizes the deleterious effects of point mutations (see Box 27-1).

6. The "standard" code is not universal; exceptions have been observed in mitochondria and a few protozoans.

Transfer RNA and Its Aminoacylation

7. Almost all known tRNAs can be arranged in a so-called cloverleaf secondary structure, which has the following key features:
   (a) A 5′-terminal phosphate group.
   (b) Three keyhole-shaped stem-and-loop structures called the D arm, the anticodon arm, and the T or TψC arm. The D arm is named for the modified base dihydrouridine (D) that frequently occurs in the loop. The anticodon arm contains the anticodon sequence in a loop. The TψC arm is named for the sequence thymidine–pseudouridine (ψ)–cytidine that the loop usually contains.
   (c) A 3′ CCA sequence containing a free 3′ hydroxyl group. This trinucleotide is either genetically specified (in prokaryotes) or enzymatically appended to the immature tRNA (in eukaryotes).
   (d) Between the anticodon arm and the TψC arm is a variable arm whose sequence and length (3–21 nucleotides) varies among tRNAs as well as among species.
   (e) Transfer RNAs have 15 invariant positions (always the same base) and 8 semivariant positions (only a purine or only a pyrimidine) that occur mostly in the loops.

8. Up to 25% of the bases of tRNA are modified posttranscriptionally. The biological significance of many of these modifications is still mysterious, since they are not required for the structural integrity of tRNA or its proper binding to the ribosome.

9. Transfer RNA has an L-shaped tertiary structure in which structural stability is maintained by extensive base stacking interactions and base pairing within and between the helical stems.

10. Attachment of an amino acid to a tRNA is carried out in a two-reaction process by an aminoacyl–tRNA synthetase (aaRS). The first reaction involves "activation" of the amino acid by its reaction with ATP to form an aminoacyl–adenylate with the release of PP_i. The mixed anhydride then reacts with tRNA to form the "charged" aminoacyl–tRNA with the release of AMP. The hydrolysis of pyrophosphate (PP_i) ensures that the overall reaction is irreversible.

$$
\begin{array}{ll}
1. & \text{Amino acid} + \text{ATP} \rightleftharpoons \text{aminoacyl–AMP} + PP_i \\
2. & \text{Aminoacyl–AMP} + \text{tRNA} \rightleftharpoons \text{aminoacyl–tRNA} + \text{AMP} \\
\hline
& \text{Amino acid} + \text{tRNA} + \text{ATP} \rightleftharpoons \text{aminoacyl–tRNA} + \text{AMP} + PP_i
\end{array}
$$

11. There are two structurally unrelated classes of aminoacyl–tRNA synthetases (Class I and Class II). These enzymes differ in
    (a) The mechanism by which they recognize their tRNA substrates.
    (b) The initial site of aminoacylation on the tRNA.
    (c) Amino acid specificity.

12. Each set of isoaccepting tRNAs is recognized by a single aminoacyl–tRNA synthetase (aaRS). Isoaccepting tRNAs have different anticodons but are conjugated to the same amino acid. aaRSs appear to recognize the unique structural features of the cognate tRNAs, not specific base sequences. Some aaRSs are capable of proofreading by a double-sieve mechanism, which requires a second active site for hydrolysis.

13. During protein synthesis, the proper tRNA is selected only through codon–anticodon interactions; the aminoacyl group is not involved. Many tRNAs bind to two or three of the codons that specify their cognate amino acids. This variation is due to non-Watson–Crick base pairing between the third codon position and the first anticodon position. For example, U in the first position of the anticodon can base pair with A or G at the third codon position. Some tRNAs have inosine at the first anticodon position, which can base pair with U, C, or A. The degeneracy in binding at this position is called wobble pairing.

Ribosomes

14. Ribosomes carry out the following functions:
    (a) They bind mRNA so that its codons are matched, with high accuracy, to tRNA anticodons.
    (b) Ribosomes contain three specific binding sites: the A site for the incoming aminoacyl–tRNA; the P site for the tRNA to which the growing polypeptide chain is attached; and the E site for the outgoing, uncharged tRNA.

(c) They mediate interactions of nonribosomal proteins that are required for polypeptide chain initiation, elongation, and termination. Many of these factors use the free energy of GTP hydrolysis to carry out their functions.

(d) Ribosomes catalyze peptide bond formation between the incoming amino acid and the growing polypeptide chain.

(e) They are responsible for the translocation of the mRNA so that codons are read sequentially.

15. Prokaryotic ribosomes are enormous ribonucleoprotein machines containing dozens of proteins and several RNA molecules. The *E. coli* ribosome has a sedimentation coefficient (which indicates its rate of sedimentation in an ultracentrifuge) of 70S and is composed of two subunits: The 30S subunit contains the 16S rRNA, while the 50S subunit contains the 23S and 5S rRNAs. The 30S subunit accommodates a single strand of mRNA (without any secondary structure); the 50S subunit harbors the growing polypeptide chain.

16. Prokaryotic ribosome structure is complex and has been pieced together through a combination of X-ray diffraction and cryoelectron microscopy. From these studies some generalizations have emerged.

(a) Both 16S and 23S rRNAs are assemblies of helices connected by loops.

(b) The four domains of 16S rRNA form distinct parts of the 30S subunit.

(c) The six domains of 23S rRNA are intertwined throughout the 50S subunit.

(d) The distribution of ribosomal proteins is not uniform. The facing surfaces of each ribosomal subunit are largely devoid of proteins; whereas the other surfaces of each subunit contains the vast majority of the ribosomal proteins.

(e) Most ribosomal proteins consist of a globular domain that is located on the surface of the ribosome and an unstructured tail that penetrates the rRNA and interacts with it via salt bridges.

17. Eukaryotic ribosomes resemble prokaryotic ribosomes in shape and function but are larger and more complex. The 80S ribosome is composed of a 40S small subunit (with 18S rRNA) and a 60S large subunit (with 5S, 5.8S, and 28S rRNAs). The 18S rRNA is homologous to the prokaryotic 16S rRNA, while the 5.8S and the 28S rRNAs are homologous to the prokaryotic 23S rRNA.

Translation

18. Polypeptide synthesis has the following key features:

(a) Ribosomes read mRNA in the 5' to 3' direction.

(b) Polypeptide synthesis proceeds from the N-terminus to the C-terminus.

(c) Chain elongation occurs by linking the growing polypeptide chain to the incoming charged tRNA.

(d) Several ribosomes can bind to a single mRNA to form a complex called a polysome or polyribosome.

19. Polypeptide chain initiation begins with Met–tRNA in eukaryotes and a modified Met–tRNA in prokaryotes, *N*-formylmethionyl–tRNA (fMet–tRNA). In prokaryotes, the AUG start codon is distinguished from other AUG codons by base pairing between the 16S

rRNA and a 5′ mRNA sequence called the Shine–Dalgarno sequence. Eukaryotes do not contain an analogous recognition sequence; initiation occurs at the first AUG of a eukaryotic mRNA.

20. Initiation of translation in *E. coli* occurs in three stages:
    (a) IF-3 binds to the 70S ribosome and promotes its dissociation into the 30S and 50S subunits.
    (b) mRNA, IF-1, and IF-2 in complex with GTP and fMet–tRNA bind to the 30S subunit.
    (c) IF-1 and IF-3 are released, allowing the 50S subunit to bind to the 30S subunit. This stimulates hydrolysis of the GTP bound to IF-2, resulting in a conformational change in the 30S subunit and the release of IF-2, which can participate in further initiation reactions.

21. Translational initiation in eukaryotes is more complex than that in prokaryotes. The three stages described for prokaryotic translational initiation also occur in eukaryotes, but up to 11 initiation factors—instead of 3—are involved. The initiation factor eIF4F binds to the 7-methylguanosine (m^7G) cap found on all eukaryotic mRNAs. This complex recruits several additional initiation factors, which then bind to the 43S preinitiation complex, forming the 48S initiation complex, which then scans the mRNA until it encounters the first AUG codon.

22. Elongation of a nascent polypeptide chain in prokaryotes proceeds as a three-reaction cycle (decoding, transpeptidation, and translocation) until a termination codon is encountered.

23. An incoming aminoacyl–tRNA, in a complex with EF-Tu·GTP, binds to the A site. This interaction is accompanied by the hydrolysis of GTP, which releases P_i and EF-Tu·GDP from the ribosome. EF-Ts replaces the GDP bound to EF-Tu, which is then replaced by GTP, thereby regenerating the EF-Tu·GTP complex. The aminoacyl–tRNA·EF-Tu·GTP undergoes a large conformational change (a 91° domain rotation!) when GTP is hydrolyzed, which moves the aminoacyl–tRNA into a position to allow the transpeptidation reaction to occur.

24. The ribosome senses proper codon–anticodon pairing through interactions with the 16S rRNA. If there is a mispairing, the aminoacyl–tRNA·EF-Tu·GTP complex dissociates from the ribosome before GTP is hydrolyzed. In addition, a proofreading step occurs after GTP hydrolysis. GTP hydrolysis is therefore the thermodynamic price of ribosomal translation accuracy.

25. The 23S rRNA of the large ribosomal subunit acts as a ribozyme to carry out the transpeptidation reaction. A peptide bond is formed through the nucleophilic displacement of the P-site tRNA by the amino group of the aminoacyl–tRNA in the A site. Hence, the nascent polypeptide is lengthened at its C-terminus by one amino acid and transferred to the A site, a process called transpeptidation. The 23S rRNA is the catalyst in peptide bond formation, although it does not participate as an acid–base catalyst. The main catalytic mechanism appears to involve positioning and orienting the substrates for reaction.

26. Translocation of the new peptidyl–tRNA occurs in several discrete steps. After transpeptidation, the acceptor end of the new peptidyl–tRNA moves into the P site, while the acceptor end of the deacylated tRNA moves into the E site. EF-G·GTP binding to the ribosome and subsequent GTP hydrolysis allows for the complete displacement of both tRNA moieties to the E and P sites, resulting in a completely vacant A site. Hence, the alternating activities of EF-Tu and EF-G keep the ribosome cycling in a unidirectional manner. In fact, EF-G structurally mimics the EF-Tu–tRNA complex.

27. Elongation in eukaryotes is similar to that in prokaryotes, except that eEF1A and eEF1B replace EF-Tu and EF-Ts, respectively, and eEF2 replaces EF-G.

28. In *E. coli*, two release factors, RF-1 and RF-2, recognize Stop codons. Binding of either RF-1 or RF-2 to the ribosome stimulates the transfer of the peptidyl group to water, thereby terminating polypeptide chain elongation. Next, RF-3 binds and, following hydrolysis of its GTP, stimulates the release of RF-1 and RF-2. Ribosome recycling factor (RRF) and EF-G·GTP then bind. GTP hydrolysis by EF-G causes RRF to move into the P site such that the remaining tRNAs are released from the P and E sites. Finally, RRF and EF-G·GDP dissociate. Eukaryotic translational termination is similarly mediated by eRF, which carries out all the activities of RF-1, RF-2, and RF-3.

29. A mutation that converts an aminoacyl-coding ("sense") codon to a Stop codon is called a nonsense mutation and results in the premature termination of translation. Such mutations can be rescued by a nonsense suppressor tRNA, which contains an anticodon to a Stop codon but carries the amino acid of its wild-type progenitor.

30. Many commonly used antibiotics interfere with the initiation or elongation steps in prokaryotic translation. Puromycin mimics tyrosyl–tRNA and aborts transpeptidation. Streptomycin inhibits translational initiation. Chloramphenicol inhibits the peptidyltransferase reaction. Tetracycline inhibits aminoacyl–tRNA binding to the ribosome. The toxin ricin inactivates the eukaryotic ribosome by excising a base from 28S rRNA.

Posttranslational Processing

31. Ribosomal proteins may recruit chaperone proteins to assist with protein folding. Trigger factor binds to the L23 ribosomal protein and recognizes short hydrophobic segments emerging from the ribosome, prior to the arrival of other chaperone proteins such as DnaK, DnaJ, and the GroEL/ES chaperonins.

32. Posttranslational modifications of newly synthesized proteins are common, especially in eukaryotes. In some cases, limited proteolysis may activate a protein, which is called a proprotein. A preproprotein also contains a signal peptide that is removed in the endoplasmic reticulum, prior to excretion of the protein. Common modifications of proteins include phosphorylation (at Ser, Thr, Tyr, and His residues), glycosylation, hydroxylation of Pro and Lys residues in collagen, and the attachment of ubiquitin or small ubiquitin-related modifier (SUMO) to Lys residues.

33. The signal recognition particle (SRP) is a ribonucleoprotein containing a 300-residue RNA (in eukaryotes) or a 114-nt RNA (in prokaryotes). When an N-terminal hydrophobic region emerges from the ribosome, the SRP binds to the nascent polypeptide and facilitates the binding of the associated ribosome to the SRP receptor complex in the endoplasmic reticulum (in eukaryotes) or plasma membrane (in prokaryotes). This results in the translocation of the newly synthesized polypeptide across the membrane.

34. Proper folding of glycoproteins in the endoplasmic reticulum (ER) relies on the activity of two homologous chaperones, membrane-bound calnexin and soluble calreticulin.

Questions

The Genetic Code

1. Why is the genetic code degenerate?

2. How many amino acid copolymers can be specified by polynucleotides consisting of two alternating nucleotides? What are the sequences of these polypeptides?

3. A cell extract from a new species of *Amoeba* is used to translate a purified mouse hemoglobin mRNA, but only a short peptide is produced. What is the simplest explanation for this result?

4. Predict the effect on protein structure and function of an A·T to G·C transition in the first codon position for lysine.

5. Identify the polypeptide encoded by the DNA sequence below, in which the lower strand serves as the template for mRNA synthesis.

```
5′  GGACCTATGATCACCTGCTCCCCGAGTGCTGTTTAGGTGGG  3′
3′  CCTGGATACTAGTGGACGAGGGGCTCACGACAAATCCACCC  5′
```

6. Describe how nucleotide substitutions at different positions in codons affect the characteristics of the encoded amino acids.

Transfer RNA and Its Aminoacylation

7. What two reactions are carried out by aminoacyl–tRNA synthetases?

8. Which of the following tRNA structural features are always involved in interactions with an aaRS?
    (a) Acceptor stem   (b) Anticodon loop   (c) TψC arm

9. A new species of yeast has just one tRNA^{Ala}, with the anticodon IGC. What does this suggest about codon usage in this organism?

10. Which of the following amino acids are likely to be linked to only one species of tRNA? What are their anticodons?
    (a) Phe   (b) Leu   (c) His

Translation

11. Reticulocyte (immature red blood cell) lysates can be made devoid of mRNA by ribonuclease treatment followed by inactivation and removal of the ribonuclease. These lysates are now capable of synthesizing protein when mRNA, GTP, and aminoacyl–tRNAs are added. How could you use this system to demonstrate that a protein is synthesized from the N-terminus to the C-terminus?

12. Match each of the functions on the left with the appropriate prokaryotic protein on the right.

    _____ Binds fMet–tRNA and GTP.   A. RF-1
    _____ Binds aminoacyl–tRNA and GTP.   B. RF-2
    _____ Recognizes Stop codons.   C. RF-3
    _____ Binds GTP and promotes RF-1 and RF-2.   D. IF-1
             release from the ribosome
    _____ Hydrolyzes GTP to GDP.   E. IF-2
    _____ Promotes the transfer of peptidyl–tRNA from   F. IF-3
             the A site to the P site.
    _____ Inhibits the interaction of the 30S and 50S subunits.   G. EF-Ts
    _____ Facilitates mRNA binding to the 30S ribosome.   H. EF-Tu
    _____ Displaces GDP from EF-Tu.   I. EF-G

13. How could you show that the inhibition of protein synthesis in a bacterial cell extract by puromycin is not irreversible?

14. You have isolated a temperature-sensitive mutant *E. coli*. At 34°C, the growth rate is nearly normal but is reduced 10-fold at 37°C. You eventually discover that EF-Tu·GDP binding to the ribosome is two-fold higher in the mutant at 37°C. What aspect of translation is affected by the mutation?

15. Which amino acid substitutions might occur in *E. coli* opal suppressors that arise by single-base mutations?

16. The error rates of translation in some rapidly growing *E. coli* strains is as high as 2 percent. What is the percentage of correctly translated protein in such bacteria (assume the protein has 100 residues)? How can these bacteria do so well?

17. Distinguish between eukaryotic and prokaryotic polypeptide initiation, elongation, and termination.

Posttranslational Processing

18. What is the adaptive significance of posttranslational cleavage of proteins such as collagen and various hormones?

19. Proteins destined for membranes or secretion have a stretch of N-terminal hydrophobic amino acids called a signal sequence. What function does this sequence serve?

Answers to Questions

1. The genetic code, which consists of 64 codons, is degenerate because the codons are used in a redundant manner. Only 21 codons would be strictly required to code for 20 amino acids and a Stop codon.

2. The six possible polynucleotides are

   AUAUAUAUAU···
   ACACACACAC···
   AGAGAGAGAG···
   CGCGCGCGCG···
   GUGUGUGUGU···
   CUCUCUCUCU···

   Each encodes a polypeptide that repeats two amino acids: Poly(AU) yields poly(Ile–Tyr) since it alternates AUA codons and UAU codons; poly(AC) yields poly(Thr–His) since it alternates ACA and CAC codons; poly(AG) yields poly(Arg–Glu) since it alternates AGA codons and GAG codons; poly(CG) yields poly(Arg–Ala) since it alternates CGC codons and GCG codons; poly(GU) yields poly(Val–Cys) since it alternates GUG codons and UGU codons; and poly(CU) yields poly(Leu–Ser) since it alternates CUC codons and UCU codons. The first amino acid in the polymer can be either of the two since the translation reading frame can begin with either codon, so technically, 12 copolymers are possible.

3. An *Amoeba* tRNA interprets a mouse sense codon as a Stop codon, thereby halting translation of the mouse RNA.

4. An A to G transition would change the lysine AAA and AAG codons to GAA and GAG, both of which encode glutamate. This would result in an acidic residue replacing a basic residue in the protein. If the change were on the protein surface, it might have no effect on protein structure or function, but if the change involved a buried ion pair, a catalytic residue, or some other essential residue, its effect might be significant.

5. The mRNA includes an AUG start codon and a UAG stop codon and has the sequence

   5′ GGACCUAUGAUCACCUGCUCCCCGAGUGCUGUUUAGGUGGG 3′

   The encoded polypeptide has the sequence

   H_3N^+–Met–Ile–Thr–Cys–Ser–Pro–Ser–Ala–Val–COO⁻

6. See Table 27-1:
    (a) Changes in the third position of a codon infrequently change the identity of the amino acid. In half of these exceptions, the amino acids are of very similar chemistry (e.g., Asp → Glu when GA(U/C) → GA(A/G)); others result in Stop codons or amino acids of very different chemical character (e.g., Ser → Arg when AG(U/C) → AG(A/G)).
    (b) Changes in the first codon position usually change the amino acid to another of similar chemistry.
    (c) Changing the second position of a codon to a pyrimidine generally specifies a hydrophobic amino acid, while changing it to a purine generally specifies a polar amino acid.

7. The first reaction "activates" an amino acid via condensation with ATP to form an aminoacyl–AMP + PP_i. The second reaction condenses this mixed anhydride with its cognate tRNA to form an aminoacyl–tRNA + AMP.

8. Only the acceptor stem is involved in all tRNA–aaRS interactions.

9. This yeast uses Ala codons GCA, GCC, and GCU but not GCG, since I can pair with A, C, or U but not G.

10. Only Phe and His are likely to be linked to one tRNA species. The anticodon of $tRNA^{Phe}$ is GAA, so it can bind both UUU and UUC by wobble pairing (see Table 27-3). Similarly, the anticodon of $tRNA^{His}$ is GUG, so it can bind codons CAC and CAU. $tRNA^{Leu}$ must have at least three anticodons (UAA, IAG, and CAG or UAG) to accommodate the six Leu codons (UUA, UUG, CUU, CUC, CUA, and CUG).

11. Add the mRNA for a known protein to the lysate. After a minute or two, add puromycin to terminate translation prematurely. Analyze the sequences of the resulting peptides to determine whether they correspond to the C- or the N-terminus of the protein. Alternatively, add to the lysate an isotopically labeled amino acid (as an aminoacyl–tRNA) that occurs only near one terminus of the protein and determine whether the label becomes incorporated into the peptides.

12.

E	Binds fMet–tRNA and GTP.
H	Binds aminoacyl–tRNA and GTP.
A, B	Recognizes Stop codons.
C	Binds GTP and promotes RF-1 and RF-2 release from the ribosome.
C, E, H, I	Hydrolyzes GTP to GDP.
I	Promotes the transfer of peptidyl–tRNA from the A site to the P site.
F	Inhibits the interaction of the 30S and 50S subunits.
E	Facilitates mRNA binding to the 30S ribosome.
G	Displaces GDP from EF-Tu.

13. Puromycin, which resembles the 3′ end of Tyr–tRNA, competes with aminoacyl–tRNAs for binding to ribosomes. Increasing the concentrations of aminoacyl–tRNAs in the extract should at least partially overcome the effects of puromycin and restore protein synthesis.

14. EF-Tu· GDP dissociation is essential for polypeptide elongation. Hence, in the mutant at 37°C, slower dissociation of the EF-Tu·GDP complex slows translation. The dramatic decrease in growth rate may be a function of kinetic proofreading in which an increase in the time of EF-Tu·GDP binding to the ribosome increases the chance of misincorporating an aminoacyl–tRNA.

15. The anticodon of an opal suppressor tRNA must be UCA. The anticodons that can mutate to UCA via a single-base change are UCC, UCG, UCU, UAA, UGA, UUA, ACA, CCA, and GCA, which correspond to amino acids Arg, Cys, Gly, Leu, Ser, and Trp (UUA corresponds to the Stop codon UAA).

16. The probability that any one amino acid is incorporated correctly is 0.98. Hence, the probability that a 100-residue protein is synthesized correctly is 0.98^{100}, or 13.26%! The bacteria survive because many translation errors do not alter protein function. In addition, protein turnover is sufficiently rapid to overcome the effects of abnormal proteins. In fact, many nonfunctional proteins have shorter-than-normal half-lives.

17. **Initiation:** Eukaryotes lack Shine–Dalgarno binding interactions, and they have many more initiation factors, including an mRNA cap–binding protein.

    Elongation: Eukaryotic eEF1 carries out the functions of prokaryotic EF-Tu and EF-Ts, while eEF2 functions similarly to EF-G.

    Termination: Eukaryotic polypeptide termination requires only one factor, eRF, which carries out all the functions of the three prokaryotic RFs.

18. Mature collagen self-assembles into large polymeric structures. Such assembly inside the cell would likely destroy the cell. Hence, the immature procollagen, which is incapable of self-assembly, exits the cell before being proteolytically converted to its mature form. In the case of peptide hormones, a mature hormone present in the cell might find its receptor and activate a signal transduction cascade that is inappropriate for that cell. This is prevented by extracellular activation, via posttranslational proteolysis, of the prohormone.

19. Signal sequences recruit signal recognition particles (SRPs), in which the resulting complex binds to an SRP receptor in the ER membrane. Binding of the signal peptide to an SRP halts translation until the SRP–ribosome–nascent polypeptide complex binds to the SRP receptor.

28
Regulation of Gene Expression

This chapter introduces you to one of the main efforts of modern biochemistry, understanding the regulation of gene expression, which defines the phenotypic properties of cells. In eukaryotes, the regulation of gene expression is more complex than in prokaryotes, largely due to the size of the eukaryotic genome, the numerous levels of DNA packaging, the complexity of the cells, and the numerous cell types in eukaryotic organisms. This chapter begins by examining genome organization in prokaryotes and eukaryotes, focusing on gene number, gene clusters, and nontranscribed DNA. The human, yeast, and *E. coli* genomes are used to illustrate the kinds of information provided by the emerging field of genomics. Included in this discussion is an overview of the organization of DNA in eukaryotes with regard to repetitive and unique DNA sequences.

The regulation of prokaryotic gene expression is discussed next, with attention focused on the *lac* operon, *trp* operon, and riboswitches, each illustrating key features common to the regulation of all prokaryotic genes. The *lac* repressor is an example of a negative regulator regulated by an allosteric effector; the *trp* operon provides an example of attenuation. Riboswitches are interesting cases of posttranscriptional regulation via the binding of allosteric effectors to mRNA.

The chapter then turns to eukaryotic gene regulation, which can occur at many points between the exposure of promoter sites within chromatin and the translation of the mRNA. Various regulatory mechanisms involve chromatin remodeling; histone modification; DNA methylation; the binding of transcriptional activators and repressors to DNA sequences upstream from the promoter; activation of signal transduction pathways; selective degradation of RNA; phosphorylation of translation initiation factors; and DNA rearrangements (as exemplified by immunoglobulin gene diversity). Progression of cells through the cell cycle is regulated by the repression and activation of key transcriptional activators and repressors. Two key regulators of cell cycle progression are p53 and pRb, mutant forms of which cause human cancers. The chapter closes with an overview of gene expression in the development of the fruit fly, *Drosophila melanogaster*. Here, a set of genes that regulate the expression of other genes orchestrates embryogenesis. *Hox* genes encode a remarkably conserved set of transcription factors that participate in embryonic patterning in *Drosophila* and a wide variety of organisms.

Essential Concepts

1. Gene expression is neither random nor fully preprogrammed. Information in an organism's genome is used in an orderly fashion during development but also responds to changes in the organism's internal or external environment.

Genome Organization

2.   The new discipline of genomics, the study of organisms' genomes, endeavors to elucidate the functions of all genes. The vast amount of data obtained from the numerous genome projects to date is stored in computerized databases, most of which are publicly available via the Internet. Genomics also explores evolutionary relationships between genes, organelles, and organisms.

3.   There is surprisingly modest correlation between the amount of DNA in an organism's genome (its C value) and its apparent morphological or metabolic complexity. This is known as the C-value paradox. However, there is a stronger correlation between gene number and morphological complexity (as measured by the number of cell types in the organism; see Figure 28-2).

4.   Only ~1.4% of the 3.2 billion base pairs of the human genome encodes proteins; this proportion is typical of most vertebrate genomes sequenced to date. So far, computer analysis indicates that there are around 23,000 protein-encoding genes in mammalian genomes, of which almost 42% are of unknown function and are referred to as orphan genes. About three-quarters of all known human genes have counterparts in other species. About one-quarter are found only in other vertebrates, while another quarter are common to all prokaryotes and eukaryotes.

5.   Approximately 60% of the human genome is transcribed into RNA:
     (a) Over 4000 genes encoding rRNA, tRNA, snRNA and other small RNAs; and
     (b) At least another 10,000 loci resulting in the transcription of noncoding RNAs (ncRNAs), most of which have no known function. Many of these RNAs are involved in RNA interference, a form of posttranscriptional gene silencing.

6.   A gene can be identified using two broad strategies:
     (a) It can identified as an open reading frame (ORF), a sequence that is not interrupted with a Stop codon; however, this method is limited due to the short size of exons and must rely on the availability of cDNA sequences from various tissues or organisms (expressed sequence tags, ESTs).
     (b) A gene can also be identified by the presence of CpG islands that are clustered upstream of many genes.

7.   In simple eukaryotes, such as yeast, it is currently possible to mutate every gene to determine its function; this information can then be applied to more complex eukaryotic organisms. In addition, human genome sequence information will make it easier to identify genes associated with disease. To this end, a catalog of single nucleotide polymorphisms (SNPs) is being compiled. A SNP occurs on average about every 1250 bp in humans, and nearly 2 million have been described so far.

8. An organism's genes are not distributed randomly throughout its genome; indeed, some genes are organized in both prokaryotic and eukaryotic genomes. Prokaryotic operons include genes related to a specific metabolic function. In eukaryotes, genes for histones and globins occur in clusters. Ribosomal RNA genes are also clustered in both prokaryotes and eukaryotes.

9. A significant portion of prokaryotic and eukaryotic genomes consists of DNA that is never transcribed. These unexpressed sequences accumulate mutations at a great rate since they experience little selective pressure; hence, they can be used to trace evolutionary relationships. In prokaryotes, these nontranscribed sequences are predominantly promoter and operator sequences adjacent to genes but also include insertion sequences and remnants of integrated viruses. In eukaryotes, nontranscribed sequences include promoters and other regulatory regions; however, the bulk of these sequences are repetitive DNA of unknown function. Some short (~10 bp) sequences are present in millions of copies, often tandemly repeated thousands of times. In humans, these short tandem repeats comprise about 3% of the genome. Eukaryotes also contain moderately repetitive sequences ($< 10^6$ copies per haploid genome) occurring in 100 to several thousand bp segments, most of which are retrotransposons. About 42% of the human genome consists of three types of retrotransposons: (a) long interspersed elements (LINEs), which are 6–8 kb long; (b) short interspersed elements (SINEs), which are 100-400 bp long; and (c) long terminal repeats (LTRs).

10. Several neurological diseases are linked to the expansion of trinucleotide repeats in certain genes. These trinucleotide repeats exhibit genetic instability: above a threshold of about 35–50 copies, they expand with successive generations (see Box 28-1). The longer the repeat, the more severe the disease and the earlier its age of onset (genetic anticipation).

Regulation of Prokaryotic Gene Expression

11. Prokaryotic gene expression is mainly controlled at the level of transcription. Translational control is probably not necessary since mRNAs have half-lives of only a few minutes.

12. Transcription of the *lac* operon is under negative control. The *lac* repressor prevents transcription initiation by binding to the operator. A metabolite of lactose, 1,6-allolactose, acts as an inducer of transcription by binding to the *lac* repressor, which causes a conformational change in the protein that dramatically reduces its affinity for the operator.

13. The *lac* repressor is a homotetramer with four functional components that interact with two DNA segments simultaneously.
    (a) The N-terminal region contains a helix–turn–helix (HTH) motif common to several other prokaryotic DNA-binding proteins. The N-terminal fragment binds to specific operator DNA sequences;
    (b) A α-helical linker region, which acts as a hinge and also binds DNA;
    (c) A two-domain core region, which binds to inducers and allosteric effectors, such as IPTG; and
    (d) A C-terminal α helix, which is required for homotetramer formation.

14. The presence of glucose prevents the expression of genes involved in the catabolism of other sugars, a phenomenon known as catabolite repression. In the presence of cAMP (which signals the absence of glucose), catabolite gene activator protein (CAP) is a positive regulator of transcription of the *lac* operon and certain other operons. CAP–cAMP binds to the promoter region of the *lac* operon via its two HTH domains and stimulates transcription.

15. Attenuation of transcription occurs in the *E. coli trp* operon, which is also controlled by the *trp* repressor (the repressor binds to the operator when tryptophan, which is a corepressor, is also present). The 5' end of the *trp* operon mRNA contains a leader sequence that encodes a short polypeptide containing two consecutive Trp residues. When Trp is limiting, a ribosome translating the leader peptide stalls at the Trp codons. Meanwhile, transcription of the leader sequence continues and the mRNA beyond the ribosome forms a hairpin secondary structure (the antiterminator) that allows transcription to proceed into the operon's five genes. When Trp (and therefore Trp–tRNA^{Trp}) is present, the ribosome proceeds rapidly through the leader sequence, which allows the formation of an alternative hairpin structure in the mRNA that terminates transcription.

16. Portions of prokaryotic mRNA act as riboswitches, such that their binding small metabolites regulates their transcription termination, translation initiation, or self-cleavage by a feedback mechanism. While this phenomenon has been most frequently described in prokaryotes, it has also been observed in fungi and plants.

Regulation of Eukaryotic Gene Expression

17. The regulation of gene expression in eukaryotes can occur at several levels, as outlined below. Keep in mind that no regulatory mechanism operates exclusively as an on–off switch and that several mechanisms may govern expression of a particular gene.
    (a) **Gene Availability.** Genes may be lost through transposition, rearranged through recombination, condensed into transcriptionally silent heterochromatin, or chemically modified in different tissues or at different stages of development.
    (b) **Transcriptional Control.** Regulation of the initiation, elongation, termination, or posttranscriptional processing phases of mRNA synthesis directly affects mRNA concentrations. Additional mechanisms for stabilizing and degrading mRNA influence the cellular half-life of the mRNA.
    (c) **Translational Control.** Regulation of the initiation, elongation, or termination phases of mRNA translation influences the concentrations of the corresponding polypeptides. Regulatory mechanisms that stabilize or degrade polypeptides also affect the rates of protein turnover.
    (d) **Protein Activation.** Posttranslational processing reactions, such as phosphorylation, acetylation, glycosylation, proteolysis, or assembly into multimers, may govern protein activity.
    (e) **Allosteric Effects.** Allosteric effectors that alter enzymatic properties or the assembly of multiprotein structures play a role in regulating gene expression.

18. Most of the DNA in a eukaryotic organism is packaged in a highly condensed form of chromatin called heterochromatin. Transcribed DNA is packaged more loosely as euchromatin and is more accessible to the transcription machinery.

19. Random X chromosome inactivation in female mammals results in equal amounts of X chromosome-encoded gene products in both males and females and is referred to as dosage compensation. Two mechanisms that appear to operate are (a) the binding of *Xist* mRNA to the chromosome from which it was transcribed, thereby inactivating that chromosome; and (b) methylation of CpG islands within many promoters on the inactive X chromosome.

20. Nearly all eukaryotic DNA is packaged in nucleosomes. For transcription to proceed, nucleosomes must be disassembled or remodeled to some extent so that regulatory proteins and RNA polymerase can bind to DNA. This remodeling is an ATP-driven process and is carried out by chromatin-remodeling complexes that create a dynamic state in which DNA maintains its overall packing but is exposed in various areas to allow interactions with the transcriptional machinery and regulatory transcription factors. The best characterized among these complexes are the yeast SWI/SWF and RSC complexes, which have homologs in higher eukaryotes. Other so-called architectural proteins that help regulate gene expression are the high mobility group (HMG) proteins, a group of small (<30 kD) proteins that get their name from their high electrophoretic mobilities in polyacrylamide gels.

21. Nucleosome structure may be altered by covalent modification, including:
    (a) Acetylation/deacetylation of certain histone Lys residues
    (b) Phosphorylation/dephosphorylation of specific Ser (and possibly Thr) residues
    (c) Methylation of certain Lys and Arg residues

    Acetylation and phosphorylation render the side chains less positively charged, leading to weakened histone–DNA interactions, whereas methylation increases the basicity and hydrophobicity of the side chains and hence tends to stabilize nucleosomal structure. These kinds of histone modifications, which have suggested the existence of a "histone code," may act in sequence or in combination to generate specific biological outcomes or cellular phenotypes.

22. The best studied chromatin modifiers are the histone acetyltransferases (HATs), which are components of multisubunit coactivators such as TAF1 (the largest subunit of the general transcription factor TFIID). The acetate donor for the HAT reaction is acetyl-CoA. Specific patterns of histone acetylation may recruit HAT-associated transcriptional coactivators, which contain ~110-residue modules known as bromodomains that bind to specific acetylated Lys residues of histones.

23. The methylation of histones and DNA leads to the inhibition of transcription, or gene silencing. *S*-adenosylmethinine is the methyl donor for these reactions.
    (a) Histone methyltransferases (HMTs) add methyl groups that are not readily removed and are most commonly seen in heterochromatin. Heterochromatin proteins (e.g., HP1) bind to methylated H3 Lys 9 via a protein module referred to as a chromodomain.

(b) DNA methyltransferases (DNA MTases) add a methyl group to cytosine residues to form 5-methylcytosine (m^5C) residues. DNA methylation appears to be self-perpetuating across cell divisions and may play a key role in genomic imprinting, which results in differential expression of maternal and paternal genes early in mammalian embryogenesis.

24. Proteins that regulate transcription are known as activators and repressors and bind to DNA targets referred to as enhancers and silencers, respectively. Since many enhancers are located upstream from the transcription start site, these regulatory proteins are also called upstream transcription factors. Many of these proteins are subject to regulation via signal transduction pathways. Certain assemblies of transcription factors that also include architectural proteins are called enhanceosomes.
    (a) Many upstream transcription factors act cooperatively to stimulate transcription in a nonstoichiometric manner such that a limited number of transcription factors can support a larger array of transcription patterns.
    (b) Mediator complexes serve as adaptors that bridge DNA-bound transcription factors with RNA polymerase II to induce or inhibit the formation of a stable preinitiation complex.
    (c) DNA sequences known as insulators limit the effects of enhancers as they define the boundaries of functional transcription units. Insulators also appear to retard the spreading of heterochromatin.

25. Several signaling systems regulate the activity of transcription factors.
    (a) Signaling pathways utilizing G protein–coupled receptors and receptor tyrosine kinases often lead to the phosphorylation of transcription factors so as to activate or inactivate them.
    (b) The sterol regulatory element binding protein (SREBP) is an inducible transcription factor that is regulated allosterically by cholesterol (see Section 20-7B).
    (c) The JAK-STAT pathway is stimulated by a family of protein growth factors called cytokines, which regulate the differentiation and proliferation of a number of cell types. Cytokine receptor dimerization, triggered by cytokine binding, allows the associated Janus tyrosine kinases (JAKs) to phosphorylate each other as well as signal transducers and activators of transcription (STAT proteins). The phosphorylated STATs dimerize and move to the nucleus, where they interact with specific DNA sequences to stimulate or repress transcription.
    (d) Members of the nuclear receptor family recognize a variety of steroid hormones and other substances and, upon ligand binding, usually dimerize and interact with specific DNA sequences called hormone response elements. The nuclear receptors consist of a ligand-binding domain, a DNA-binding domain, and a transcription activation domain.

26. Alternative mRNA splicing, which generates tissue-specific isoforms of a gene product, is a posttranscriptional regulatory mechanism discussed in detail in Section 26-3A.

27.   A variety of factors can modulate the half-life of an mRNA:
    (a)   Progressive deadenylation of the 3′ end by deadenylases
    (b)   Removal of the 5′ cap by decapping enzyme
    (c)   Protein binding to AU-rich sequences in the 3′ untranslated region of mRNA
    (d)   Formation of secondary structures, usually at the 5′ or 3′ end
    (e)   Presence of certain RNA-binding proteins
    (f)   Presence of premature Stop codons (see Box 28-3)

28.   Noncoding RNA plays an important role in the inhibition of gene expression (gene silencing). Naturally occurring RNAs of this kind are referred to as short interfering RNAs (siRNAs) or micro RNAs (miRNAs). Several hundred miRNAs have been identified in mammals, although their targets are still largely unknown.

29.   Gene silencing by short noncoding RNA is referred to as RNA interference (RNAi), which was first characterized by experiments in which added RNA interfered with gene expression. Inhibition is due to base pairing between the siRNA and a complementary mRNA. The mechanism of RNAi involves (a) an ATP-dependent protein (Dicer) that cleaves dsRNA precursors into 21- to 23-nt siRNA fragments; (b) the transfer of siRNA to an RNA-induced silencing complex (RISC); and (c) the cleavage by RISC of mRNA targets that form duplexes with the siRNA. RNAi has been extremely useful in "knocking out" gene expression in numerous model systems.

30.   Translational control occurs in embryonic cells, which contain stockpiled maternal mRNA. Globin synthesis is also controlled at the level of translation, via heme-regulated inhibitor (HRI), so that the protein is synthesized only when heme is also available.

31.   Somatic recombination and hypermutation serve as a unique mechanism to generate antibody diversity during the development of lymphocytes. Both immunoglobulin heavy and light chains contain four segments encoded by an exon (a leader segment, L, a variable region segment, V, a joining segment, J, and a constant region segment, C). The heavy chain also contains a diversity segment, D, which is also encoded by one exon. The various segments occur in tandem arrays in the genome in germline DNA. During their development, clones of B lymphocytes engage in somatic recombination events that eliminate all but one randomly selected exon of each segment. As activated lymphocytes undergo cell division, the $V/D/J$ exons also undergo hypermutation at a rate of 10^{-3} base changes per nucleotide per cell division. Alternative splicing allows B lymphocytes to produce membrane-bound and secreted isoforms of specific immunoglobulins.

The Cell Cycle, Cancer, and Apoptosis

32.   The eukaryotic cell cycle is divided into four phases:
    (a)   M phase (for mitosis), when mitosis and cell division occur.
    (b) G_1 phase (for gap), a variable but long part of the cell cycle.
    (c)   S phase (for synthesis), when DNA replication occurs.
    (d) G_2 phase, a relatively short period of preparation for mitosis.

Some cells enter a quiescent phase (G_0) after M phase in which they cease dividing.

33. Progression through the cell cycle requires the expression of certain genes, including the cyclins (whose levels rise and fall in the cell cycle) and the cyclin-dependent kinases (CDKs). CDKs are positively regulated by cyclin-dependent phosphorylation and negatively regulated by CDK inhibitors (CDKIs). Activated CDKs phosphorylate several nuclear proteins, including histone H1 and proteins that disassemble the nucleus and rearrange the cytoskeleton.

34. The tumor suppressor p53 accumulates in response to DNA damage and leads to inhibition of CDKs, thereby arresting cell division to allow DNA repair or, if repair is unsuccessful, triggering cell suicide (called programmed cell death or apoptosis). p53 is a transcriptional activator of a CDKI ($p21^{Cip1}$) that inhibits both the G_1/S and G_2/M transitions. The accumulation of excessive levels of p53 due to irreparable DNA lesions leads to apoptosis. p53 is also a sensor that integrates information from signaling pathways such as the Ras signaling cascade, which leads to the stabilization of p53.

35. Apoptosis is a phenomenon associated with a variety of developmental events including digit formation, development of the immune system, immune responses, and organ development and maintenance. Apoptosis is mediated by a cascade of proteases (caspases) that is triggered by either the extrinsic pathway or the intrinsic pathway. The extrinsic pathway activates the caspase cascade via the activation of so-called death receptors (Fas), while the intrinsic pathway (often initiated by the loss of key cell–cell interactions or the absence of hormones) begins with the activation of cytosolic Bcl-2 proteins. Bcl-2 associates with mitochondria to release cytochrome c into the cytosol, which then binds to Apaf-1 to form a scaffold for several procaspase-9 molecules. This interaction results in the autoactivation of procaspase-9, which then activates procaspase-3 to initiate programmed cell death.

36. Another key cell cycle regulator is pRb, a tumor suppressor, which binds and inhibits a transcription factor (E2F) that stimulates the transcription of numerous S phase genes. Phosphorylation of pRb by various CDKs leads to the dissociation of pRb from E2F, allowing the transcription factor to activate S phase genes.

37. Embryonic development in *Drosophila melanogaster* begins with rapid nuclear divisions in the fertilized egg followed by a cellularization process that leads to formation of a cell monolayer called the blastoderm. During this development, the embryo begins a segmentation process that ultimately specifies the differentiation of adult body structures.

38. A hierarchy of gene interactions orchestrates early patterning of the *Drosophila* embryo:
    (a) Maternal-effect genes define the embryo's anteroposterior and dorsoventral axes, even before embryogenesis begins, via the deposition of mRNA in different portions of the egg. The resulting protein gradients subsequently influence the expression of embryonic genes.
    (b) Segmentation genes, which specify the correct number and polarity of embryonic body segments, include gap genes, pair-rule genes, and segment polarity genes. The products of these genes activate and repress other genes, producing sharp stripes of proteins that define the embryonic body segments.

(c)   Homeotic selector genes specify segment identity; their mutation changes one body part into another. The developmental fate of a body segment depends on its position in a gradient of homeotic gene products.

39.   Many homeotic genes contain a conserved 60-residue polypeptide sequence called the homeodomain or homeobox. Homeodomains occur in many developmentally important genes, called *Hox* genes, in vertebrates and invertebrates. The homeodomain encodes an HTH DNA-binding motif, and *Hox* proteins are transcription factors that regulate the expression of other genes usually involved in defining key positions along the anteroposterior axis of the embryo.

Questions

Genome Organization

1.   From Figure 28-2, select the two types of organisms that most dramatically illustrate the C-value paradox.

2.   Based on the information in Figure 28-3, what percentage of known human genes are involved in signal transduction?

3.   Can the *D. melanogaster* histone genes be considered a type of moderately repetitive DNA?

4.   How might recombination during meiosis result in the expansion of trinucleotide repeats?

Regulation of Prokaryotic Gene Expression

5.   What observation indicates that transcription of the *lac* operon is never completely repressed?

6.   Match the molecules or DNA sequences on the left with the terms on the right.
     _____ Attenuator   A.   Tryptophan
     _____ Inducer   B.   TPP
     _____ Allosteric inhibitor C.   3·4 terminator hairpin
     _____ Leader   D.   IPTG
     _____ Co-repressor   E.   TrpL

7.   A mutation in which element of the *lac* operon (*P, O, Z, Y, A*) and/or the *lac* repressor gene (*I*) will lead to constitutive expression of the structural genes of the operon? Explain.

8.   Imagine a culture of *E. coli* grown for many generations in the presence of glucose and absence of lactose. How do these bacteria "turn on" the *lac* operon if the permease gene has been "off" for so long?

9.   What mechanisms have been observed in riboswitches for the inhibition of gene expression?

Regulation of Eukaryotic Gene Expression

10.   Match the structural elements on the left with their functions on the right.

____ Promoter	A. Transcription factor binding site
____ Silencer	B. Transcription terminator
____ Operator	C. Protein coding region
____ Attenuator	D. Inducer binding site
____ Enhancer	E. RNA polymerase binding site
____ Structural gene	F. Repressor binding site

11.   Which of the following DNA structures are likely to be transcriptionally active: Barr body, euchromatin, heterochromatin, DNase I–sensitive regions, and highly methylated DNA.

12.   Match the following proteins or protein domains on the left with the functions on the right.

____ Enhanceosome	A. Binds acetylated Lys residues on histones
____ Bromodomain	B. Acidic region found in upstream transcription factors
____ Mediators	C. Defines the boundaries of functional transcription units
____ Chromodomain	D. Part of chromatin proteins that bind methylated histones
____ Insulator	E. Complex of architectural proteins containing coactivators and/or corepressors
____ Activation domain	F. Adaptor proteins that bind upstream transcription factors and RNAPII

13.   Given its mechanisms, explain how RNAi might act catalytically.

14.   Not counting variation due to hypermutation and the addition and deletion of nucleotides at the junctions between V, D, and J segments, what is the total possible number of different antibodies that a human can produce? Assume that the number of possible λ chains is the same as the number of possible κ chains.

The Cell Cycle, Cancer, and Apoptosis

15.   Which phase of the cell cycle is likely to be missing in the early embryonic cells of the frog *Xenopus laevis*, which divide about every 20–30 minutes?

16.   Why is a mutated p53 gene present in so many cancers?

17.   How do mutations resulting in nonfunctional ATM predispose individuals to cancer?

18. Inactivation of the eukaryotic proto-oncogene *c-myc* (see Box 13-3 on p. 421) involves premature termination of its RNA transcripts near the promoter. What feature of this phenomenon is likely to be similar to attenuation in prokaryotes? What feature must be different?

19. Why is it important that cells other than reticulocytes not synthesize heme-controlled repressor?

20. Match the following genes involved in *Drosophila* development with the statements below.

    A. Homeotic genes   B. Pair-rule genes
    C. Gap genes   D. Segment polarity genes

    _____ Gradients of these gene products define the polarity of body segments.
    _____ A deletion in one of these genes converts an antenna into a leg.
    _____ The spatial expression of these genes is regulated by the distribution of a maternal mRNA.
    _____ Mutations in these genes result in deletion of portions of every second body segment.
    _____ The products of these genes contain helix–turn–helix motifs.

21. How do Bicoid and Nanos proteins determine the differential expression of the *giant, Krüppel*, and *knirps* genes?

22. Do homeotic genes encode morphogens? Explain.

Answers to Questions

1. Some algae (with less than five cell types) have more than 1000 times more DNA than some arthropods (with dozens of different cell types).

2. At least 12.2% of the human genome appears to encode genes involved in signal transduction (1.2% are signaling molecules, 5% receptors, 2.8% kinases, 3.2% regulatory molecules). However, other genes important to signal transduction are not apparent in this pie chart, notably phosphatases.

3. The *Drosophila* genome contains ~100 copies of the histone genes, which can therefore be classified as moderately repetitive sequences.

4. A recombination event that was not in register would result in two daughter cells with differing numbers of trinucleotide repeats, as diagrammed below for a CAG repeat (only one DNA strand is shown). The longer the repeat, the more out-of-register the recombination is likely to be.

$$\cdots \cdots \text{CAGCAGCAGCAGCAGCAGCAGCAG} \cdots \cdots$$
$$\text{| | |}$$
$$\cdots \cdots \text{CAGCAGCAGCAGCAGCAGCAGCAG} \cdots \cdots$$

meiotic recombination and
cell division

daughter 1

$\cdots \cdots$ CAGCAGCAGCAGCAGCAGCAGCAGCAGCAGCAGCAG $\cdots \cdots$

daughter 2

$\cdots \cdots$ CAGCAGCAGCAG $\cdots \cdots$

5. The natural inducer of the *lac* operon is 1,6-allolactose, which is produced from lactose by the action of β-galactosidase. The *lac* operon must be transcribed at a low level in order to generate a small amount of β-galactosidase as well as galactoside permease, which allows lactose to enter the cell.

6.

  C Attenuator
  D Inducer
  B Allosteric inhibitor
  E Leader
  A Co-repressor

7. Mutations in *I* or *O* can lead to constitutive expression of the *Z*, *Y*, and *A* genes of the *lac* operon. A mutation in *I* (encoding the *lac* repressor) that prevents binding to the operator will allow expression of the *Z*, *Y*, and *A* genes in the absence of inducer. Similarly, mutations in the operator that prevent binding of the *lac* repressor will also lead to constitutive expression of the *Z*, *Y*, and *A* genes.

8. Rarely is the activation or repression of transcription an all-or-none affair. In order for the bacteria to be responsive to lactose, the bacteria must occasionally transcribe the *lac* operon and produce a few molecules of the permease to allow a relatively rapid response to the presence of lactose. The presence of lactose allows for a rapid and high level production of the proteins encoded in the *lac* operon.

9. Riboswitches with bound metabolites have been found to inhibit gene expression via (a) the masking of the Shine–Dalgarno sequence, (b) formation of a premature transcriptional termination structure, and (c) activation of a self-cleaving reaction.

10.
    E Promoter
    A Silencer
    F Operator
    B Attenuator
    A Enhancer
    C Structural gene

11. The transcriptionally active structure is euchromatin. DNase I–sensitive regions only mark potentially active loci. The Barr body, heterochromatin, and highly methylated DNA are transcriptionally inactive.

12.
    E Enhanceosome
    A Bromodomain
    F Mediators
    D Chromodomain
    C Insulator
    B Activation domain

13. After the RISC–siRNA complex has bound an mRNA and degraded it, the complex can interact with another mRNA, and so on, so that a single RISC–siRNA can degrade multiple mRNA molecules.

14. Antibodies combine a heavy chain and a light chain. Since there are ~2000 different κ and λ light chains, the number of different combinations is

    (2×10^3 possible κ light chains)(8.4×10^6 possible heavy chains) + (2×10^3 possible λ light chains)(8.4×10^6 possible heavy chains) = 3.36×10^{10} different antibody molecules

    (When one factors in hypermutation and the addition and deletion of nucleotides at the junctions between V, D, and J segments, this number increases to an estimated 10^{18} possible antibodies, although this is far greater than the number of antibodies that an individual produces in a lifetime.)

15. These cells progress directly from M phase to S phase, essentially skipping G_1 phase.

16. p53 normally prevents cell division when DNA is damaged. Loss of this function leads to mutations that contribute to uncontrolled cell proliferation.

17. Mutations resulting in nonfunctional ATM do not allow cells to stabilize p53 via phosphorylation initiated by DNA damage, which normally stimulates ATM-mediated phosphorylation of p53. Unphosphorylated p53 binds to Mdm2 and so becomes targeted for proteolytic degradation by proteosomes; hence, the steady-state level of p53 cannot rise sufficiently to stimulate DNA repair.

18. The eukaryotic transcript likely forms a transcription termination structure similar to that in prokaryotic attenuation. The obvious difference is that in eukaryotes, ribosomal binding and translation cannot be involved in transcription termination.

19. Heme-controlled repressor phosphorylates eIF2 so that eIF2·GTP cannot be regenerated. This prevents translation initiation and would inhibit protein synthesis in all cells. The inhibition of protein synthesis does not harm reticulocytes, which synthesize little protein other than hemoglobin.

20.
    _D_ Gradients of these gene products define the polarity of body segments.
    _A_ A deletion in one of these genes converts an antenna into a leg.
    _C_ The spatial expression of these genes is regulated by the distribution of a maternal mRNA.
    _B_ Mutations in these genes result in deletion of portions of every second body segment.
    _A_ The products of these genes contain helix–turn–helix motifs.

21. Bicoid and Nanos proteins are translated from mRNAs that are maternally deposited in the anterior and posterior poles, respectively, of the unfertilized egg. They are therefore present at different concentrations along the length of the early embryo (see Fig. 28-49). The gradients of these proteins establish a gradient of expression of Hunchback protein that decreases nonlinearly from anterior to posterior. Specific concentrations of Hunchback protein regulate the level of transcription of the *giant, Krüppel,* and *knirps* genes so that these proteins form bands across the embryonic body.

22. Homeotic genes encode transcription factors, which cannot be morphogens because they do not diffuse between cells or between nuclei in a syncytium.